exploring ENERGY
Sources/Applications/Alternatives

by

HOWARD BUD SMITH
Managing Editor—Technology
Goodheart-Willcox Co., Inc.

Former editor-in-chief
Industrial Education Magazine

South Holland, Illinois
THE GOODHEART-WILLCOX COMPANY, INC.
Publishers

Library of Congress Catalog Card Number 84-21105
International Standard Book Number 0-87006-477-0

3456789-85-9876

Library of Congress Cataloging in Publication Data

Smith, Howard S. (Howard Sylvester)
Exploring energy.

Includes index.
1. Power resources. 2. Energy conservation.
I. Title.
TJ163.2.S63 1985 621.042 84-21105
ISBN 0-87006-477-0

INTRODUCTION

Whatever moves or changes in our world is made to do so by the use of a force we call energy. Even the human body must depend upon energy for the ability to talk, walk, and work. Indeed, without it, our marvelous technological society would not exist.

Even so, until the energy shortages of 1973, little thought was given to the importance of a ready supply of energy. Nor was much thought given to the finite nature of our fossil fuels—that one day they would be gone, never to be replaced.

EXPLORING ENERGY: Sources, Applications, Alternatives deals with many aspects of our energy supplies. It explains how we extract, process, convert, and use them to power our mobile society. It also deals with the alternative sources—where they come from and how we can use them to conserve dwindling fossil fuel reserves.

Further, this text examines how and where we use our energy, where demand is the heaviest and what the outlook is into the year 2000 and beyond. It explains the various "use sectors" and what demands each makes on available energy supplies.

From a base of knowledge which includes an understanding of the various alternate energy sources, this book then deals with the need and the tactics of conservation from one sector of society to another. A final chapter presents a view of the impact the changing energy picture will have on energy careers of the future.

Each chapter includes a list of words to know, test questions, and suggested outside activities. An instructors/activity guide, designed for the instructor's use, includes drawings and directions for various projects and experiments in alternate energy and energy conservation.

Howard Bud Smith

CONTENTS

Fig. 1-1. The sun is the greatest source of energy we know. Top. Every day, winter or summer, it lights up the earth. The trees grow because of sunshine and snow falls because of it. (Cascade Pioneer-Advertiser). Bottom. When the sun sets, the earth grows dark.

1 NATURE OF ENERGY

The information given in this chapter will enable you to:

- Define energy.
- Give the types of energy.
- List and describe the six forms of energy.
- Describe how light and heat travel.
- Define the term inertia.
- List two forms of energy conversion that occur in nature.
- Give a simple explanation of the laws of thermodynamics.
- Explain entropy and give an example.
- List the sources of energy available for use today.

Energy is the ability to do work. We are all familiar with the idea of work. We could say that anything that causes movement has energy. That would be true, but there is much more that is caused by energy than just movement. We really need to define what work is.

Work is that which causes change. This would account for objects being moved, for cold objects becoming warm, or for darkness to be lighted. However, it would not account for the energy used, for example, to prevent objects from falling or scattering. Thus, we can say that work involves change, movement, or support.

Looking around us we can see a great deal of energy being used and changes occurring. To give only a few examples:

1. The sun lights and heats the earth, Fig. 1-1.
2. Winds cause trees to bend and flags to flutter, Fig. 1-2.

Fig. 1-2. Air in motion is called wind. We know it can move things. We see it moving a flag.

Fig. 1-3. Flowing water has energy as it moves from high ground to lower ground. Even slow-moving water will carry a leaf or a stick or a boat downstream.

3. Water, Fig. 1-3, moves from higher ground to lower ground. It can move so slowly that it hardly disturbs a floating leaf. It may move rapidly so that waves can change shorelines by moving sand about, Fig. 1-4.

Fig. 1-4. Waves such as these can move sand and change the shoreline of a lake or ocean.

HUMAN ENERGY

People exert energy too. Leaping, running, and pulling all require use of energy within us, Fig. 1-5. Our actions create an energy that all moving bodies have.

Power is sometimes confused with energy. It is related to energy but is not the same. A later chapter will define this term.

Fig. 1-5. We say that people have energy and do work. They can run, jump, throw, and pull. (Cascade Pioneer-Advertiser)

EFFECT OF ENERGY

You can easily see the effects of energy when it causes motion such as a windmill turning or an automobile moving down the street. It is even quite easy to realize that energy causes light and heat. You can see the light and you can feel the heat. It is not as easy to realize that energy is causing plants and trees to grow. See Fig. 1-6.

Fig. 1-6. Energy from the sun is stored in plants and fruit. We use plants as food and we sometimes use them as fuel.

Fig. 1-7. Clouds are made up of moisture drawn into the atmosphere by the work of the sun.

While you may know it is happening, you cannot see the sun lifting water into the air as vapor. You may see the effects of it in clouds or rain, Fig. 1-7.

Nature sometimes releases great forces. Sometimes the forces cause great damage. A high wind will destroy buildings and trees, Fig. 1-8. Lightning starts fires, Fig. 1-9.

Fig. 1-8. Natural forces get out of control. Wind destroys property and trees. (Cascade Pioneer-Advertiser)

As you can see, energy is available in different ways. It produces many different effects. Since this book will be concerned with the single subject of energy it is important that you become aware of the types and forms.

Fig. 1-9. Occasionally, lightning causes fires.

TYPES OF ENERGY

Energy can be classified in two ways:

1. Potential energy.
2. Kinetic energy.

Potential energy is energy that is waiting to make changes or cause motion. It is at rest, in storage, inactive. There are two kinds.

1. Potential energy of position.
2. Potential energy of form.

POTENTIAL ENERGY

A rock sitting at the top of a hill, or the weights on a cuckoo clock represent potential energy of position. Both are higher than their surroundings. They are said to have energy of gravity. Gravity is the force which attracts bodies toward the earth. The hammer of a pile driver, Fig. 1-10, lifted in the air and ready to be dropped, has this type of potential energy too.

It has another type as it drops. We shall discuss this type later.

POTENTIAL ENERGY OF FORM

Potential energy of form is the result of twisting, compressing, stretching, or bending a material out of its natural shape. Metal and rubber are examples of materials that can be

Fig. 1-10. The hammer of a pile driver has potential energy because it is in a position to fall and move something. (Cascade Pioneer-Advertiser)

Fig. 1-11. When you stretch a rubber band and place it around something you "load" it with energy. In trying to return to its original size it "works" at holding the cards together.

deformed to make them do work. When they are deformed (compressed, stretched, or bent) they try to return to their original shape or length, Fig. 1-11. In doing so they exert force if attached to a load.

KINETIC ENERGY

Kinetic energy is the ability of objects that are moving to do work. The word "kinetic" comes from the Greek word "kinema" which means motion. This energy can be found in wind, water falling over a dam, or an automobile rolling downhill. Any body in motion has kinetic energy. Fig. 1-12 shows a type of kinetic energy.

Fig. 1-12. A girl leaping into a pool has kinetic energy as she falls toward the water. The kinetic energy is absorbed by the water which is pushed away as her body enters the pool.

FORMS OF ENERGY

Potential and kinetic energy are also related to the form in which energy is found. The forms of energy include:

1. Light energy.
2. Heat (or thermal) energy.
3. Mechanical energy.
4. Chemical energy.
5. Electrical energy.
6. Nuclear energy.

All forms are important since they can be made to do work for us. Each will be defined and discussed.

LIGHT ENERGY

It is not easy to understand light energy. However, scientists have theories about it.

Another name for it is RADIANT energy. This is energy that travels as a wave motion. Its name suggests that it comes from a type of radiation. This kind of energy travels through outer space and earth's atmosphere at a high rate of speed.

Fig. 1-13. Heat and light are usually found together. There was enough light for the camera to record this campfire scene.

The scientific term for this kind of energy is ELECTROMAGNETIC RADIATION. Light is only the small part of electromagnetic radiation which can be seen. We will talk later about its causes and how it is thought to work. The sun is the greatest source of energy in the known universe. It sends out enormous quantities of electromagnetic radiation.

HEAT ENERGY

Heat energy is very closely related to light energy. It, too, comes from electromagnetic radiation. It is to be expected, then, that heat and light are usually found together. Did you ever sit before a campfire that did not give off both heat and light, Fig. 1-13? The heat is an invisible part of electromagnetic radiation. Like light, the invisible rays which cause heat travel through space at a high rate of speed.

Electromagnetic radiation

Just what is this electromagnetic radiation, aside from its being responsible for both light and heat? Scientists have two theories about how it is caused and how it acts. One is called the WAVE theory. This theory is based upon the activity of electrons in atoms of matter. The other explanation is called the QUANTUM or "particles of energy" theory.

Wave theory

The wave theory explains radiation as the effect of electrons jumping out of their orbits (circular paths) around the nucleus (center) of an atom. The electrons get into a different level as they continue to whirl around the nucleus, Fig. 1-14. They may even whirl right away from their own atom and join another atom.

For some reason not known, this releases a great deal of energy. The energy does not stand still. It travels 186,000 miles a second through space. You will learn more about this theory as you study solar energy in Chapter 8.

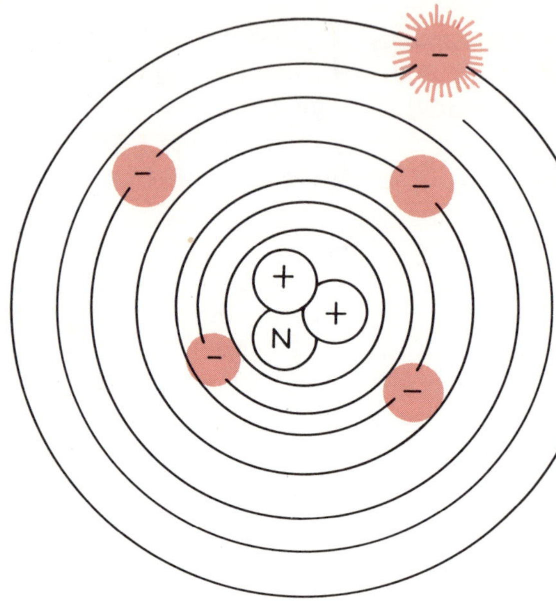

Fig. 1-14. According to the wave theory of radiation, electrons jumping out of one orbit into another orbit around an atom release huge amounts of energy.

Quantum theory

Not all behavior of electromagnetic radiation can be explained by the wave theory. In such cases, scientists find it easier to consider radiation as particles of energy.

These particles or PHOTONS, are bundles of energy that travel through space and give up their energy upon striking a surface through which they cannot pass. It is believed now that the wave and quantum theories complement (help) each other.

Heat/light differences

Though both light and heat are the result of radiant energy, heat energy acts differently from light energy. Light exists outside of matter and does not enter into matter that is not transparent or translucent. (Transparent means you can see through it; translucent means it transmits light but you cannot see through it.) Heat exists outside of matter, too, but it will enter matter and cause its temperature to rise. The atoms of matter that heat has entered begin to move about more rapidly. The higher the temperature of the material, the more rapid the motion of the atom.

Also, heated atoms must have more room to move about. This accounts for the well-known fact that heated liquids or gases expand. For example, if water or air is heated in a closed container, expansion may cause the container to bulge or burst, Fig. 1-15.

Fig. 1-15. Fluids in closed containers will expand as they are heated and may even burst their containers.

Fig. 1-16. Convection taking place in a liquid. Warm water is less dense since its molecules are farther apart. Being lighter, it rises while cooler water sinks.

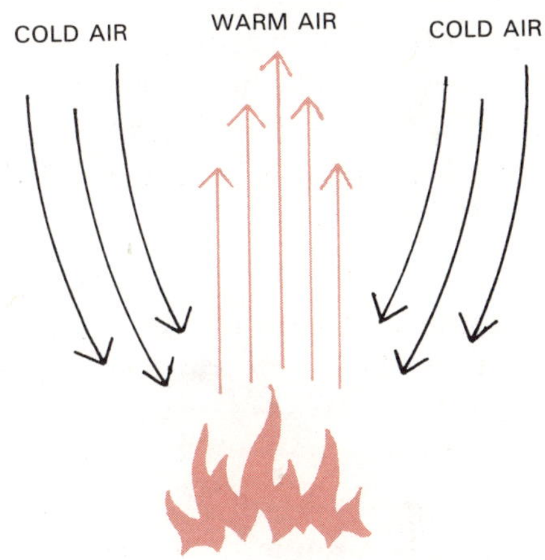

Fig. 1-17. Convection occurs in the air above a fire. Warm air rises and cooler air descends to take its place.

HOW HEAT TRAVELS

While light travels by way of radiation only, heat can move from place to place by three methods:

1. Radiation which was just discussed under "electromagnetic radiation."
2. Convection.
3. Conduction.

Convection

Convection is movement within liquids or gases. The heat travels by circulation of a heated liquid or air. Circulation will continue until all of the material is the same temperature.

Water illustrates this action. Because of expansion, hot water is lighter than cold water. Therefore, cold water sinks to take up space vacated by warmer water which tends to rise, Fig. 1-16.

Likewise, when air is heated, it becomes lighter and rises. Cooler air sinks and takes its place. See Fig. 1-17. A room may be heated on this principle. Warm air from the furnace rises as it enters the room. Cold air sinks to the floor level and is drawn back to the furnace through the cold air return. See Fig. 1-18.

Conduction

Conduction is heat travel produced by molecular action. Heated molecules of a substance move about so rapidly that they collide with one another. Warmer molecules bump into cooler ones and give some of their heat to the slower moving cool molecules. This action is repeated until all molecules in the substance are the same

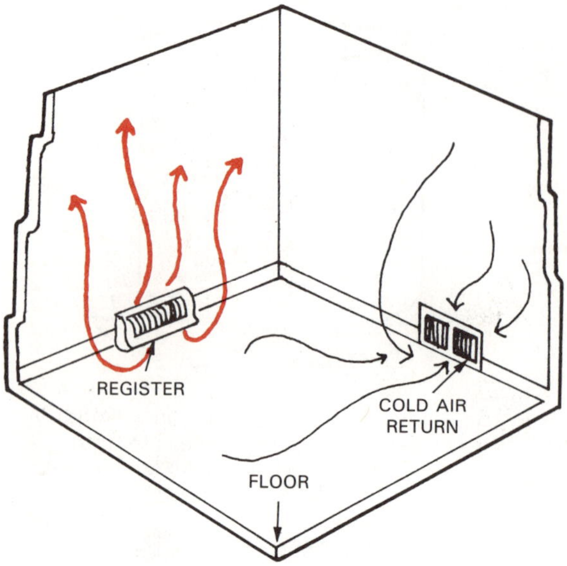

Fig. 1-18. Home heating uses the principles of convection. Warm air comes out of a floor-level register and rises, heating the room. Cool air enters an opening that will take it back to the furnace to be heated.

Fig. 1-19. Heat will travel along a solid material, such as an iron, as heated molecules jostle (bump into) cooler molecules. They pass along some of their heat and, thus, the metal gradually becomes hot all over.

temperature. See Fig. 1-19. This principle is applied to cool an automobile engine and to cook food.

MECHANICAL ENERGY

Of all the forms of energy, MECHANICAL ENERGY is the one we usually think of as "doing work." Mechanical energy is the same as kinetic energy which you learned about earlier. However, the term "mechanical" is generally used wherever the energy is harnessed to do work.

Though its name suggests machines, mechanical energy is not limited to the work output of machines. All moving bodies have it. Bodies in motion do work because they resist forces exerted to stop them. This tendency to stay in motion is called INERTIA. The heavier an object is the more inertia it has.

CHEMICAL ENERGY

Chemical energy is locked away in the molecules of many kinds of substances. Growing plants collect radiant energy from the sun, carbon dioxide from the atmosphere, and minerals and water from the soil. They combine all these chemically to form new molecules that make up the pulp and fiber of the plant. The stored energy can be released by burning. For example, wood is burned to create heat energy which is used to run a steam engine, Fig. 1-20. The burning rearranges the molecules, releasing heat and light in the form of radiant energy. The ashes left behind are an entirely different form of matter.

Gasoline, oil, and natural gas are other examples of chemically stored energy. A battery also stores energy as a chemical and releases it as electrical energy, Fig. 1-21.

ELECTRICAL ENERGY

Electrical energy results when negatively charged particles are attracted to positively charged particles. The negatively charged particles are called ELECTRONS. The positively charged particles are PROTONS. Attraction between them causes a flow of energy. Electrons can be made to produce light and heat when they flow through fine wires in a light bulb. They will also create a magnetic field and cause an electric motor to rotate and provide power for turning machines. Electric current can also produce heat

Fig. 1-20. Burning releases the energy stored in wood which was once part of a living plant. It provides both light and energy to heat coffee or run an old steam engine. (Cascade Pioneer-Advertiser)

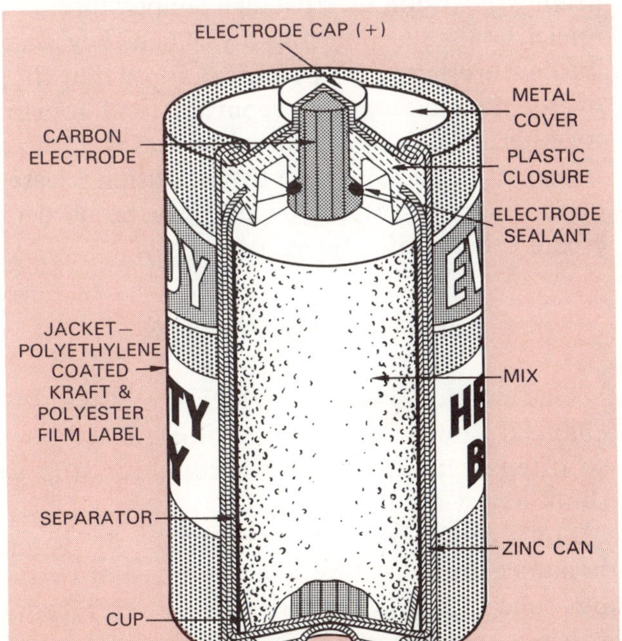

Fig. 1-21. Cutaway of a small battery such as is used in flashlights and toys. It can store electrical energy in a chemical paste which is called a "mix." (Union Carbide Corp.)

when traveling through certain types of wires, Fig. 1-22.

Another fact which makes electrical energy so popular is the ease with which it moves from one

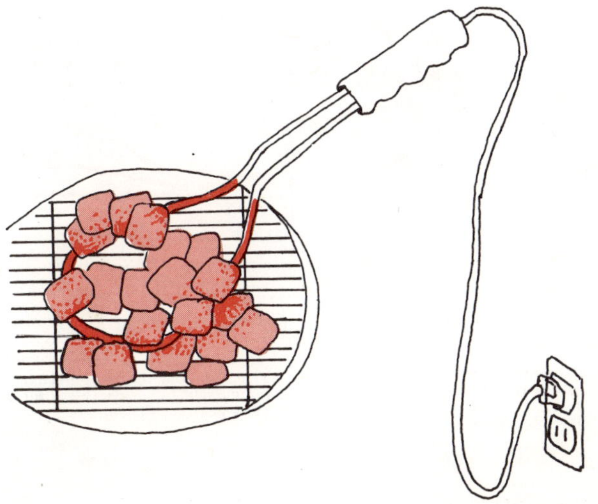

Fig. 1-22. An electric charcoal lighter heats up and starts charcoal because wires in the lighter resist the electric current passing through them.

place to another. It can be produced at one location and moved long distances before it is used.

NUCLEAR ENERGY

Atomic or nuclear energy is released when certain kinds of atoms are split. The release of the energy can be swift as in the explosion of an atomic or hydrogen bomb. It can also be controlled or slowed down to produce heat to rotate a turbine for making electric power or for propelling a ship.

A nucleus is the center of an atom. All nuclei (plural for nucleus) are made up of positive particles called PROTONS and neutral particles called NEUTRONS. Nuclear energy is produced when the nucleus of an atom of uranium is changed. Uranium is used because its atoms have many protons and neutrons in their nuclei.

Fission

One way of changing an atom is to bombard it with a particle such as a neutron. This causes the nucleus to split into two parts each with the same number of protons and neutrons. This changes the mass (weight) of the nucleus and causes it to give up tremendous amounts of heat energy. See Fig. 1-23. The splitting of the nucleus is called FISSION.

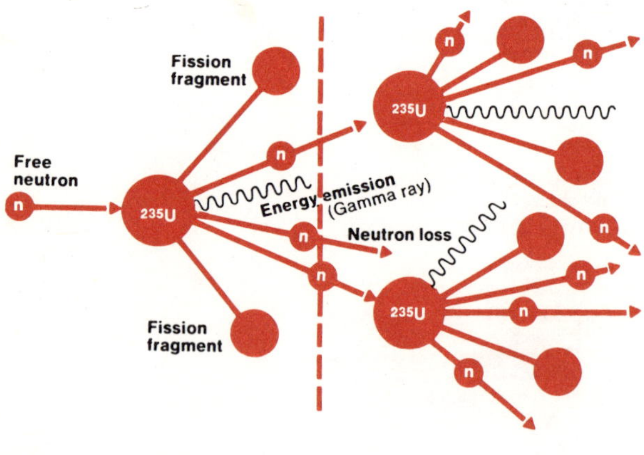

A free neutron entering a ^{235}U nucleus, causes it to fission.

Fission divides nucleus into smaller fragments, releases energy and 2 or 3 neutrons.

In a chain reaction, neutrons cause more nuclei to fission, releasing more energy and neutrons.

Fig. 1-23. Nuclear energy is caused by breaking up an atom of a certain kind of uranium. It creates other kinds of matter and releases great quantities of heat. (Atomic Industrial Forum, Inc.)

If other nuclei are struck by neutrons flying out from the first fission, the splitting or fission will continue. This is called a CHAIN REACTION. When controlled inside a reactor, the release of heat takes place at a slower rate than the exploding of an atomic bomb. It is, therefore, possible to put the heat energy to use. Usually it heats water for steam energy.

Fusion

Another way of tapping the energy of the atom is to combine two light atoms into one heavy atom. This process is called FUSION. It propels two nuclei together. This is difficult because the nuclei are both positively charged and tend to repel one another. This tendency is overcome by heating the fusion fuel to a high temperature. The nuclei bounce around rapidly. Then they slam into each other with such great force that they overcome their natural repulsion and remain together. Fig. 1-24 is a simple drawing of how fusion takes place. As with fission, fusion releases great quantities of heat which can be used to create other useful forms of energy.

ENERGY CONVERSION

One characteristic of energy is that it is easily changed from one form to another. It happens so often in nature that we hardly ever stop to think about the conversion.

Some of the common conversions occurring in nature are caused by solar energy. For example, solar radiation is responsible for several conversions which occur quietly all around us.

1. The sun, like a giant pump, lifts water from the earth's surface. Some is collected from lakes, oceans, and rivers. Part of it comes from the evaporation of moisture from other surfaces. Every year, 100,000 cubic miles of

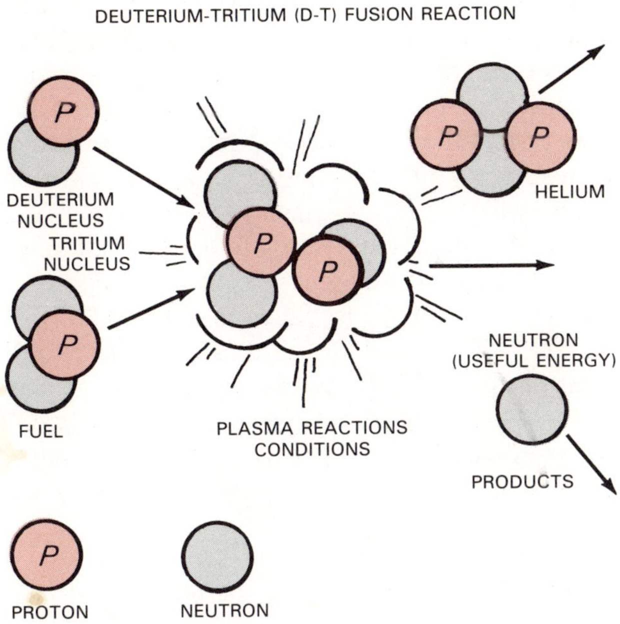

Fig. 1-24. Fusion is a form of nuclear energy. Two light atoms are combined with a heavier one. As they join, energy is released in the form of heat.

moisture are drawn into the atmosphere by the sun. It is later released as rain, sleet, dew, or snow to replenish the flow of rivers. This flow is mechanical energy which can be captured and put to use producing other forms of energy.

2. Plants collect solar radiation and store it in matter which makes up leaves, stems, and stalks. This process, called PHOTOSYNTHESIS, was discussed earlier. The energy will remain locked in the plant as chemical energy until some outside agent transforms it into still another form of energy.
   a. Plants eaten by animals or humans supply the energy needed to maintain body temperature and nourish muscles which control the movements of the body.
   b. Burning the plant changes its chemical energy into heat energy.
   c. Plant life buried and subjected to heavy pressure, forms fossil fuels, Fig. 1-25. These stored forms of energy are rich in carbon and are useful for providing heat when burned.

Fig. 1-25. A chunk of bituminous coal. It was once plant life which became buried and changed to carbon under pressure.

LAWS OF ENERGY CONSERVATION

Energy follows certain basic behavior characteristics known as the LAWS OF THERMODYNAMICS. The first of these laws is that: Energy cannot be destroyed. It can be changed from one form to another, but it never ceases to exist.

At first glance, this seems like a wonderful arrangement. It it cannot be destroyed, can we continue to use it over and over again?

The answer is no because of the second law of thermodynamics. It is impossible to make heat flow from a cold body to a warmer one without using another source of energy. The effect of the law is that, once used, energy loses its ability to do work even though it is not destroyed.

A child on a swing, Fig. 1-26, shows how this principle works. Mechanical energy was used to set the swing in motion. When no more energy is put into pushing the swing, it will gradually stop. It would seem that all the energy is gone. However, if you had a sensitive thermometer, you could measure that the mechanical energy had changed to a low heat. It would show an increase in the temperature:

Fig. 1-26. Child on a swing. Eventually the swing will stop because no energy is put into making it go.

1. Of the swing.
2. Of the hinges holding the chain to the supports.
3. Of the air around the swing.

The heat is not of a high enough temperature to be useful. It is passed off to the atmosphere. In such cases the energy is said to be DEGRADED (weakened) to a point where we are not able to make use of it.

ENTROPY

The child on the swing is an example of what entropy means. The swing slowed down and came to a stop. The energy was not all gone, but it was so scattered that it could do no more work. The principle of entropy is this: eventually all energy, as it is used, becomes so random (scattered) that it loses the ability to do work.

Let us consider another example. Suppose that you were to heat up water to operate a small model of a steam engine. When the water becomes steam, it will run the engine as long as you continue to heat the water. However, when you turn off the burner supplying the heat, the steam will cool. The engine will stop running. The water is still hot but it no longer has the energy to run the steam engine.

SOURCES OF ENERGY

What does it take for a form of energy to be useful as a source? Several things are required.

1. It must be plentiful and easy to acquire.
2. It must be easily converted into a form which our machines can use.
3. It must be easy to transport.
4. It must be easy to store.

Up to now, only a few energy sources have been developed that fit all these requirements:

1. Hydrocarbons or fossil fuels. They include: natural gas, coal, and petroleum.
2. Nuclear energy.
3. Hydroelectric power.

Other sources are available which may not now meet all the requirements as well as those just listed. However, they could be developed to the point of meeting the requirements to a higher degree. This group includes:

1. Solar energy. The radiation of the sun, discussed earlier, can be used directly to produce power and heat buildings.
2. Forms of kinetic energy, including tidal, wave, and wind energy.
3. Thermal energy. This includes geothermal (heat from the center of the earth) and the heat stored in ocean waters.
4. Biofuels. These include a group of gases and liquids which can be extracted from organic wastes, seaweed, grains, and a variety of plant life.

RENEWABLE AND NONRENEWABLE

Fossil fuels, since the process which formed them took millions of years, are considered nonrenewable. It means that once used, they cannot be replaced.

Energy sources which can be replaced are called RENEWABLE. Because these sources will not run out, they need to be studied and developed.

All of these energy sources, renewable and nonrenewable, will be discussed more fully in later chapters.

DO YOU KNOW THESE TERMS?

Biofuels, chain reaction, chemical energy, conduction, convection, degraded, electrical energy, electromagnetic radiation, electron, energy, entropy, fission, fossil fuels, fusion, gravity, kinetic energy, mechanical energy, molecules, nuclear energy, nucleus, neutron, orbit, photon, photosynthesis, potential energy, proton, radiant energy, translucent, transparent, work.

SUGGESTED ACTIVITIES

1. Secure books from your library and research the quantum theory of energy. Prepare a written report for your class.
2. Build or bring in a toy or device which demonstrates that heated fluid, such as air or water, rises.
3. Construct and/or demonstrate a toy or device which illustrates a principle learned in this chapter. Ask your instructor for suggestions and help.
4. Place an open topped glass container of water in direct sunlight. Cover the top with a piece of clear plastic or glass. Observe at intervals. What happens to the underside of the cover? Report your observations to the instructor or class. Discuss what happens and why.

TEST YOUR KNOWLEDGE

1. What is energy?
2. Energy and energy sources are able to produce (check correct answers):
   a. Heat.
   b. Light.
   c. Motion.
   d. None of the above.
   e. All of the above.
3. ______________ energy is the energy that is waiting to do work.
4. Define kinetic energy and given an example.
5. List the six forms of energy and briefly describe each one.
6. Heat moves from one place to another in three ways: ____________, ____________, and ____________.
7. Chemical energy is locked away in the ____________ of many kind of substances.
8. How is nuclear energy produced?
9. Give two examples of energy conversion which take place in nature.
10. The first Law of Thermodynamics states that energy ____________ be ____________.
11. List three fossil fuels.
12. Fuels made from plant life or organic wastes are said to come from a (renewable, nonrenewable) source.
13. Entropy means that eventually all energy, as it is used, becomes so scattered that it ________________.

Two examples of energy are shown in this photo. Can you pick them out?

2 HARNESSING ENERGY

The information given in this chapter will enable you to:

- Trace development of energy conversion from primitive to modern times.
- List and describe a number of energy converters including windmills, water wheels, turbines, internal and external combustion engines, electric motors and generators, and fuel cells.

Our ancestors did not know much about using energy. At first they depended upon the natural warmth of caves to shelter them from the cold and to keep them safe. Then they learned to use fire. They found that it helped to keep them warm, Fig. 2-1. During thousands of years of simple existence, early humans had to depend on their own muscle power to travel, find food, and protect themselves.

Fig. 2-1. Primitive humans used fire to provide warmth and to cook food. Wood was the first fuel.

FIRST ENERGY USED

In that time they discovered that muscle power could hurtle a rock at an animal, bringing it down to provide a bigger supply of food. Then they learned to form the rocks into better weapons. The fire was found to be useful in shaping spears, knives, and arrows. Cold water, dropped carefully on the edges of heated stones would break away small chips until the first cutting edges were formed. See Fig. 2-2. Thus, our ancestors could make their muscle (mechanical) power work better. When they used these improved tools, they once more enlarged the supply of food and got better protection against animals and enemies.

The first communities were developed as families and tribes began to take up more or less permanent possession of lands around the valleys of the Nile, the Tigris, and the Euphrates Rivers in Egypt. There is also evidence that humans living in southern California 70,000 to 100,000 years ago were fashioning crude tools. Their newly acquired tool-making skills allowed them to grow food from the soil and kill wild game. Animals, which they had tamed, provided both a source of food and mechanical energy to help them power tools and work the soil, Fig. 2-3.

As civilizations developed the need for travel, rivers were used to carry rafts downstream. Later

Fig. 2-2. Later the heat of the fire was used in toolmaking. Top. Cold water dashed on a hot stone, broke it into jagged chips. Bottom. The chips were used as arrow or spear heads and as crude knives.

Fig. 2-3. Animals were tamed and made to provide the power to pull farming implements.

Fig. 2-4. Sails were found to be better than muscle power to drive boats.

still, fixed sails were added, Fig. 2-4, to capture the mechanical energy of the wind.

LABOR-SAVING MACHINES

The first attempts at applying engineering principles to labor-saving machines probably came around 300 B.C. Greek scientists developed a number of these but regarded them as little more than toys and curiosities for the amusement of the rich. However, Archimedes, who lived from 287 to 212 B.C., is said to have put one of the inventions to use. To defend a Greek city from a Roman invasion, he used large curved mirrors to reflect concentrated sunlight that burned up the sails of the Roman ships.

The inventors of the Middle Ages were much better at making machines that were more efficient as well as labor-saving. Both animal and human muscle were better used. See Fig. 2-5 and Fig. 2-6. For the first time, human muscle and even horses were replaced with another source of mechanical energy. The first water wheels captured the energy of falling or flowing water.

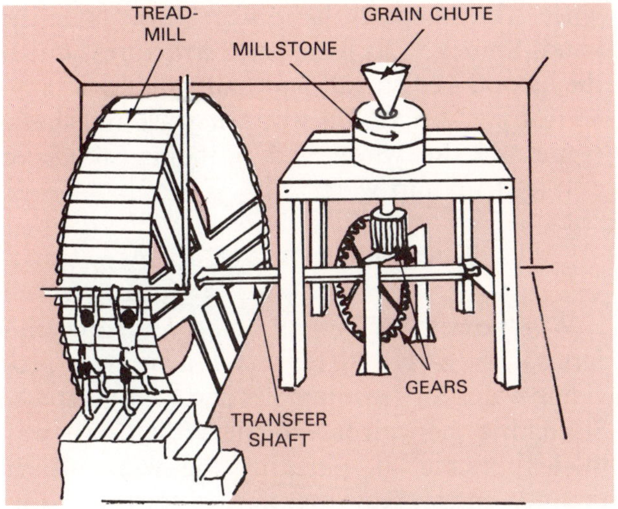

Fig. 2-5. Human power was multiplied and used to drive huge grain mills.

Fig. 2-6. Animals were trained to power mills.

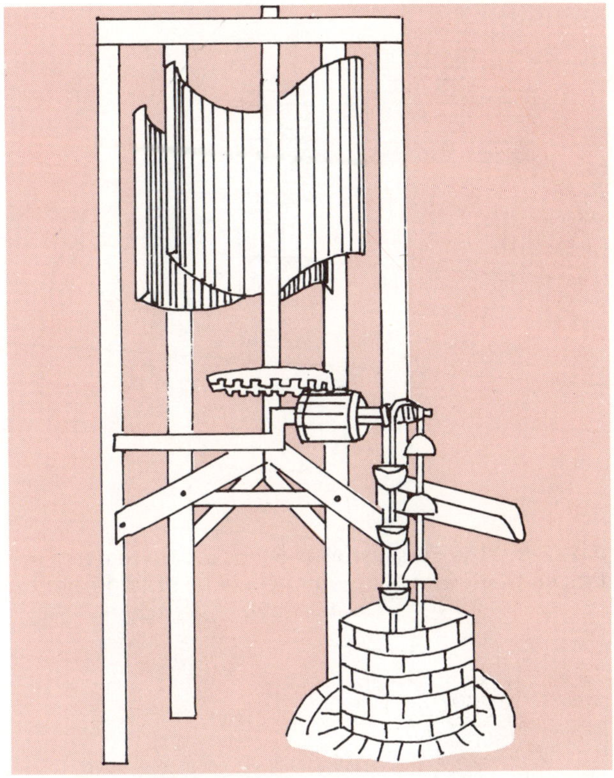

Fig. 2-7. A sketch showing an early Persian windmill. It could catch the wind from any direction.

EARLY WINDMILLS

The first windmills were thought to be a Persian invention. They were popular in the desert where heavy winds blow most of the time. Fig. 2-7 shows an early design. The vanes were mounted on a vertical shaft. It looked somewhat like a modern revolving door. The vanes were made of wood.

The shaft was connected to a wooden disc which had pegged teeth suspended along its outer edge. These meshed with a slotted drum. The drum operated a long, endless rope ladder which had dippers attached to it. The ladder, called an elevator, extended down into the well. As the drum turned, the dippers picked up water which was brought to the surface and dumped into a trough, Fig. 2-8. The top of the tower was open so that the vanes might catch the winds from every direction.

EUROPEAN WINDMILLS

The windmill first began to appear in Europe in the 12th century. The first ones were called post mills and looked like the sketches in Fig. 2-9. The vanes looked like four huge paddles. They were wooden frames with canvas stretched over them. The vanes were supported by a post and an open structure like a treehouse. This allowed the mill to turn into the wind. Sometimes the open structure was closed in for storage.

The smock mill, which came later, was an improvement over the post mill. It had a second

Fig. 2-8. The Persian mill was hooked up to gears and lifts so that buckets of water could be drawn from the well below.

Fig. 2-9. Later European designs improved upon the Persian mill. Top. Smock windmill. Bottom. Post windmill.

wheel which kept the large wheel turned into the wind. Smock mills were huge structures. Often they stood 120 ft. tall and had vanes or "sails" 80 ft. long. A series of wooden gears and shafts connected the wheel with grinding stones or devices to pump water.

WINDMILLS IN AMERICA

Windpower became popular on American farms in the early 19th century as a means of providing water for farm families and their livestock. Stretching the railroad to the West Coast was made possible, in part, by the windmill. It pumped and stored water needed for the steam-powered locomotives. Earlier designs had wooden towers and large wooden wheels, Fig. 2-10. A vane (fan-like piece), extending behind the wheel, kept the wheel turned directly into the wind. Later, towers and wheels of galvanized

Fig. 2-10. Early American mills were made of wood.

steel replaced the wooden mills. Though often abandoned due to the coming of cheap electric power to the farms, many of the towers still dot the midwestern and western farmlands, Fig. 2-11. The wheel rotated an offset shaft or crank which lifted a long shaft in a pumping action. The shaft was connected to the piston of a lift pump. As the wheel revolved, the piston of the pump was

Fig. 2-11. Later designs were of metal. Many of them still stand although few are used for pumping water anymore.

made to move up and down to bring up water from the well.

RISE AND DECLINE OF WIND POWER

Though relatively cheap, wind power was never highly dependable, nor was the windmill as powerful as the water wheel. It was more popular, of course, in lowlands which did not have water power. By 1810 it had become an efficient source of power and was used for irrigation, drainage, and, to some extent, a power source for grinding corn and small grain. In 1890, Denmark built the first wind generator. Eventually other European countries adapted them for generating electric power. The excess electricity was stored in batteries for use during calm days.

HARNESSING WATER POWER

While primitive travelers might have been contented to let the power of moving water carry their boats downstream, later generations found another way to use it. The water wheel, which was invented about 200 B.C., captured the energy of flowing or falling water and converted it to power for turning millstones. This was a big improvement over old methods of grinding corn with a stone bowl and a smooth rock.

The first water wheels were designed to let the water flow underneath where it pushed against paddles. This is known as an UNDERSHOT WATER WHEEL, Fig. 2-12. During the Middle Ages, the water wheel had been perfected to the point where a single wheel could develop 40 to 60 horsepower.

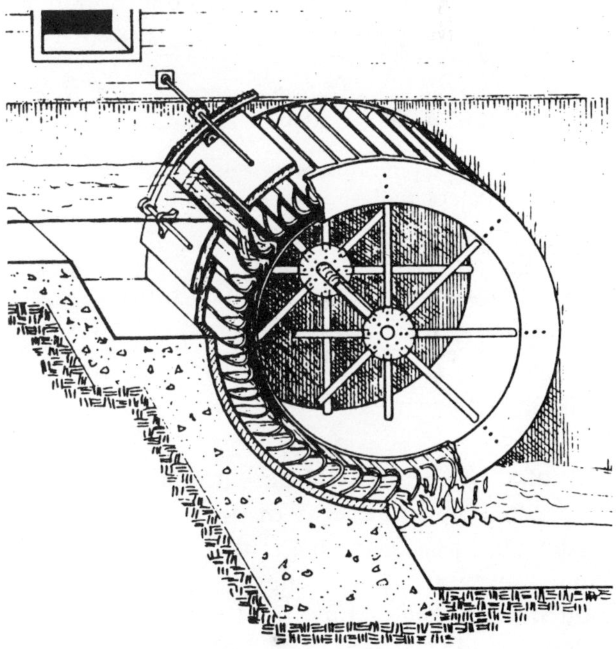

Fig. 2-12. Undershot water wheel. In this early design, the water flowed under the wheel. It pushed against the paddles to turn the wheel.

In 1750, John Smeaton, a British civil engineer, designed the OVERSHOT WATER WHEEL. This was a big improvement over the undershot wheel. The water poured over the top of the wheel on its downward side. Thus, the weight of the water, added to its motion, made the wheel

Fig. 2-13. Model of an overshot water wheel. It was more efficient than the earlier undershot wheel.

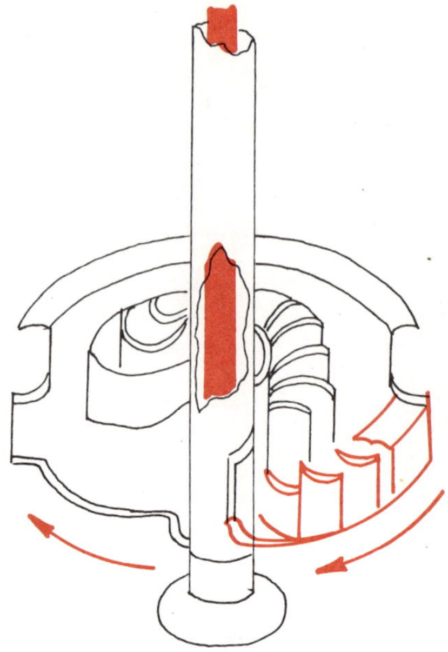

Fig. 2-14. Fourneyron's water turbine. Cutaway shows stationary vanes in black and movable vanes in color. Water coming down the hollow outer shaft flowed out through the stationary (not moving) vanes and struck the movable vanes.

do more work. That is, it had more power. See Fig. 2-13.

TURBINES

Water wheels did not get enough power out of the mechanical energy of the water. This led to development of the water turbine. One of the early successes was designed by Benoit Fourneyron, Fig. 2-14. It had curved vanes and was suspended on a vertical shaft. The shaft was hollow and the water entered the turbine through the shaft. As it entered the turbine it was forced to move outward, turning the blades before it ran off the edge.

The modern water turbine is mainly used to generate electricity. Power is supplied by dammed up water. The kinetic energy is supplied by the height of its fall to the turbine below. This is called "head." The water is channeled through a sluice (a special waterway or tunnel). It strikes the turbine and causes it to spin rapidly as it passes through the fins or blades. Being connected by a shaft to a generator, the turbine causes electricity to be generated. Fig. 2-15 is a cross section of a modern hydroelectric power plant.

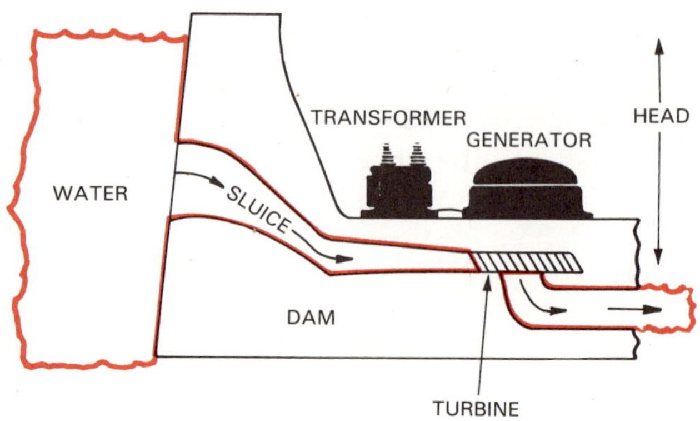

Fig. 2-15. Modern water turbine. Kinetic energy is supplied by the "head." This is the distance the water falls before it hits the turbine.

USING HEAT TO PRODUCE MOTION

Cave dwellers built fires to create more comfortable surroundings. Early metal workers burned wood and coal to melt their materials for casting and forging. Heat, however, was not used for creating motion until someone invented a machine which could contain and convert the heat energy. This invention was the heat engine.

HEAT ENGINES

Devices which can contain the energy of heat and make it do work are called HEAT ENGINES. The first one was developed about 130 B.C. by Hero, a scientist, writer, and mathematician of Alexandria—part of present-day Greece. His machine, Fig. 2-16, hardly resembled a steam engine or steam turbine as we know them today. It looked like a pressure cooker with a large ball mounted on top of it. The lower section of the "kettle" held water. It had a tight-fitting cover. Two pipes extended upward from the flat top and were connected to a hollow ball. The ball, which could rotate on the pipes, had two nozzles. These were angled to face in opposite directions. When water was boiled in the kettle, the steam escaped through the pipes into the hollow sphere. As pressure built up in the sphere, the steam escaped through the nozzles. This caused the ball to rotate.

You can easily see the force of the steam created by boiling water if you observe a covered pot in which food is being prepared. The pressure will actually lift the lid until some steam can escape and relieve the pressure.

What do you suppose would happen if you were to attach a whistle so that the steam would have to travel through it to escape? This principle is used on the whistling teakettle, Fig. 2-17. Is the steam doing useful work by blowing the whistle?

Hero had no idea of the importance of his invention. If further attempts were made to try to harness steam they were not recorded until Leonardo da Vinci invented a gun which used steam to fire bullets. That was in 1495.

Fig. 2-16. The first heat engine was a Greek toy. Steam escaped from the kettle-like container below. It traveled through the pipes to the hollow sphere. As the pressure built up, the steam escaped through the spigots. This caused the sphere to spin.

Fig. 2-17. Boiling kettle demonstrates the power of expanding gases. The steam will blow the whistle.

STEAM POWERED WATER PUMPS

Little came of da Vinci's invention and little more was done with steam as a power source until English miners needed a better way to pump water out of the peat mines. In 1698 Thomas Savery designed and built a steam operated pump that worked on pressures above 150 psi (pounds per square inch). It marked the beginning of the modern era of power development which saw the employment of fossil fuels (coal, petroleum, or natural gas) as chief source of energy.

Savery's engine was further improved when he went into partnership with Thomas Newcomen. The improved engine shot steam into a hollow cylinder to move a piston. The piston was connected to one end of a rocking beam. The other end was connected to a pump rod. As the steam worked the piston, the beam rocked up and down like a seesaw. See Fig. 2-18.

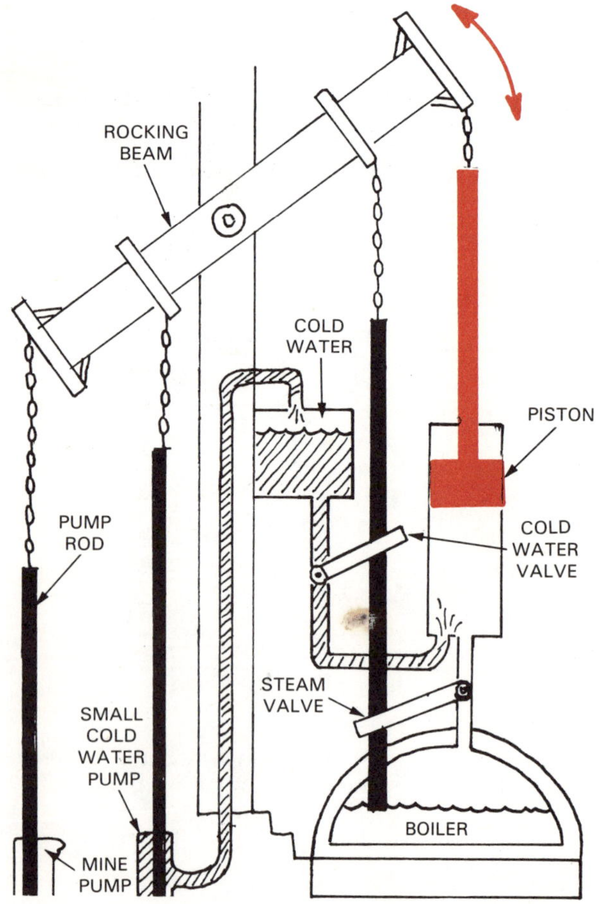

Fig. 2-18. Newcomen's engine was used to pump water from English coal mines. The steam piston rocked the beam up and down to move the pump piston in a pumping motion.

HARNESSING THE FOSSIL FUELS

The earliest practical steam engines were designed to burn fossil fuels. A Scottish instrument maker, James Watt, developed the first successful reciprocating steam engine, Fig. 2-19. Other designs based on the same principle would drive boats and locomotives. See Fig. 2-20.

HOW THE STEAM ENGINE OPERATES

The steam engine depends on the pressure of super hot steam. The steam is let into a large cylindrical chamber that has been fitted with a movable cap called a piston. The pressure of the expanding steam pushes against the piston causing it to move along the cylinder. The piston is attached to a long rod which turns a crank. The crank will turn whatever is connected to it.

STEAM AND GAS TURBINES

The turbine is a more efficient mechanism for converting heat energy and kinetic energy of

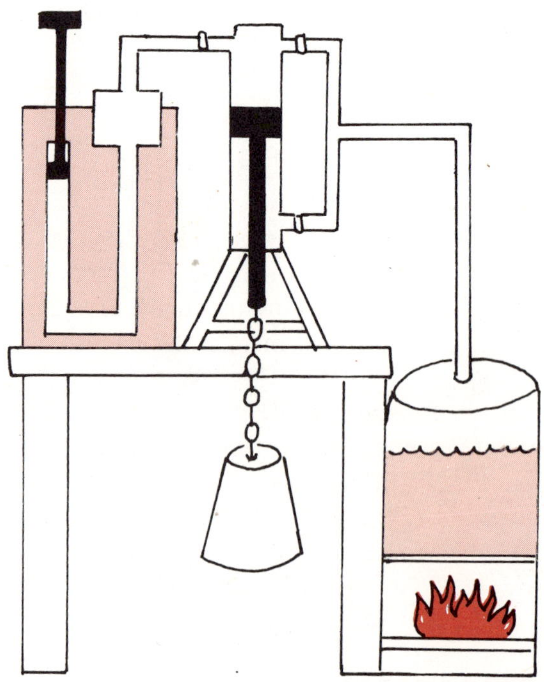

Fig. 2-19. Simplified drawing of Watt's steam engine. It did not work too well and used huge quantities of fuel. The problem was that the heat was wasted by having to cool and reheat the walls of the cylinder at each cycle.

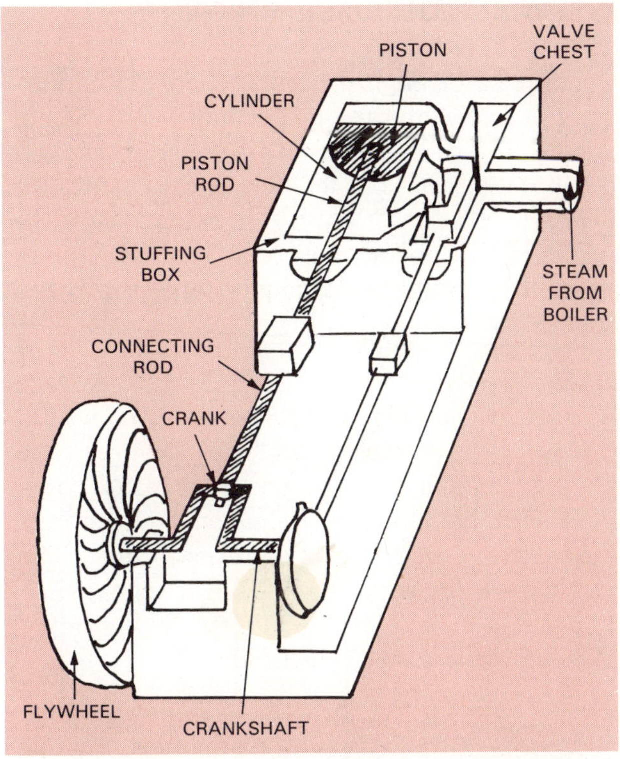

Fig. 2-20. Later version of Watt's steam engine. He learned to pack insulation around the cylinder. This is the area called the "stuffing box." The engine was adapted successfully to boats, tractors, and locomotives.

steam or gas to mechanical energy. It is able to use the pressurized steam and convert it directly to rotary motion. The steam engine has to convert linear (straight-line) motion to rotary motion. In the process, some mechanical energy is wasted in friction and inertia.

Another reason for the efficiency of the turbine is that energy can be applied to it continuously. While the piston of a reciprocating steam engine returns to the top of the cylinder it cannot use the energy of the steam. That is, its conversion of energy is only intermittent (off and on).

TYPES OF STEAM TURBINES

In the IMPULSE turbine, high pressure steam is piped to the turbine blades. It escapes through

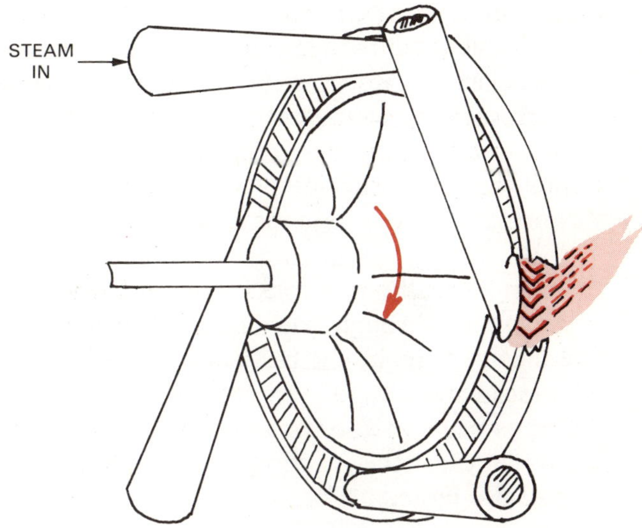

Fig. 2-21. The de Laval steam turbine is an example of the impulse type turbine. High-pressure steam is piped to the turbine blades.

nozzles and is moving at very high speed as it strikes the blades. See Fig. 2-21. In modern turbines the speed may be faster than 4000 ft. per second.

REACTION TURBINE

In a REACTION type turbine, Fig. 2-22, the steam passes through several sets of blades. The result is a stronger "push." All of the blades are

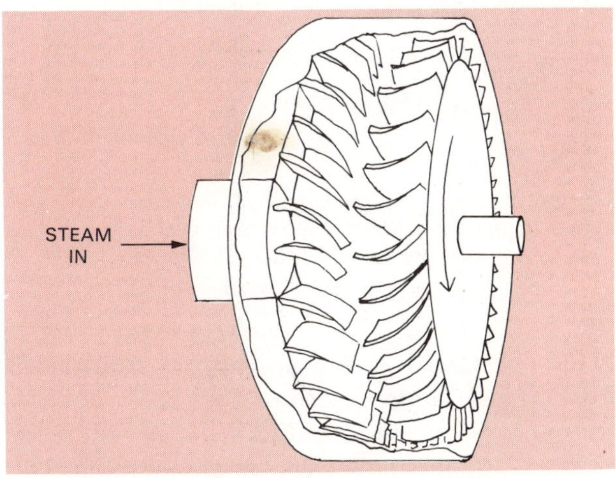

Fig. 2-22. Single element of a Parsons steam turbine. Steam had to pass through several sets of blades.

on the same shaft. As the steam leaves one set of blades it is redirected so it strikes the next set at an angle that causes great pressure even though the pressure of the steam has been reduced.

COMPOUND TURBINE

A third type is known as the COMPOUND turbine. Actually, it is a series of turbines. The steam passes through one after the other. Each is designed to make the best use of the steam as its pressure and temperature drop, Fig. 2-23. This design is used on very large turbines where the large blades of the reaction type would be impractically large.

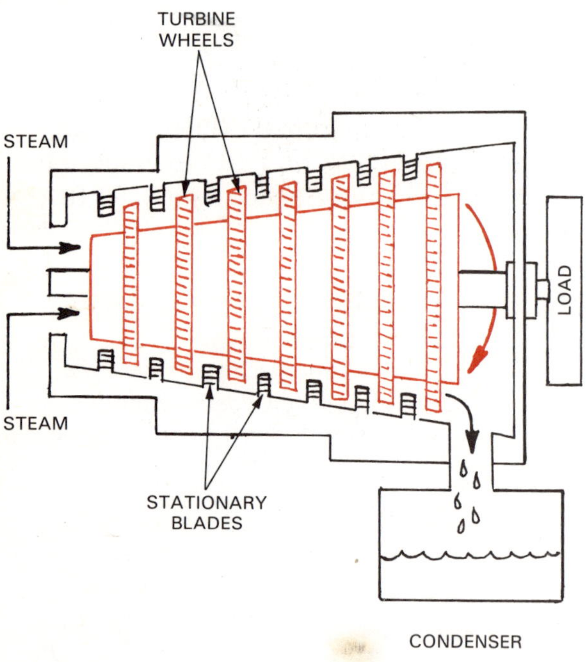

Fig. 2-23. Compound turbine is actually a series of turbines. It makes the best use of steam as temperature and pressure of the steam drop.

Improvements and modifications continued. These included Charles Curtis' velocity-compound turbine introduced in New York in 1897 and Zoelly's multistage impulse turbine. The latter was introduced in 1900. It had fewer stages than the Curtis turbine and was very simple.

INTERNAL COMBUSTION ENGINES

Combustion engines harness energy by capturing the force of rapidly burning fuels. Steam engines are called external combustion engines because the fuel is burned outside the engine. See Fig. 2-24. Internal combustion engines burn the fuel inside the engine, Fig. 2-25. The two popular types are the gasoline engine and the diesel engine.

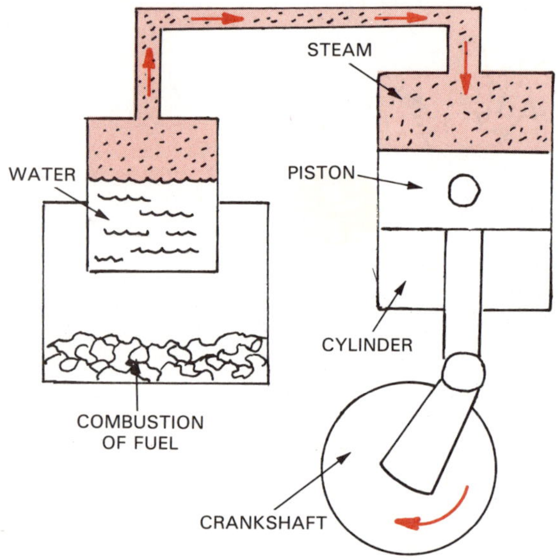

Fig. 2-24. Steam engines are external combustion engines because the fuel supplying the energy is burned outside the engine.

HOW THE INTERNAL COMBUSTION ENGINE DEVELOPED

Experimental models of four-stroke internal combustion engines were developed in both England and Europe around 1800. Fuel was a mixture of coal, gas, and air. About 60 years later, The French engineer, Etienne Lenoir, built a two-stroke engine that ran smoothly but did not develop very much power. Even so, several hundred were sold over the next few years.

Alphonse Beau de Rochas, another French inventor, developed the engineering principles that led to the development of the first successful four-

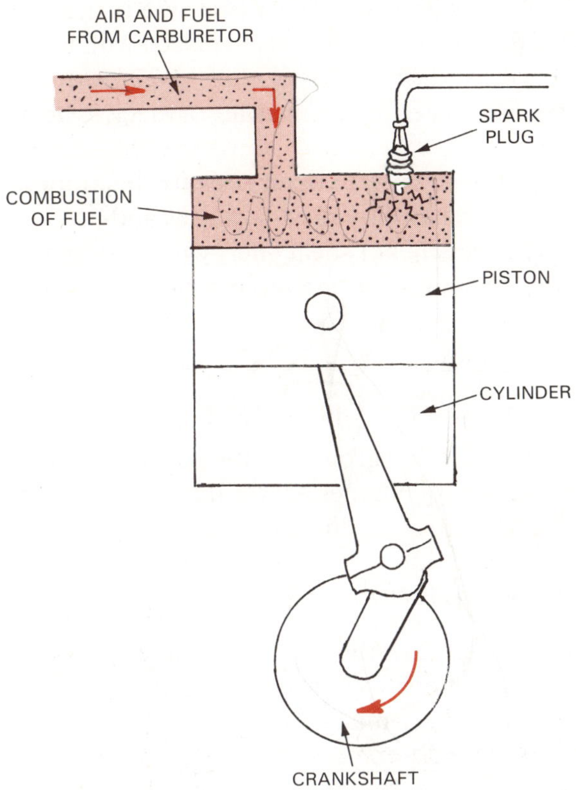

Fig. 2-25. Internal combustion engine is basically a free-moving piston being pushed up and down in a cylinder. Power is provided by burning fuel.

stroke engines. He listed the sequence of operations:

1. "Suction during the first outstroke to bring in the fuel-air mixture."
2. "Compression of the mixture during the following instroke."
3. "Ignition of the compressed charge. . .by means. . .of a spark from an induction coil, with the expansion of the combustion products and the performance of useful work during the next outstroke."
4. "Expulsion of the burned gases during the next instroke."

Fourteen years later, Nikolaus A. Otto, a German engineer, applied these principles in the first four-stroke engine. Even today, most internal combustion engines operate on these basic principles. It is often called the "Otto Cycle."

Otto's engine was manufactured in Germany and in the United States. Nearly 50,000 of the engines were sold between 1877 and 1894.

Improvements in the internal combustion continued. In 1855, Gottlieb Daimler, the German inventor and engineer, designed and built an engine that burned gasoline. Another German engineer, Rudolph Diesel, built the first diesel engine. It was designed to operate on low-grade fuels. See Fig. 2-26.

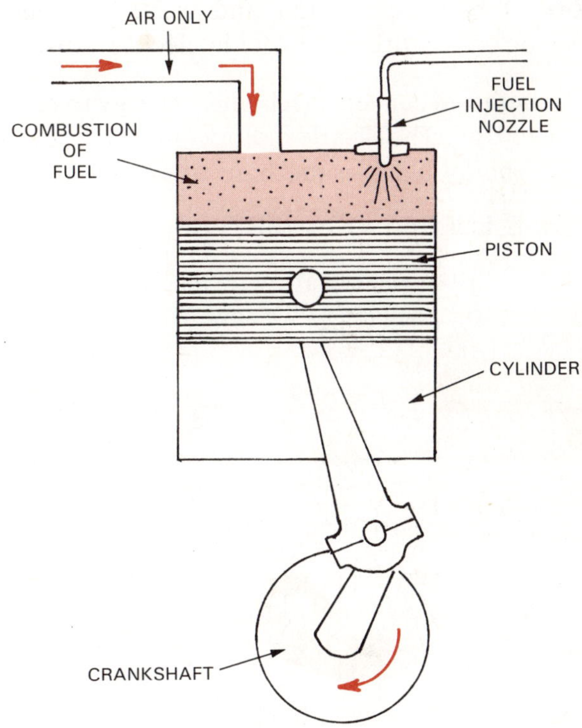

Fig. 2-26. Diesel engine burns low-grade fuel inside the cylinder.

HARNESSING IS CONVERSION

You have just studied the history of human effort to shift the burden of work from themselves to machines. First they were able to train beasts to do the work. Then they designed and adapted machines for this purpose.

One common element in these efforts was the use of some agent that could convert energy. One form of energy was usually converted to a different form so it could perform the task that needed doing. Thus, the devices or beasts that were harnessed to do the work came to be called ENERGY CONVERTERS and the process, ENERGY CONVERSION.

MODERN CONVERSION SYSTEMS

Many of the ancient conversion devices are still with us in improved designs. The gasoline and diesel engines, together with the steam turbine, supply most of the mechanical energy used in the world for transportation and electric power. Hydroelectric turbines provide electric power.

Generators convert mechanical energy to electrical energy. The electric motor converts the electrical energy back to mechanical energy.

OTHER ENERGY CONVERTERS

Use of fossil fuels to provide heat for cooking and comfort, began, as we saw, with campfires to heat caves and cook meat that was turned on a spit. The open fire soon was replaced with crude fireplaces and then stoves, Fig. 2-27. Modern homes are heated with furnaces or electric heat.

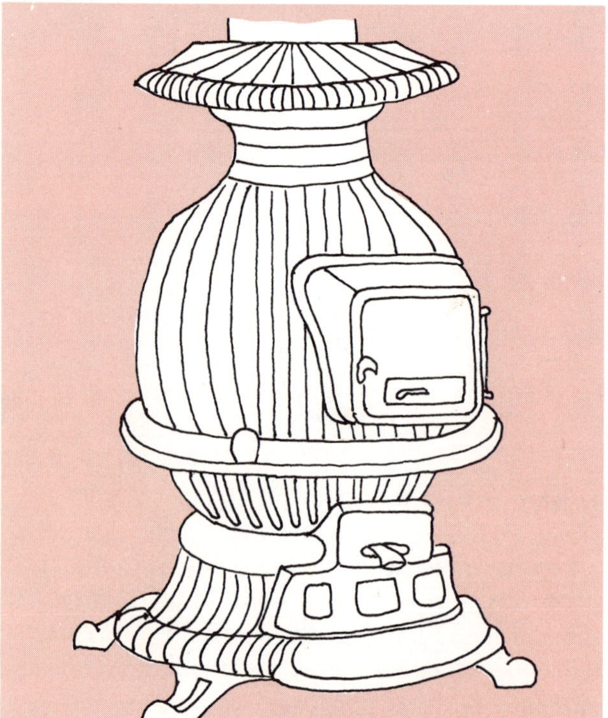

Fig. 2-27. The stove is a more efficient heating system than an open fire or a fireplace.

Likewise, the use of open fires for fashioning tools was replaced by blast furnaces and process heat. (Process heat is heat energy used to manufacture products of metal, plastic, and other materials.) Factories today use electric power to run machines.

In the 20th century, jet engines and rocket engines joined the growing list of prime movers. These powerful engines are designed for the aircraft and aerospace industry.

LOOKING AHEAD

At some time in the future we can expect that more efficient energy converters will be perfected. More than likely many of these will be devices that can convert energy directly from one form to another with few or no moving parts. Some of these inventions are already in the development stages. Because power systems and energy converters must always work together so that energy may be put to work, understanding them is important for an understanding of energy. For this reason, Chapter 3 will be devoted to power systems and Chapter 4 to modern energy conversion.

DO YOU KNOW THESE TERMS?

Compound turbine, diesel engine, generator, heat engine, impulse turbine, internal combustion engine, overshot water wheel, post mill, reaction turbine, smock mill, turbine, undershot water wheel, windmill.

SUGGESTED ACTIVITIES

1. Prepare a report on an invention of the last 100 years which harnessed energy or improved upon the usage of energy.
2. Construct a working model of a water wheel and demonstrate it to your class.

3. Build a model of an early windmill, working with metal or wood. Use descriptions and illustrations in the text to assist in the design. Explain its operation to your class.
4. Sketch a device which can convert steam into motion. Suggest how it might be built.

TEST YOUR KNOWLEDGE

1. Name two early forms of mechanical energy.
2. The __________ __________ was invented to make use of the energy in falling or moving water.
3. The __________ was the first invention that captured wind for pumping water and grinding grain.
4. The invention of the machine in No. 3 was credited to:
   a. The Dutch.
   b. Settler in America.
   c. The English.
   d. The Persians.
5. Describe the importance of the windmill to American farms.
6. The water wheel gave way to the __________ because the latter made better use of the energy of the falling water.
7. What is the meaning of the word "head" in relation to the energy of falling water?
8. Devices which can contain the energy of heat and make it do work are called __________ __________.
9. Make a sketch of Newcomen's steam engine.
10. The steam __________ uses the energy of steam and changes it directly to rotary motion.
11. Describe the difference between an external combustion engine and an internal combustion engine. Give examples of each.
12. When a device or machine is designed to change one form of energy to another which can be used to do work, it is called:
    a. A machine.
    b. An energy converter.

Space shuttle at launch. Several carefully engineered power systems make this space travel system possible. Can you suggest how many separate power systems are involved?
(NASA and Rockwell International)

3 POWER SYSTEMS

The information given in this chapter will enable you to:

- Define a power system.
- List four basic power systems.
- List and describe the role of each essential element of a power system.
- Define the following energy/power terms: work, power, torque, velocity, head, pressure, Btu, efficiency, and quad.

In Chapters 1 and 2 you learned about the nature and sources of energy and how they can be harnessed to do work. In this chapter you will study power systems and their relationships. You will also learn how energy is measured.

POWER SYSTEMS

A power system is important to the harnessing of energy. It converts, transmits, and controls energy in such a way that the energy performs useful work. A lawn mower, Fig. 3-1, is a simple power system. The mower is small and has few parts. On the other hand, some power systems are very large. They take up a great deal of space and have many parts. An electric power plant would be a good example of a large power system. See Fig. 3-2.

Fig. 3-1. A simple power system. Lawn mower converts a form of energy to do useful work.

BASIC POWER SYSTEM

There are four basic power systems:

1. Electrical systems.
2. Mechanical systems.
3. Fluid systems.
4. Heat systems.

No matter how small or big, all power systems have certain basic elements that have the same function. The elements are:

1. A source of energy. Often, this is a fuel of some type, but it could be another form of energy, such as water power or wind.
2. A conversion method. There must be a way of

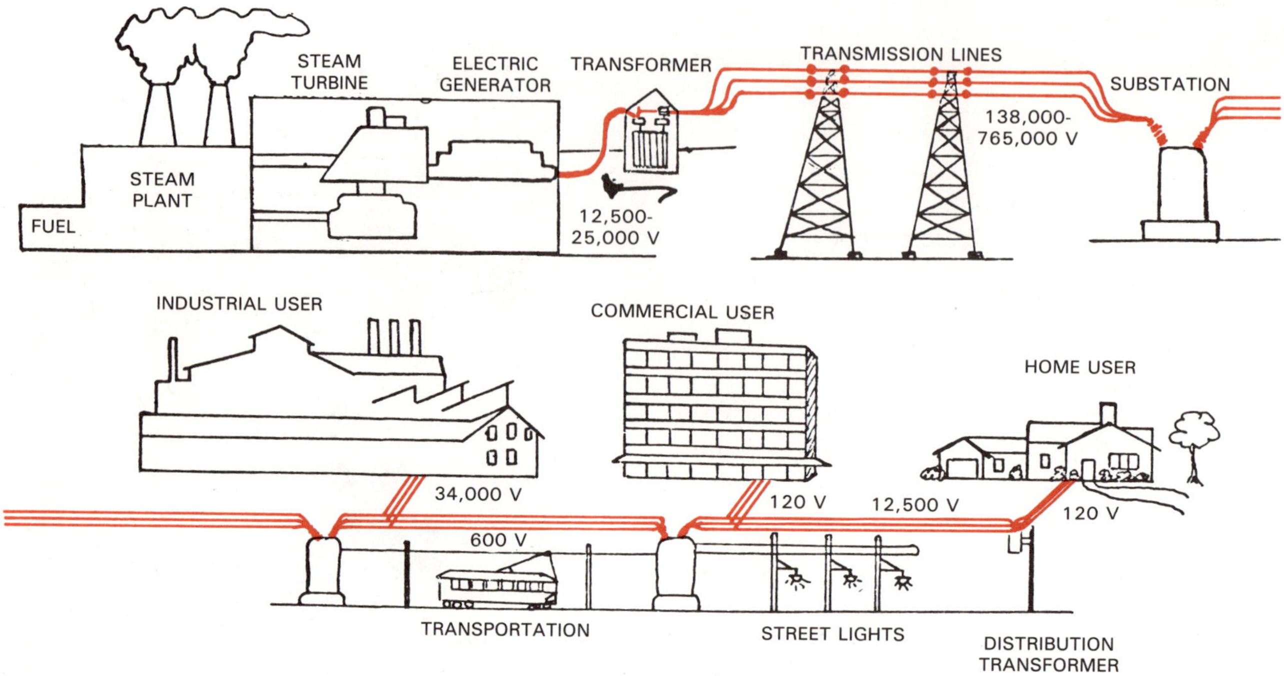

Fig. 3-2. Electric power stations are a complex power system. They are large, take up much space, and have many parts.

COMPARING POWER SYSTEMS

ELEMENT	LAWN MOWER	POWER PLANT	EXPLANATION
Energy Source	Fuel and tank to contain it.	Piles of coal, natural gas pipeline, nuclear rods, force of falling water.	Gasoline is fed continuously to engine; power plants are supplied fuel on continuous basis—it is available night and day.
Conversion Method	Engine burns fuel and uses expanding gases created by combustion.	Fuels and fission create steam to rotate turbine. Falling water also spins turbine which operates electrical generator.	Gasoline engine converts chemical energy into heat energy so engine can create spinning motion (kinetic energy). Power station uses chemical, nuclear, or kinetic energy to create electrical energy.
Transmission Path	Piston and crankshaft	Electric power lines	Kinetic or mechanical energy travels from expanding gases to piston, to crankshaft to cutting blade. Electrical energy travels through wires to run electric motors and lights.
Control System	Throttle lever handle for steering	Switches and transformers	Mower's speed and direction are controlled. Transformers control amount of power coming through power lines. Switches turn electric power on and off.
Measuring Devices	Oil pressure gauge Gasoline tank gauge	Electric meters	In both systems, gauges tell operator operating condition of power system. Fuel, current, and voltage can be measured.
Load	Cutting blade	Electric motors Electric lights Electric heating	Load is the work done by systems; mower cuts grass; electric power runs machine, provides light.

Fig. 3-3. Comparison of large and small power systems. Each has all the requirements of a system.

changing the energy so it can do work.

3. A transmission path. This is a way of moving the energy to the load where it can do the work from which it is designed.
4. A control system. Like a leash on a dog, a control keeps the energy in check so that it does its work and no more.
5. Measuring devices. These may be meters or indicators. Their purpose is to give the user information about the power system that is needed to keep it working properly.
6. An output or load.
7. Losses. Not all the energy released is used. If you compare a lawn mower with a power plant you will see that both meet all the characteristics of a power system. See Fig. 3-3.

ENERGY SOURCE

From previous chapters you already know that an energy source is a supply of energy which can be tapped when there is work to do. Some energy sources are CONSTANT; others are called CHANGEABLE sources because availability is not certain. It is subject to change.

CONSTANT ENERGY SOURCE

A constant energy source is one that is always on hand and ready for use when it is needed. Nearly all constant energy sources are the result of storage. The storage may have been accidental or purposeful. See Figs. 3-4 and 3-5. Fuels such as coal, crude oil, and natural gas resulted from natural or accidental storage. Natural processes caused them to be trapped in beds and pools. This accidental storage makes them a constant source.

Water stored behind a dam to provide a continuous source of mechanical energy is purposeful storage. Electricity, however, is one constant energy source which is hard to store. Small quantities can be captured in batteries but, usually, electricity must be generated as it is used.

Most power systems are designed to use a constant source of energy. Systems built around the use of electric power, natural gas, and petroleum are the most common. Electric power operates lights, electric motors, and appliances, Fig. 3-6. Sometimes it is used for heating and cooling.

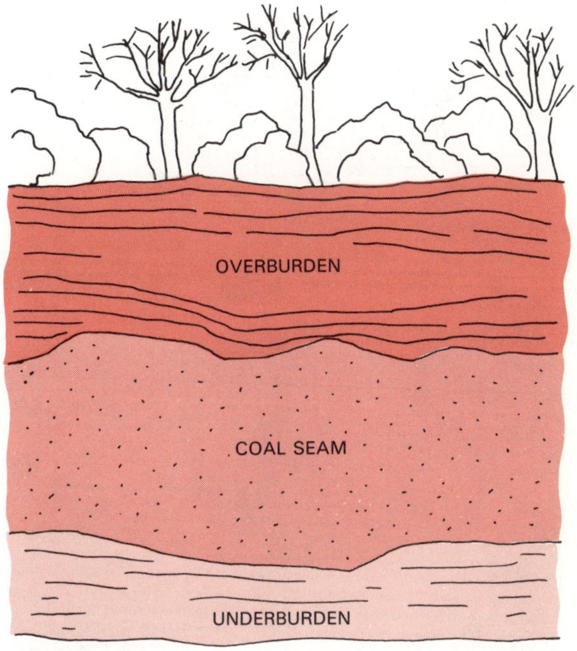

Fig. 3-4. Accidental storage of fuel source. Coal bed formed by nature lies stored until it is mined.

Fig. 3-5. Gasoline is stored in a container until it is needed.

Fig. 3-6. Electricity is a constant source of power. It can be used to produce light, heat, or mechanical energy. Top. Modern kitchen stove uses electricity for heat. Bottom. Dishwasher has electric motor to power water pump. (Westinghouse)

Natural gas provides a fuel source for cooking, heating homes, office buildings, and factories. Petroleum provides gasoline to fuel internal combustion engines.

CHANGEABLE SOURCES

Some energy sources are available only some of the time. These types are called changeable sources.

The wind is a good example of such a source. It can run many machines cheaply. However, you cannot depend on it day in and day out. Sometimes there is no wind.

Solar energy is also a changeable source. The sun delivers large quantities of energy to the earth. It can be made to do a great deal of work. In fact, it helps to produce many other sources of energy. Still, solar radiation cannot be depended upon entirely. On cloudy days there is little sunlight and at night none at all. At best, it only shines about 12 hours out of the day in most parts of the world, Fig. 3-7.

Tides and ocean waves are another form of changeable energy. Like the wind and sun, they are not always working.

As constant sources are used up, changeable sources will become more and more important. Scientists are, therefore, working on ways to convert and store the energy from these sources so that the stored energy will work while the source is not available.

CONVERSION OF ENERGY

As we know from Chapter 2, energy may need to be converted from one form to another before it can do work. Some power systems may make more than one conversion. Lawn mowers and automobiles cannot use fuel directly. Fuel must be converted to hot gases. Fig. 3-8 shows transfer of energy from expanding hot gases through a piston to a crank.

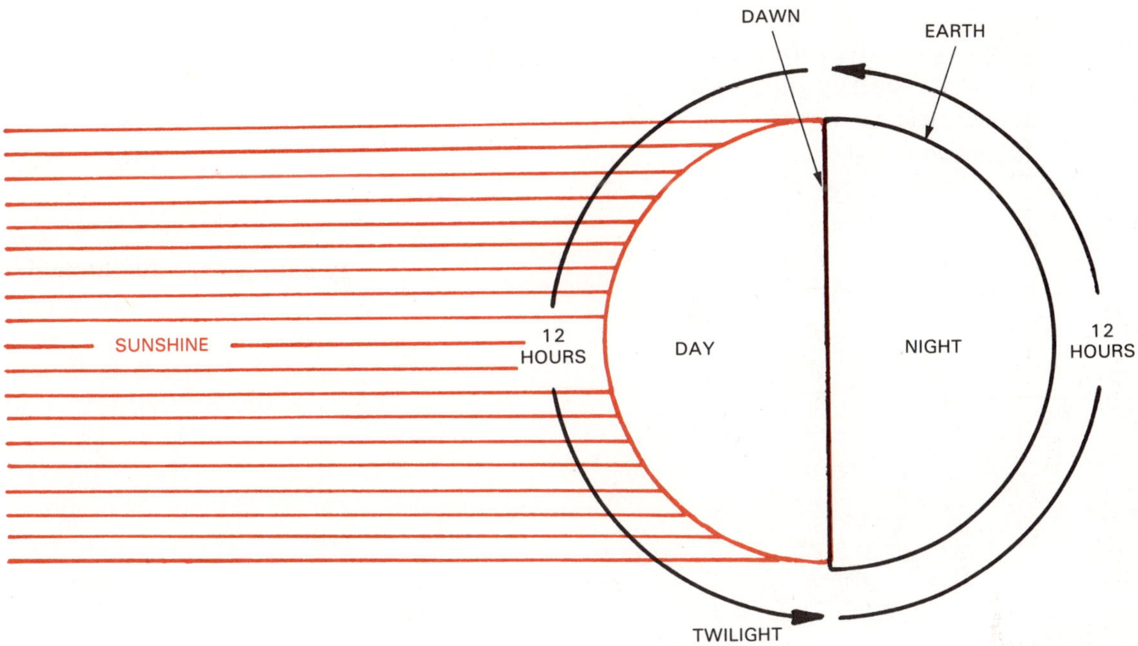

Fig. 3-7. Because of the earth's rotation, sunlight is available only about half of every 24 hour period. The drawing shows why this is so.

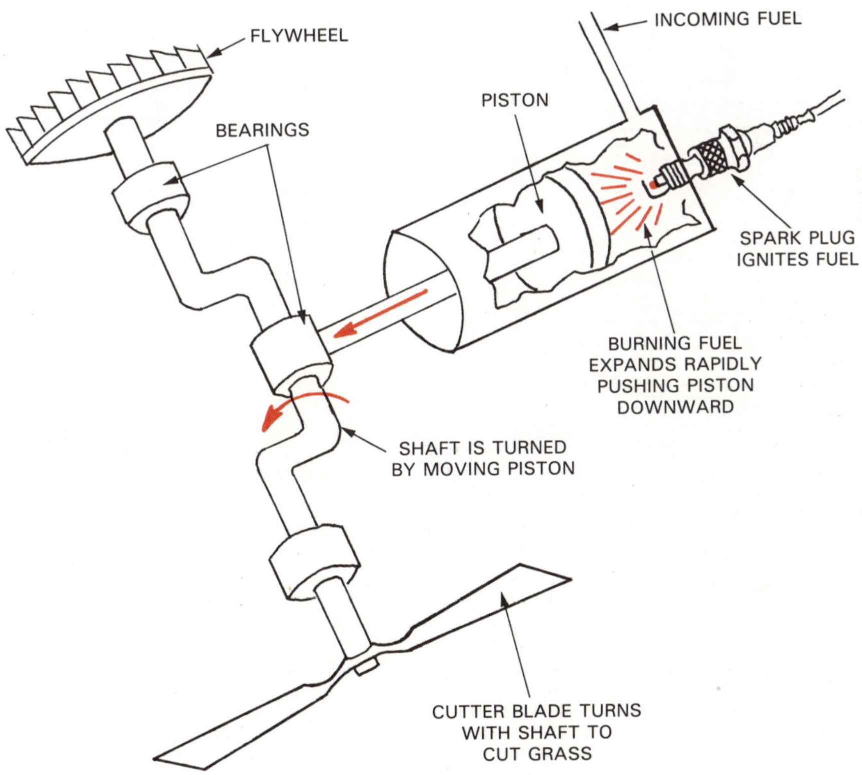

Fig. 3-8. Lawn mower engine converts fuel (chemical energy) to hot gases which create mechanical force to turn the mower blade so it will cut grass.

A coal or oil-fired power generating station makes several energy conversions. First it converts stored chemical energy into heat energy. Then the heat energy is converted to mechanical energy. Finally, mechanical energy is changed into electrical energy. See Fig. 3-9.

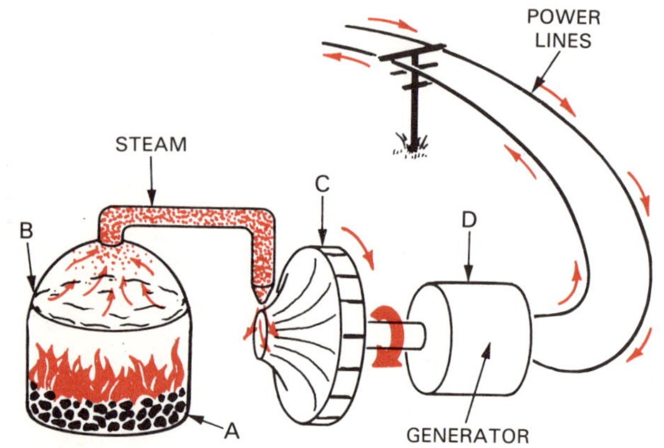

Fig. 3-9. Converting the energy of coal to electricity takes several steps or conversions. A—Coal is burned to produce heat energy. B—Heat raises water temperature until it turns to steam. C—Heat energy in steam is converted to mechanical energy as it spins the turbine. D—Energy of whirling turbine drives generator. The generator converts mechanical energy to electrical energy.

TRANSMISSION OF ENERGY

The third element of a power system, transmission, is the act of moving power from the source to the load. Three basic methods are employed for this purpose:

1. Fluid linkage. Fluids or gases are forced through pipes or tubes to the load. The fluids are driven by the force of the converted energy. The brake system on an automobile operates this way. Pressure from the driver's foot or a pump driven by the engine forces brake fluid through tubes to push the brake shoe against the brake drum. See Fig. 3-10. Hoists are another example of fluids transmitting power from one point to another, Fig. 3-11.

   Industry often uses compressed air to move switches and operate controls on assembly lines. The air is under pressure and moves through tubes and channels in metal parts called gates.
2. Electrical linkage. Moving through wires, electric current carries the energy of the whirling turbine to homes and factories. There it can be used for lighting and to operate appliances. The wires, known as transmission lines, Fig. 3-12, allow the generating station to be placed near its source of energy.
3. Direct mechanical linkage. A shaft or a belt

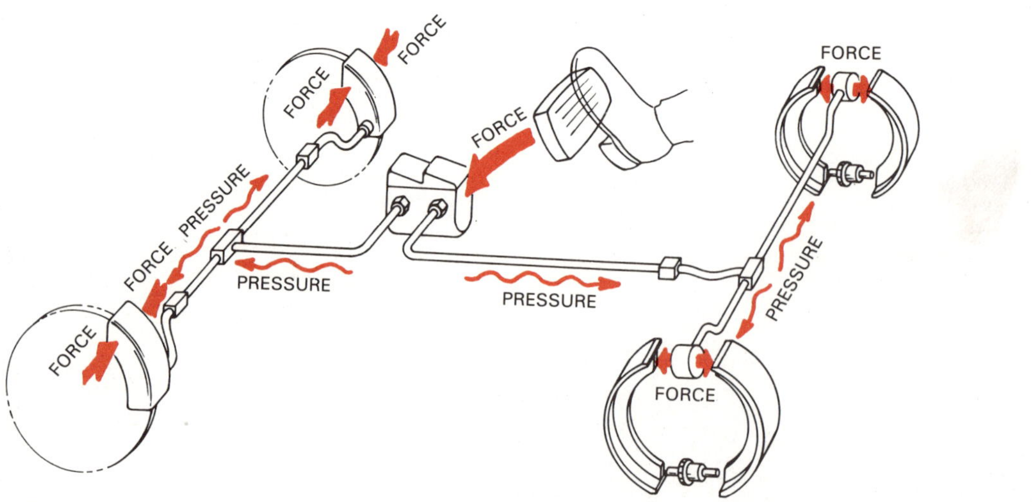

Fig. 3-10. Brakes on an automobile use fluids to transmit stopping power. The foot on the brake pedal is the power applied. The force is moved from the pedal to all four wheels by fluid. (EIS Div., Parker-Hannifin)

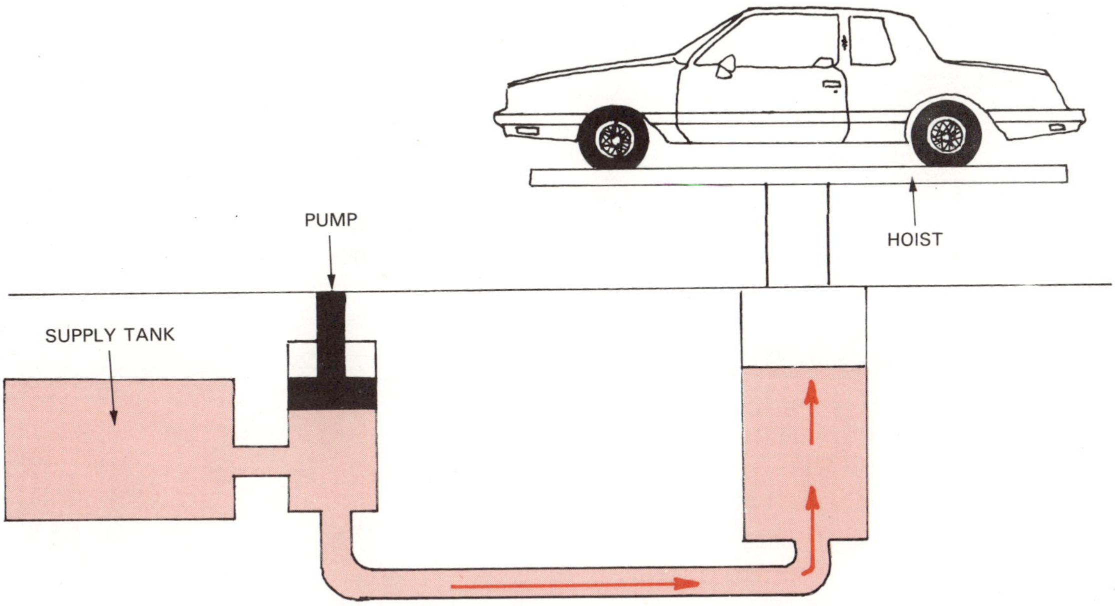

Fig. 3-11. Power to lift an automobile into the air is transmitted to the hoist by the mechanical action of a pump. Fluids are forced through pipes and cylinders. The fluid, in turn, forces a piston to move up a cylinder, lifting the automobile.

Fig. 3-12. Electric power can be transmitted long distances by power lines such as these.

moves power from one spot to another. The driveshaft of an automobile takes the rotational motion of the engine and carries it to the wheels. Ropes, chains, or cables also can be used to transmit mechanical energy, Fig. 3-13.

Fig. 3-13. Transmission system, Bicycle chain transfers power from pedal to back wheel. Pumping action of rider (arrow) is changed to rotating action by pedals and chain.

CONTROL OF ENERGY

Controls in power systems do three things:
1. Turn power on and off.
2. Regulate speed or cause some other change in the way power is applied.
3. Control direction.

Switches control the electric lights and motors in your home and school. See Fig. 3-14. The automobile has a switch for turning on electric current for starting and operating the engine. Other switches may be used to control lights, windows, locks, and windshield wipers.

Electric power plants have switches and transformers which control the electric current going out to customers, Fig. 3-15.

Fig. 3-14. Switches control electric lights.

TYPES OF CONTROLS

Controls can be divided into two basic types:
1. Total control. A switch is of this type. When it operates, power is either on or off. The switch on an automobile turns current from the battery on or off.

Fig. 3-15. Power plants require many kinds of controls. Top. Main control panel in power plant turns power on and off. Bottom. Power plant switchyard with huge circuit breakers. (Power Authority of New York)

2. Partial control. Controls of this type regulate direction and speed.

As important as "off and on" control is, can you imagine driving an automobile using only the ignition switch? You actually need several partial controls to drive an automobile. The shift lever controls transfer of power from the engine to the drive wheels. The steering wheel keeps the vehicle on the road. The accelerator produces changes in speed. A brake pedal slows or stops the vehicle.

CONTROL OF UNWANTED OUTPUT

Some of the output of power systems cannot be used. Sometimes it must be regulated to make the system safe or to protect it from damage. Outputs which must be controlled include:

1. Friction. This is the resistance between two bodies rubbing against each other, Fig. 3-16.

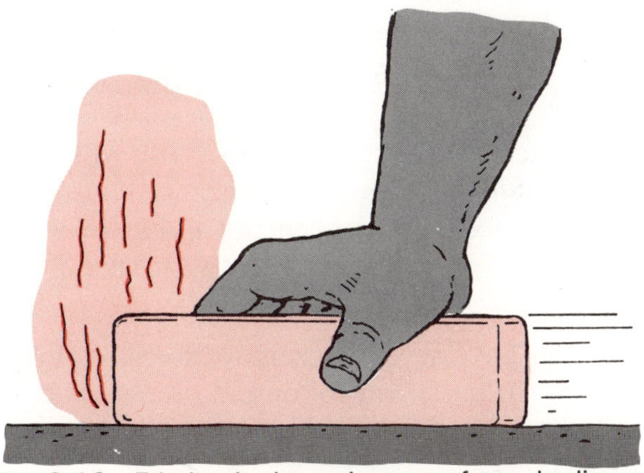

Fig. 3-16. Friction is the resistance of two bodies or surfaces rubbing against each other. (EIS Div. Parker-Hannifin Corp.)

2. Waste heat. Electric power plants and factories cannot use all the heat they generate. Part of it must be carried away and eventually released to the air. See Fig. 3-17.

MEASUREMENT

Measurement is important in a power system for more than one reason:

Fig. 3-17. Cooling towers are used by power plants to carry away waste heat.

1. It determines the amount of energy in the system.
2. It may determine how much work the system is doing.
3. It may also determine how fast work is being done.

METERS AND GAUGES

It is necessary to know about the operating condition of parts of a power system. Indicators such as gauges give this information. The indicators for some systems are very simple. Others are complicated.

A simple gauge to show how much fuel remains in the fuel tank may be all that is needed on a lawn mower. An automobile will have measuring instruments such as:

1. Fuel gauge.
2. Ammeter for measuring current output of the alternator.
3. Heat indicator for the engine coolant temperature.

4. A speed indicator to gauge how fast the automobile is traveling.
5. Oil pressure gauge to indicate whether the engine is getting proper lubrication.

Fig. 3-18 shows the dashboard of an automobile. The operator needs the gauges to be sure the vehicle is operating properly.

Power stations have many gauges to show how much electrical pressure (voltage) and how much electricity (current) is moving through power lines. Special meters measure how much electricity is being used by customers.

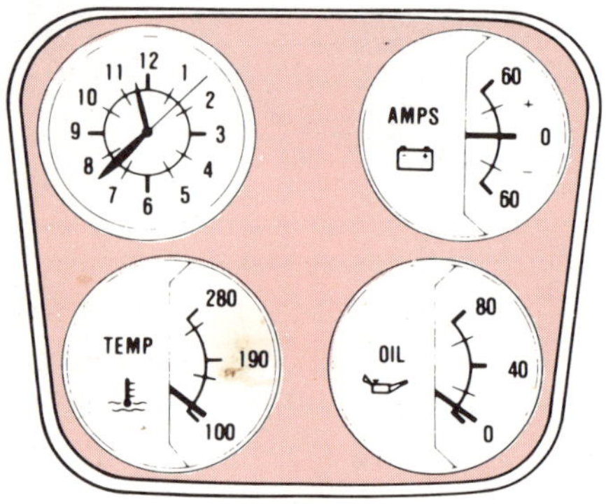

Fig. 3-18. Instruments are necessary for monitoring the operation of a power system. Instrument panel of automobile tells the operator the operating condition of the charging, cooling, lubrication, fuel supply, and, sometimes, the braking system. (AMC)

LOAD

The load of a power system is the useful output. A bicycle is a good example of a simple system. The energy source is the mechanical force exerted on the pedals. The crank changes the pumping motion to circular or rotating motion. The chain transmits the motion to the rear wheel. The weight of the person and the weight of the bicycle are the load.

LOSSES

Losses are an unwanted portion of power systems. As explained earlier, it is an unwanted element of energy usage. You will learn about the efficiency of power systems later when you study conservation methods. Power losses are one of the reasons why efficiency is low in some conversions of energy and high in others.

Two important elements mentioned earlier contribute to loss of energy. They are friction and waste heat. Extra energy must be used to overcome the tendency of parts moving past each other to stick. Even with the use of lubricants such as oil and wax, friction cannot be entirely stopped.

Waste heat is simply a part of the energy that has lost its ability to do any more work. It must be passed off to the air, water, or other matter around it. In the automobile, for example, the heat is passed from the engine to the water in the cooling system. The radiator and the heater pass the heat off to the atmosphere.

POWER/ENERGY TERMS AND MEASUREMENT

As you study energy and power you will begin to use many new terms. These terms are used whenever people discuss energy and its relationship to power, work, and energy sources.

Loads have several terms to describe them. Work is the basic unit for output or system load.

Work is force times the distance through which the force acts. A force is anything that causes an object to move.

It is easy to see some types of work being done. For example, when the rider pumps the pedals, a bicycle moves forward, Fig. 3-19. It is not as easy to see electricity doing work. Current travels unseen through wire conductors. While we cannot see it move, we can see the effect. We see it in the lighted electric bulb and the rotating of the electric motor. These effects tell us that the electricity is at work.

Fig. 3-19. Load is the useful output of any power system. It is what is moved or changed. Bicycle's load is the weight of the rider and the weight of the bicycle itself.

MEASURING WORK

Work is measured in foot-pounds for conventional U.S. measure. A foot-pound is the amount of force necessary to move a 1 lb. load a distance of one foot.

The mathematical formula is:

$$\text{WORK} = \text{DISTANCE} \times \text{FORCE or}$$
$$W = D \times F$$

To see how this formula works, suppose that the weight of the rider of a bicycle is 150 lb. How much work is done if the bicycle is pedaled 500 ft.?

$W = D \times F$
$W = 500 \times 150$
$W = 75{,}000$ ft.-lb.

In SI metric, which is being adopted by most countries, work is measured in joules. The unit is named after James Prescott Joule, an English scientist. The joule (J) is the work done when a force of 1 newton moves an object 1 meter. The formula is:

$$\text{Work} = \text{Distance (in meters)} \times \text{Force (in newtons)}$$

One newton is the force used to accelerate 1 kilogram of mass to 1 meter per second squared.

How many joules of work are done when a force of 300 newtons (N) is applied for a distance of 150 meters?

Using the formula:

$W = 150 \times 300 = 45{,}000$ joules

POWER

Work and power are often confused. Power is used when work is meant and vice versa. It is easier to separate them if you only remember that work does not concern itself with how long it takes. Only distance and force matter.

Power is the amount of work done in a given amount of time.

To measure power we use the following formula:

$$\text{Power} = \frac{\text{work}}{\text{time}} \text{ or } P = \frac{w}{t}$$

It can also be written:

$$P = \frac{D \times F}{t}$$

If a rider and bicycle weighing 150 lb. travel 500 ft. in 20 seconds, how much power is developed?

$$P = \frac{D \times F}{t}$$
$$P = \frac{500 \times 150}{20} = \frac{75{,}000}{20}$$
$$= 3750 \text{ lb.-ft. per sec.}$$

HORSEPOWER

In U.S. conventional measure, the standard unit of power is the horsepower, Fig. 3-20.

One horsepower (hp) is the energy needed to lift 33,000 lb. 1 foot in 1 minute.

If 200 lb. are lifted 165 ft. in 1 minute, how much horsepower is developed?

To find the answer:

1. Find out how much work was done.
   Work = D × F
   Work = 165 ft. × 200 lb. = 33,000 ft.-lb.
2. Now use the formula to find horsepower.
   $$\text{Hp} = \frac{\text{work}}{\text{time in minutes} \times 33{,}000} = \frac{33{,}000}{1 \times 33{,}000} = 1$$
3. The answer is 1 hp.

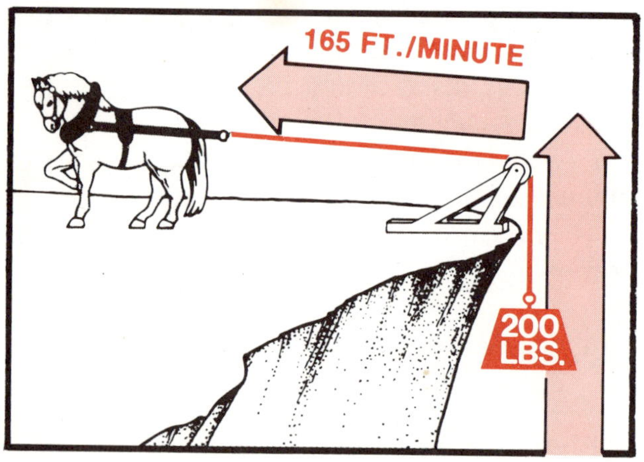

Fig. 3-20. Horsepower is the rate of doing work.

In SI metric the unit of power is the watt (W). One watt is the power needed to produce work at the rate of 1 joule (J) per second. A joule is the force needed to move 1 kg of mass 1 meter. Since the watt is a small unit, power is usually measured in kilowatts which is 1000 watts. A kilowatt equals about 1.34 hp. The formula for determining power is the same:

Watts = work (in joules or kilojoules) ÷ time

$$\text{Power (in watts)} = \frac{\text{distance (in meters)} \times \text{force (in newtons)}}{\text{time}}$$

or:

$$\text{Power} = \frac{\text{work (in joules)}}{\text{time}}$$

If a bicycle and rider with a mass of 70 kg move 300 meters in 20 seconds, how much power is developed?

$$\text{Power} = \frac{\text{distance} \times \text{force}}{\text{time}} = \frac{300 \times 70}{20} = 1050 \text{ W}$$

TORQUE

Power has just been described in terms of a force being exerted in one direction. However, not all motion is linear (in a straight line). An engine, for example, or a crank both produce a TURNING or ROTATIONAL force, Fig. 3-21. This twisting force is called TORQUE.

Torque is measured the same way as work, by multiplying force (in pounds) times distance (in

Fig. 3-21. Linear motion is converted to circular motion (torque) as the rider presses down on the pedal. The wheels roll along the ground again converting the torque to linear motion.

feet). There is one slight difference. Work is measured distance times force: *torque is always measured force times distance.* The answer is always given in pound-foot (lb.-ft.) units.

If a force of 50 lb. is applied to move a load 10 ft., how much torque is developed?

Torque = Force times Distance (F × D)

T = F × D

T = 50 lb. × 10 ft.

T = 500 lb.-ft.

In SI metric, torque is called MOMENTS OF FORCE. The unit of measure is the newton meter (N·m). The formula is the same one used for U.S. conventional measure:

$$\text{Moments of Force} = \text{force} \times \text{distance}$$

An engine uses a force of 10 newtons to move a load 2 meters. Compute the Moments of Force.

Moments of Force = 10 × 2

Moments of Force = 20 N·m

VELOCITY

Velocity is the distance traveled divided by the time of travel in one continuous direction. It is used to measure speeds of things moved by some form of energy.

For example, gasoline is used to propel an automobile at velocities that are measured in miles or kilometers per hour. Measurement can also be in meters per minute, feet per second, or feet per minute.

The formula for velocity measurement is:

$$\text{Velocity} = \frac{\text{distance in the same direction}}{\text{time}}$$

ACCELERATION

Acceleration is related to velocity. It is a change in velocity over a period of time. It is expressed in feet per minute, miles per hour, meters per minute, and kilometers per hour. The formula for determining acceleration is:

$$\text{Acceleration} = \frac{\text{change in velocity}}{\text{time}}$$

An important example of acceleration has to do with gravity. A body falling toward the earth will always be accelerated at 32.8 ft. (10 m) per second. That is, its velocity will increase 32.8 ft. per second *every second.*

CALCULATING POTENTIAL ENERGY

An important aspect of acceleration for engineers is calculating the energy (kinetic) which exists in an object at rest (but which might fall or could be made to fall). The energy resulting from falling matter is related to acceleration. The formula is:

p = mgh where:

p = potential energy.

m = mass or weight.

g = rate of free fall acceleration (32.8 ft. per sec.).

h = the distance the object or body falls.

To see how it works, try this problem:

A 10 lb. rock sits at the edge of a cliff. It is 100 feet to the bottom of the cliff. What is the gravitational potential energy of the rock?

Use the formula p = mgh

$$p = 10 \times 32.8 \times 100 = 32{,}800 \text{ foot-pounds}$$

HEAD

Head is related to potential energy of gravity. It is the difference between the level of the water inlet at a dam or reservoir and the outlets. See Fig. 3-22.

EFFICIENCY

Efficiency is a ratio (percentage) between the amount of energy put into a power system mea-

Fig. 3-22. Head is the difference in height between the inlet and outlet of water stored behind a dam or in a tower. The greater the head, the greater the force at the outlet.

sured against the amount of work the system does. The efficiency of an engine can be computed using the formula:

$$\text{Efficiency} = \frac{\text{energy out}}{\text{energy in}} \times 100 \text{ Percent}$$

PRESSURE

Pressure is the force exerted by a contained gas or liquid. Its formula is:

$$\text{Pressure} = \frac{\text{force}}{\text{area}}$$

Pressure is an important force in energy systems. Its applications include:

1. Systems which depend upon air for operation: Compressed air may be used to operate some types of brakes; to inflate tires, Fig. 3-23; and to control machines in manufacturing operations.
2. Systems which operate with fluids under pressure:
   a. Steam for driving turbines for generation of electricity.
   b. Liquids under pressure in hydraulic systems where they lift heavy loads and exert extremely high forces. An example of such a system is the hydraulic brake system on an automobile. Refer again to Fig. 3-10.

The U.S. conventional unit of measure for pressure is pounds per square inch. It is abbreviated "psi." The metric unit of pressure is the pascal (Pa). A pascal is a force of 1 newton per square meter.

BRITISH THERMAL UNIT (Btu)

When heat energy moves from one body to another you can measure it. The British thermal

Fig. 3-23. Air pressure is an important force in our energy conscious world.

unit (Btu) is the U.S. conventional unit of heat measurement. It is the amount of thermal (heat) energy required to raise the temperature of 1 lb. (about a pint) of water 1 degree Fahrenheit. The Btu is used in:

1. Calculating the amount of heat in fuels.
2. Determining the heat output of mechanisms such as furnaces and boilers.
3. Calculating the heat requirements for spaces to be heated.
4. Measuring the cooling capacity of an air conditioning unit.
5. Measuring heat loss in a home.

   A chart for converting various quantities to Btus is shown in Fig. 3-24.
6. Measuring solar radiation. This is expressed as Btu per sq. ft. per hour.

ENERGY CONVERSION TABLE

To convert from	to	Multiply by
kilowatt-hour	Btu	3412.8
1 ton bituminous coal	Btu	26,200,000
1 bbl. crude oil	Btu	5,600,000
1 bbl. residual oil	Btu	6,290,000
1 gal. gasoline	Btu	125,000
1 gal. #2 fuel oil	Btu	138,800
1 cu. ft. natural gas	Btu	1031
1 mcf natural gas	Btu	1,031,000
1 therm natural gas	Btu	100,000
1 Btu	kWh	0.000293

Fig. 3-24. Factors for converting various energy units to British thermal units.

BARREL

The barrel is a unit of measure used in the petroleum industry. It is equal to 42 gallons. The term is left over from the early days of the oil industry when crude oil was shipped from the oil fields to refineries in wooden barrels, Fig. 3-25. Since then, railroad tank cars, huge seagoing tankers, oil barges, and pipelines have replaced the barrel. The term is used throughout the industry particularly in discussing resources, supply, and demand.

Fig. 3-25. Wooden barrels, once used to ship oil, are long gone, but oil is still measured by the barrel.

QUAD

A quad is a measurement of energy consumption equal to 10^{15} Btu. This amounts to 1 quadrillion Btu. This is about the same amount of energy released by the burning of a half million barrels of oil daily for a year.

DO YOU KNOW THESE TERMS?

Acceleration, ammeter, barrel, Btu, changeable energy source, constant energy source, efficiency, foot-pound, head, horsepower, joule, moments of force, pascal, power, pressure, quad, torque, velocity, watt, work.

SUGGESTED ACTIVITIES

1. Visit a power station where electricity is generated. Ask about job opportunities and the kinds of work done there. Take part in a class discussion on what you heard and observed.
2. Prepare a poster showing the different parts of an electric power system. Label each as to

whether it is: energy source, conversion, transmission, control, measurement, or load.

3. Secure a photograph of a small, complete power system from a magazine or old automotive shop manual. Label different visible parts of the system according to the part's function in the power system.
4. Draw up a list of all the different kinds of power systems in the community where you live. List the source of energy for each.
5. Construct a device which is capable of moving about with an energy source carried with the device.

TEST YOUR KNOWLEDGE

1. Define the word "power system."
2. Name the six basic elements or parts of a power system.
3. A ____________ energy source is one that is always at hand when needed.
4. A changeable energy source is one which:
   a. Can readily be changed from one form of energy to another.
   b. Is available some of the time but not all of the time.
   c. Neither of the above.
5. List the conversions made by a power station in producing energy.
6. What three things do controls do in power systems?
7. Friction is the ____________ between two bodies rubbing against each other.
8. A gas gauge is part of ____________ in a power system.
9. Work is measured in ____________ ____________ in the U.S. conventional measuring system.
10. The ____________ is the metric term for measurement of work.
11. What is torque? What term is used in its measurement?
12. Velocity is the distance traveled divided by the ____________ of travel.

4 ENERGY CONVERTERS

The information given in this chapter will enable you to:

- ☐ Define energy conversion.
- ☐ Describe what happens in energy conversion.
- ☐ Explain the difference between indirect and direct energy converters.
- ☐ List and describe several indirect energy converters.
- ☐ List and describe several direct energy converters.

While you were studying how energy is harnessed you became familiar with the term ENERGY CONVERSION. It means the changing of one form of energy to another form. This conversion is an important step in any use of energy. It accomplishes two things: it unlocks the energy and/or changes it into a form which is appropriate for the job to be done. Sometimes a series of conversions are made before the energy is finally used, Fig. 4-1.

TYPES OF CONVERSION

There are many different ways of converting energy. It is possible to separate the converters into different types.

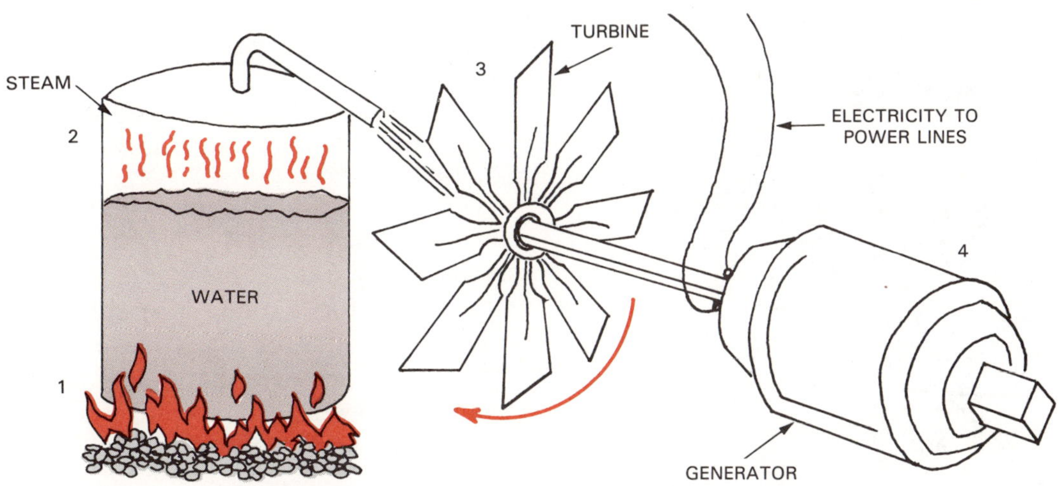

Fig. 4-1. Simplified diagram of an electrical power plant. Many conversions take place before the electric power is produced. First, chemical energy is turned to heat energy; then the heat energy creates steam. The mechanical energy of the steam pressure drives the turbine. Finally the mechanical energy is converted to electrical energy by the generator.

One way is to group them by the number of conversions that must take place before the energy can be used. If there are several conversions before the energy is in the right form to do work, the method is called INDIRECT CONVERSION. But, if the energy is used after one step, the method is called DIRECT CONVERSION.

INDIRECT CONVERSION METHODS

Because our society has become so dependent upon fossil fuels, most converters, up to now, have been designed to harness the energy of hydrocarbons (FOSSIL FUELS). The converters use combustion to produce heat from substances such as gasoline, coal, natural gas, and other materials containing carbon. For this reason they are called combustion heat converters or engines.

INTERNAL COMBUSTION HEAT ENGINES

One type of combustion engine, Fig. 4-2, is designed to vaporize and burn a mixture of air and gasoline inside a closed chamber called a cylinder. It changes chemical energy to heat energy and then converts the heat energy to mechanical energy. Being similar to the steam engine discussed in Chapter 2, it consists of one or more cylinders within a heavy metal casting. The top of the cylinder is sealed except for two small openings covered by valves. One valve controls the intake of fuel. The other opens to get rid of exhaust (spent fuel).

Fig. 4-2. An automotive engine is a heat engine. It burns fuel inside its cylinders to provide power for the automobile. (Ford Motor Co.)

Each cylinder has a piston which is fastened to a crank. The pistons are free to move up and down inside the cylinders.

A piston on the downward stroke draws in an explosive vapor (mixture of fuel and air). The mixture is compressed on the upward stroke. As the piston reaches the top of its stroke, a spark ignites the vapor. It burns rapidly causing the gases to expand with explosive force. The piston is driven downward in a power stroke.

On the upward stroke, the spent mixture is pushed out of the cylinder through one of the valves at the top. When the piston reaches the top, the exhaust valve closes and the intake valve opens once more. The cycle, consisting of intake, compression, power, and exhaust, begins again.

This is known as the four-stroke cycle engine. See Fig. 4-3. Other designs require only two strokes to complete the combustion cycle. They are known as two-stroke cycle engines. See Fig. 4-4.

Diesel engines, Fig. 4-5, like the gasoline engine, have cylinders and pistons and burn a liquid fuel.

They also may be either two or four-stroke cycle. The major differences are:

1. The air and fuel are mixed inside the cylinder.
2. The diesel fuel is not as explosive as gasoline.
3. There is no electrical spark to ignite the fuel. Ignition depends on temperature and pressure of the fuel vapors.

ROTARY INTERNAL COMBUSTION ENGINES

Rotary combustion engines, Fig. 4-6, develop power through a triangular shaped rotor which

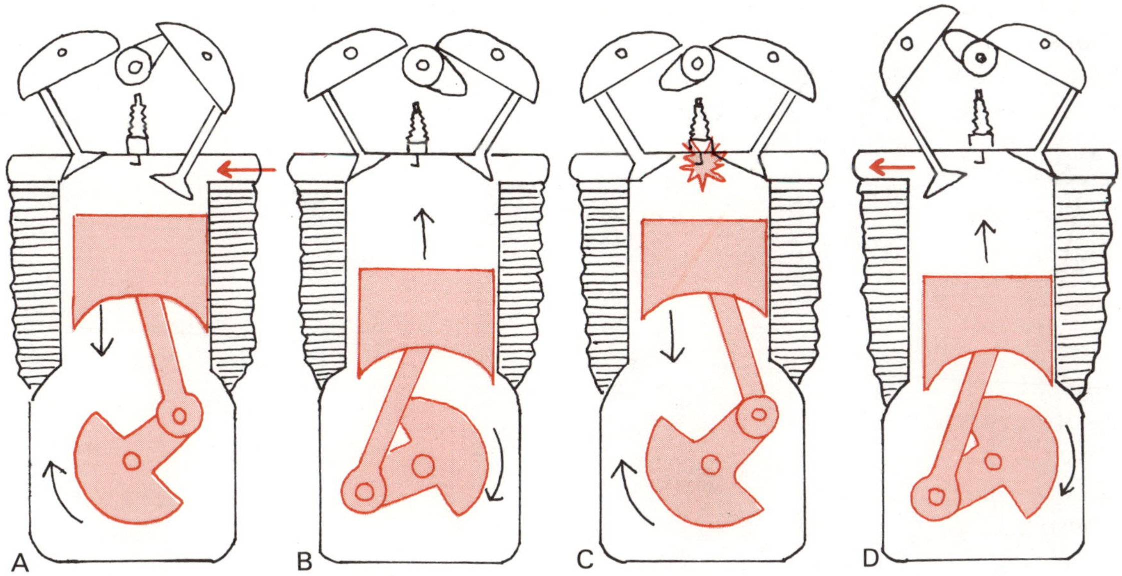

Fig. 4-3. Operation of a four-stroke cycle internal combustion engine. They are also called reciprocating engines because the pistons move up and down in a back and forth motion. A—First stroke of the cycle. Piston moves down to draw in fuel vapor through open valve at upper right. B—Next stroke is upward to compress the gas vapor. Valve is closed. C—Spark ignites fuel and drives piston down in power stroke. D—Exhaust stroke. Piston moves upward driving out spent fuel through open exhaust valve, top left.

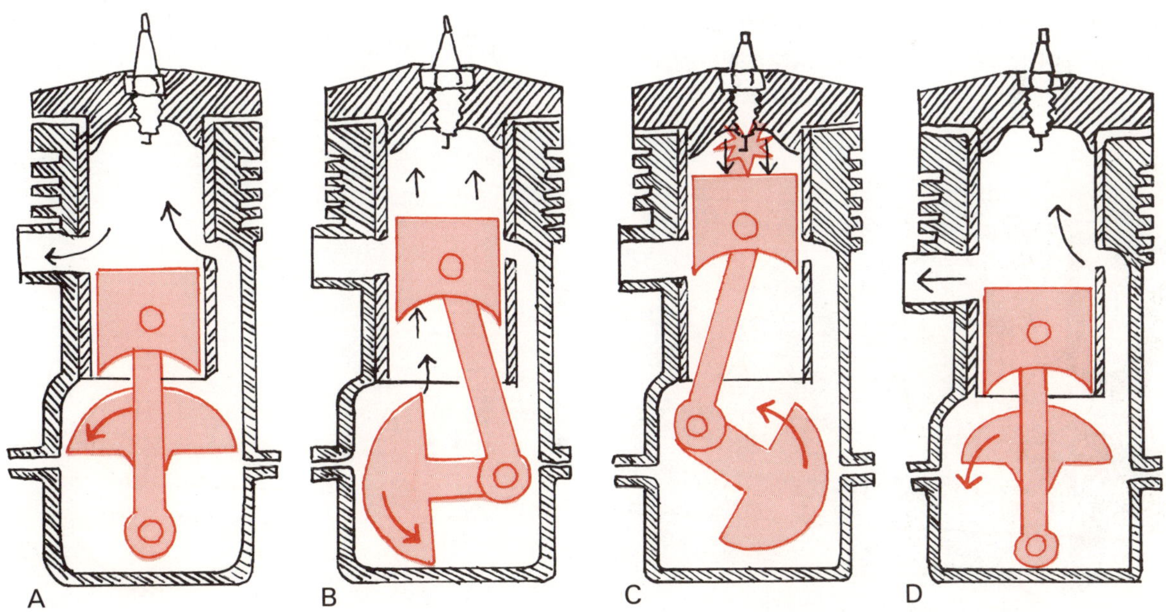

Fig. 4-4. Two-stroke cycle engine operation. A—At bottom of first downstroke, spent fuel is exhausted and new charge of fuel vapor enters cylinder. B—On upstroke fuel is compressed. C—Fuel ignites and piston moves downward in power stroke. D—At end of downstroke spent fuel is exhausted and new charge of fuel enters.

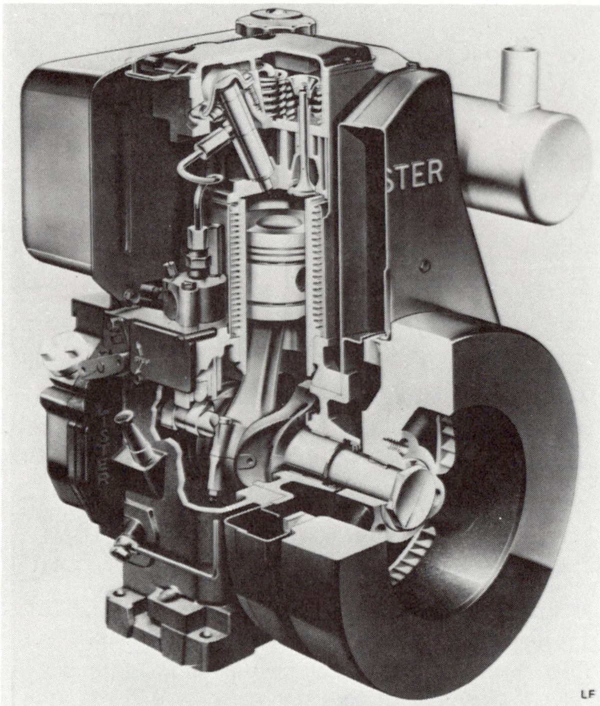

Fig. 4-5. Diesel engine is another type of internal combustion heat engine. (Lister Diesel Inc.)

spins inside a chamber. At three different points during its rotation, it compresses a vaporized fuel-air mixture which is then ignited. As with

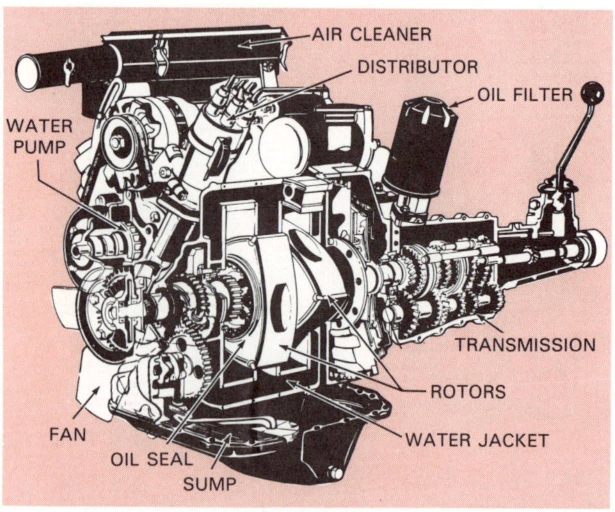

Fig. 4-6. Rotary engine is also known as a Wankel engine. It has three power thrusts during every revolution. (Sun/Mazda)

the conventional gasoline engine, the burning gases expand and drive the rotor.

Usually referred to as the Wankel engine, after its German inventor, Felix Wankel, the rotary engine has several advantages over the reciprocating internal combustion engine.

1. It is smaller and quieter.
2. It is more efficient and develops more torque for its size. This is simple to explain. It has three power impulses during every revolution whereas the reciprocating engine has but one power stroke every two or four revolutions. Further, the reciprocating engine's piston must reverse direction at the end of every stroke. The rotor of the Wankel travels only in one direction. Fig. 4-7 shows the operation of one chamber of the Wankel engine.

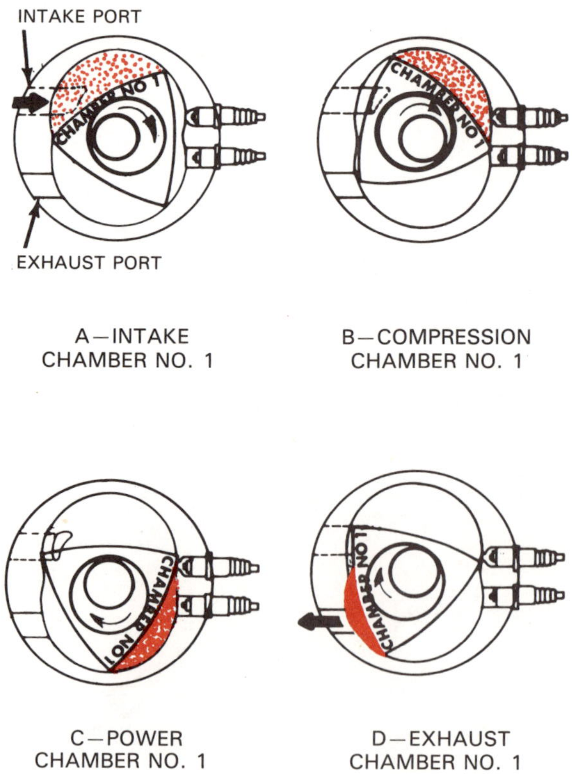

Fig. 4-7. Combustion cycle of a Wankel engine shows one chamber during one 360 deg. rotation. Each of the other chambers is going through the same cycle but at different times.

GAS TURBINE ENGINES

Like steam turbines (Fig. 2-22, pg. 29), gas turbines have large, finned rotors which absorb the high energy of pressurized gas and spin at a high rate of speed. Where the steam turbine is designed primarily for stationary operation, the gas turbine is found in mobile (moving) vehicles such as aircraft, boats, locomotives, and automobiles. Fig. 4-8 shows a simple sketch and a photograph of a gas turbine designed for use in an automobile. It consists chiefly of a compressor, combustion chamber, a turbine, and an exhaust nozzle.

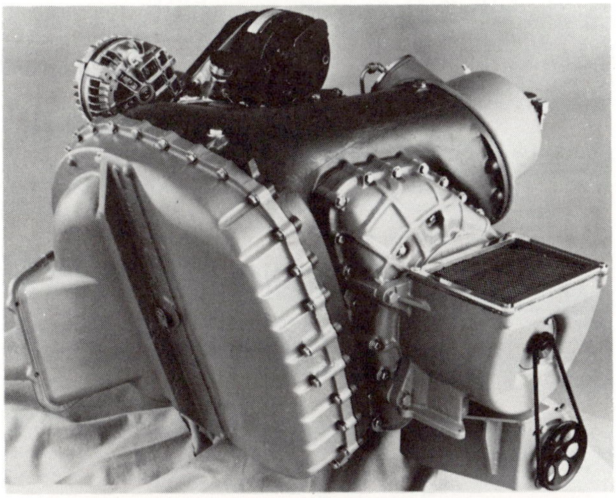

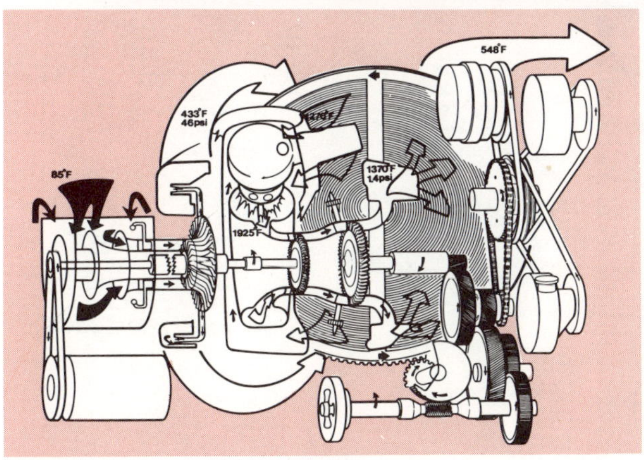

Fig. 4-8. Gas turbine engine. Top. Photo of engine designed for use in an automobile. It is compact and powerful for its size. Bottom. Schematic diagram of same turbine shows airflow and temperatures. (Chrysler Corp.)

Compressed air is forced into the combustion chamber. Fuel is added and ignited. Expanding gases, at 1300° to 1700°F (104° to 927°C) move to the turbine. There they expand to atmospheric pressure and transfer their energy to the turbine. It rotates at high speed to drive the compressor and a propeller or drive wheels.

After the hot gases have passed through the turbine they are used in one of two ways. In the turbojet and turbofan engines, the gases provide jet propulsion. In the turboprop and turboshaft engines, they are used to drive rotors or propellers.

JET AND ROCKET ENGINES

The jet and rocket engines work much like the air escaping from an inflated balloon. If you were to blow up a balloon and then release it, it would be propelled through the air for a few seconds by the escaping air, Fig. 4-9.

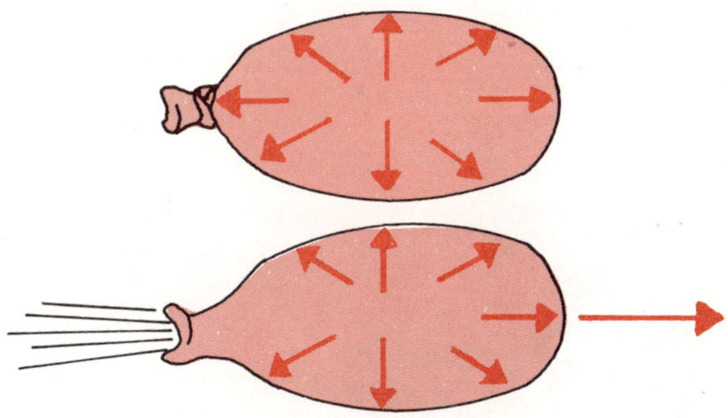

Fig. 4-9. Balloon demonstrates physical principle that "for every action there is an opposite and equal reaction." Top. As long as air cannot escape from balloon, pressure inside is equal on all sides. Bottom. As soon as air is allowed to escape, pressure at front of balloon is greater and propels balloon in that direction.

The balloon is moved by the pressure of the air inside not by the air escaping. In the same way, the jet engine moves in the direction of the greater pressure inside of it, Fig. 4-10.

The jet engine and the rocket engine differ in one respect. The jet draws its air for combustion

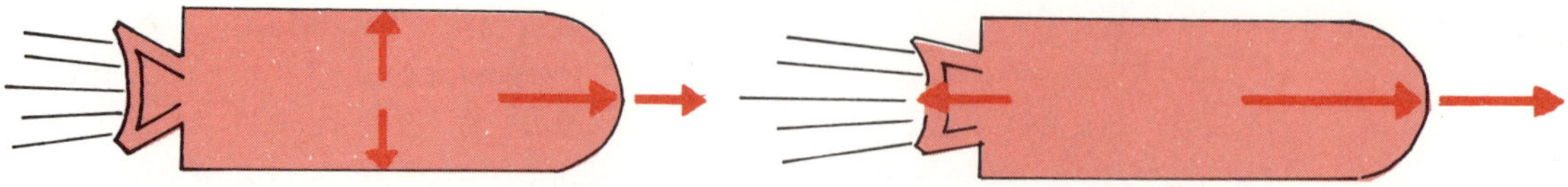

Fig. 4-10. How jet engine operates. Left. Like the balloon, jet moves in direction of the greatest pressure. Right. Action and reaction in a jet.

from the outside. The rocket carries its own supply. Jets are suitable for flights within the earth's atmosphere. Rockets, since they carry their own supply of oxygen, are suitable for outer space travel. See Fig. 4-11.

Jet propulsion is possible because of a natural principle known as Newton's Third Law of Motion. According to this law, for every action there is an equal and opposite reaction. In the balloon the action is the air rushing out of the neck. In the jet engine the action is the hot gases rushing out the rear of the engine. The reaction in both cases is the unequal pressure inside.

The most widely used jet engine today is the gas turbine jet. Jets have replaced piston engines in all military and most commercial aircraft.

EXTERNAL COMBUSTION ENERGY CONVERTERS

External combustion converters burn their fuel outside of the cylinder or chamber where the conversion from heat to motion takes place. The external combustion engines have never become major power sources for transportation. However, a great deal of experimentation is being done today to perfect them for automotive use.

There are several types:

1. Steam engine. It was discussed in Chapter 2.
2. The Stirling cycle engine.
3. The free piston engine.

STIRLING CYCLE ENGINE

The Stirling cycle engine, Fig. 4-12, was invented in 1916 by Rev. Robert Stirling. Though

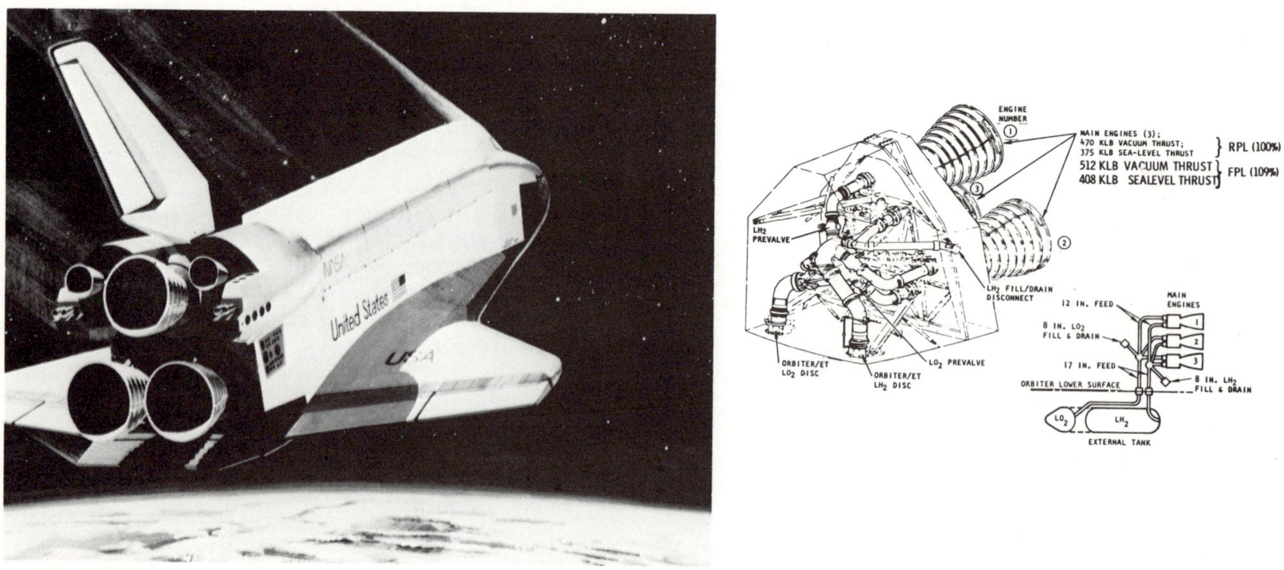

Fig. 4-11. Rocket engine is designed to propel vehicle in space. The air and fuel it consumes is carried in the spaceship. Left. Space shuttle orbiter Columbia. Right. Schematic of one of the engines. (NASA)

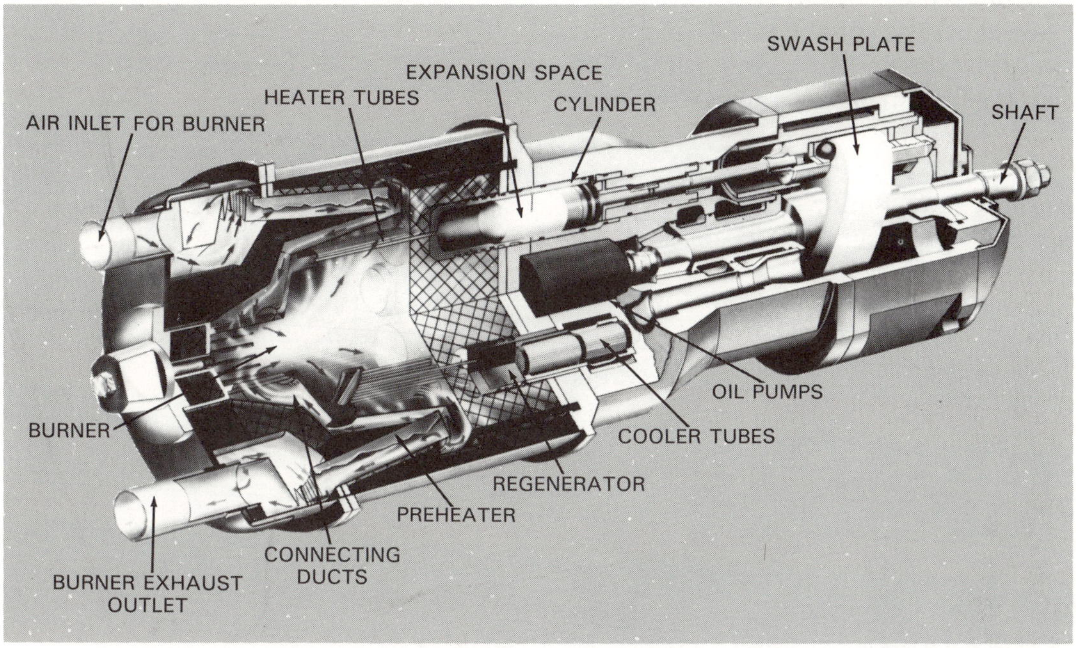

Fig. 4-12. Stirling cycle engine. More efficient than the reciprocating engine it may be the automotive power plant of the future if some of its problems can be solved through engineering. (Ford Motor Co.)

fossil fuels are the usual combustion material used, any heat source can be used, including solar.

The combustion process takes place in a chamber called the burner. The heat is transferred to a fluid such as gas or air through an exchanger. This is a device that can collect the heat of combustion and transfer it without taking on any of the matter which is carrying the heat.

The exchanger heats up the compressed gas or air, which, in turn, expands. As pressure increases, the gas pushes on pistons to create mechanical energy. A crank or swash plate, Fig. 4-13, changes the linear (straight-line) motion to rotary (circular) motion.

Then the gas in the cylinder is cooled by an outside cooler. The gas contracts (takes up less space) and pressure drops.

Current development has added refinements to the Stirling cycle engine. However, the basic

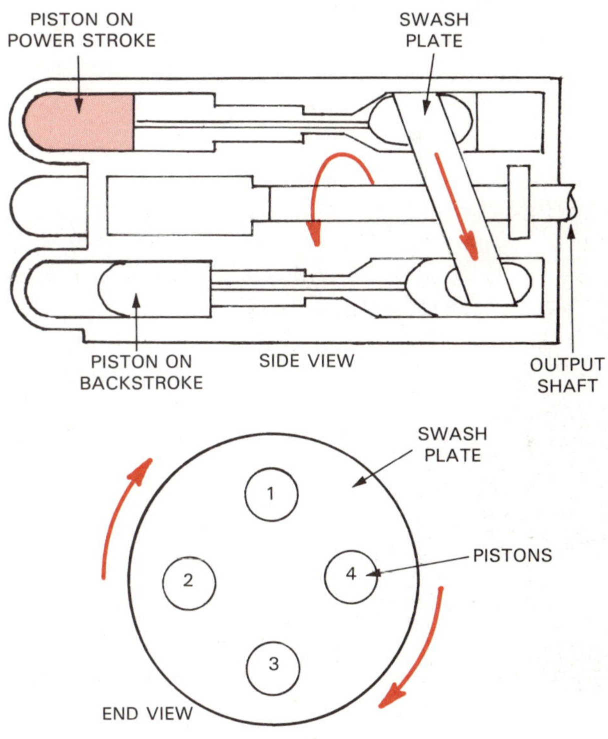

Fig. 4-13. Cutaway of the pistons of a Stirling cycle engine. As they move back and forth, they rotate the swash plate.

operation is the same as described. The engine has two power strokes of the piston for every revolution of the crank. This makes it more efficient than the two-stroke or four-stroke internal combustion engines.

Efficiency of today's models is about 30 percent. This is better than that of modern gas engines, Fig. 4-14.

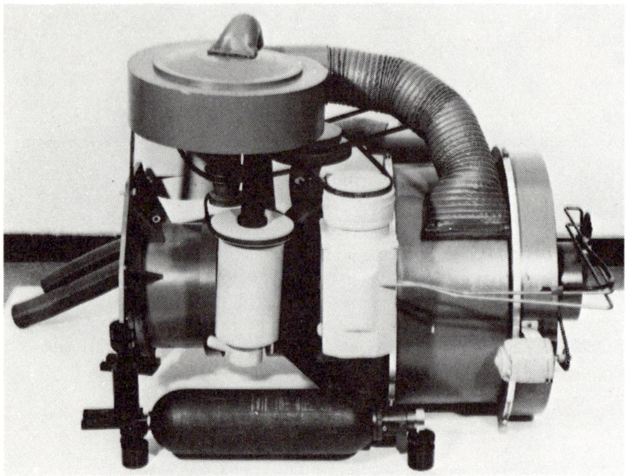

Fig. 4-14. An experimental Stirling engine developed for automotive use. (Ford Motor Co.)

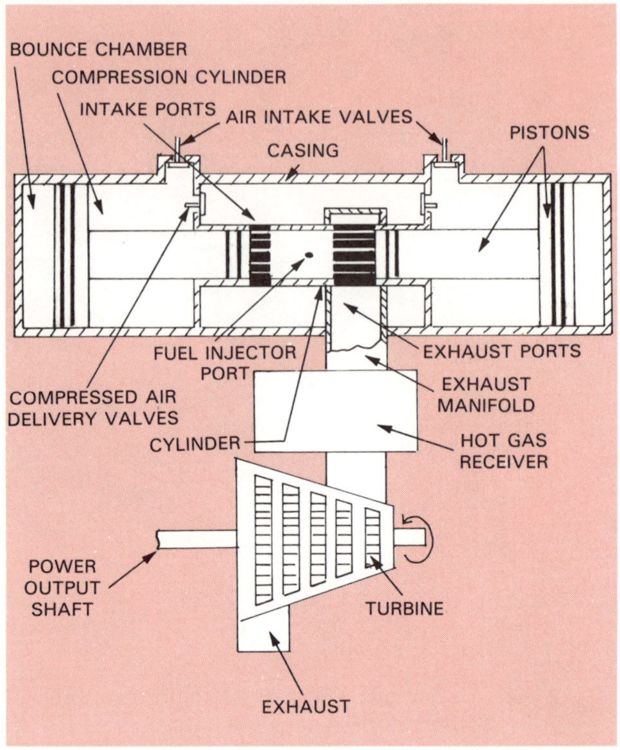

Fig. 4-15. Simple diagram of the free piston engine. It combines a turbine with parts of a reciprocating engine.

PROBLEMS WITH THE STIRLING CYCLE ENGINE

The Stirling cycle engine is not used in modern vehicles because it has a few problems. A solution has not been found to the proper cooling of the engine. Since it runs at higher temperatures, it has to get rid of more heat than a conventional gasoline engine.

A second problem is the working gases that operate the pistons. Helium and hydrogen work best but are difficult to contain because they are under such high pressure (200 atmospheres or 2940 psi). Hydrogen could be a safety hazard if a faulty seal leaked this highly flammable gas.

FREE PISTON ENGINE

Another energy converter engineers are trying to perfect is the free piston engine, Fig. 4-15. Somewhat of a hybrid, it combines elements of both reciprocating and gas turbine engines.

The pistons are called "free" because they are not connected to a rod or crankshaft and they have no ignition system to ignite fuel. See Fig. 4-16. The cylinders in which the pistons move are stepped so that there are two diameters. The large end of the pistons are called the "bounce" pistons. As the engine starts, air pressure is forced into the bounce chamber and two opposing pistons move toward each other. When the pistons are very close to each other, fuel is injected into the cylinder between them. The fuel is ignited by the high pressure and the pistons are forced apart violently. This causes air in the bounce chamber to be compressed providing energy for the return stroke. At a certain point, ports open in the cylinder and allow the hot, expanding gases to move into the turbine and turn it.

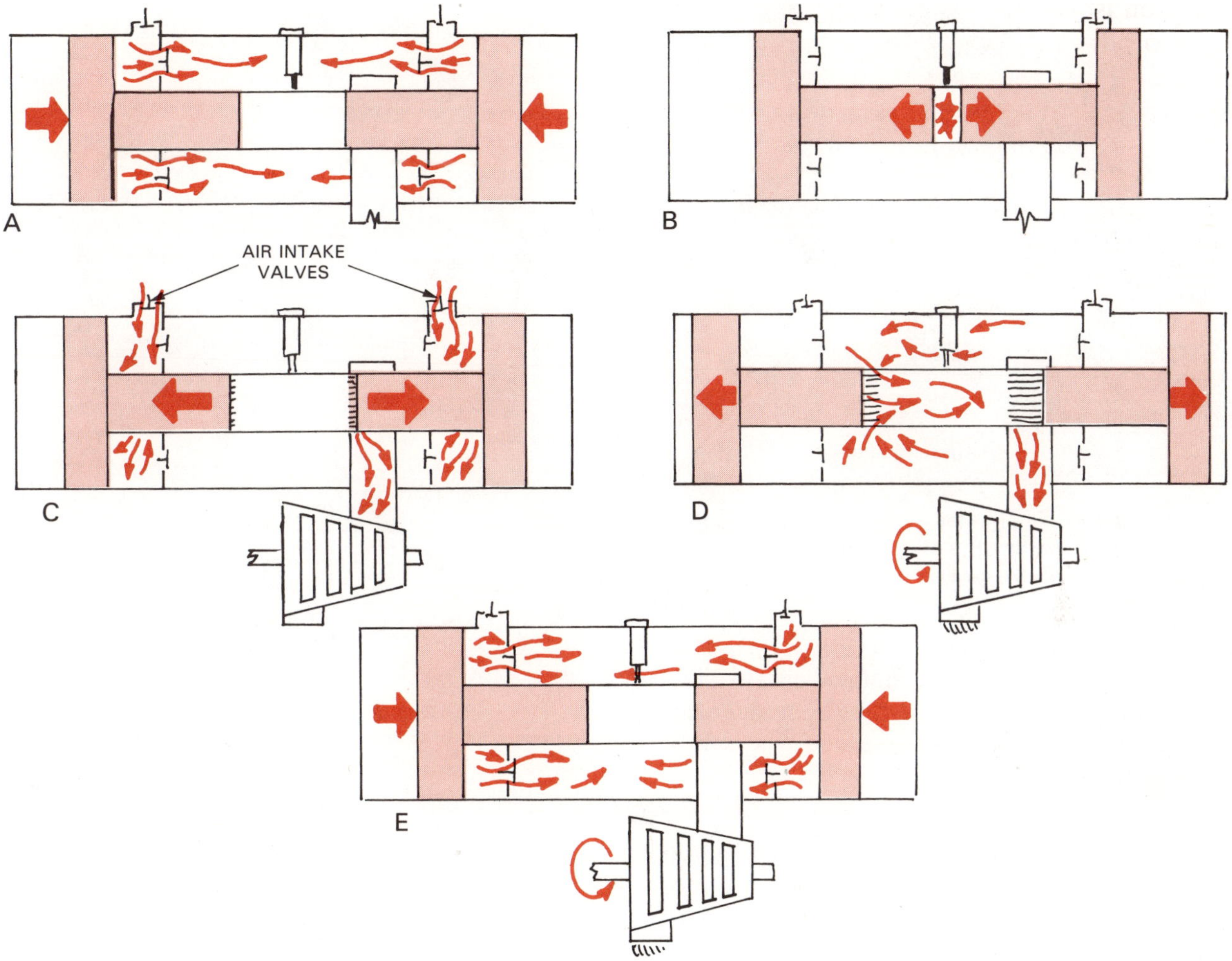

Fig. 4-16. Operating sequence of free piston engine. A—Air pressure from the starting system feeds into the bounce chambers at either end. B—When the ends of the pistons are close to each other, a spark ignites fuel fed into the chamber. C—Pistons, moving outward, compress air in the bounce chambers. This provides energy for the return stroke. Intake valves open to let air into the combustion chamber. Exhaust ports open and allow air into the turbine. D—Intake ports open at the end of the power stroke. Then, compressed air in the enclosure around the cylinder rushes into the power chamber pushing the remaining exhaust gases out to the turbine. E—Compressed air in the bounce chambers move the pistons back together to start the next cycle.

Even though the free piston engine will run on almost any fuel, it is not in commercial use. Various problems exist with the design. It is noisy, has low power, is hard to start, and violent forces of combustion soon cause breakdown.

GENERATORS AND MOTORS

Generators and motors are also indirect converters. Generators convert mechanical energy to electrical energy. The electric motor converts the electrical energy back to mechanical energy.

Generators, Fig. 4-17, produce electricity by passing a wire through a magnetic field. This act causes electrons to flow through the wire. If the wire is connected to a load by a conductor (metal wire) the electricity will drive the load (a light or an electric motor for example).

The electric motor, Fig. 4-18, converts the

Fig. 4-17. An oil-fired turbine generator uses fossil fuels to create heat, to generate electricity. (Niagara Mohawk Power Corp.)

electrical energy back to mechanical energy. The light bulb converts the electricity to light and heat.

BOILERS AND FURNACES

Another type of indirect energy conversion device is the boilers and furnaces used to produce steam and hot air for space heating and in-

Fig. 4-18. Electric motor converts electricity back to mechanical energy.

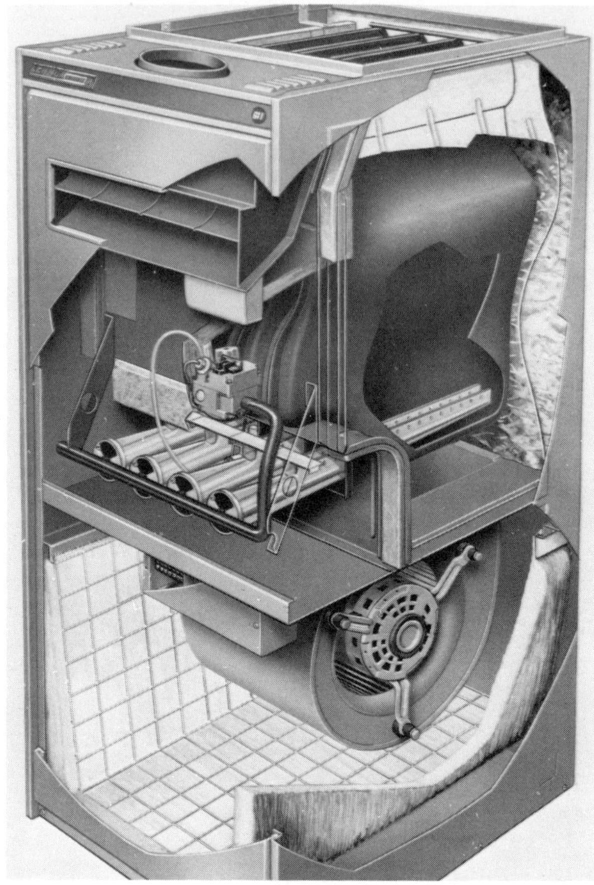

Fig. 4-19. A hot air furnace is a type of heat engine. (Lennox)

dustrial processes. These systems are a type of heat engine. They use a fossil fuel or solar energy to produce heat. See Fig. 4-19.

The Rankine engine is a reversible heat engine. It uses heated water to heat up a volatile (turns to vapor easily) fluid such as Freon.

The Rankine engine is generally used to drive air conditioners or generators. One of its uses is to produce electric power in ocean thermal energy conversion (OTEC). This is covered in Chapter 12 on Thermal Energy. See Fig. 4-20.

DIRECT CONVERSION

The battery is a familiar example of a device which makes a direct conversion of energy. It uses

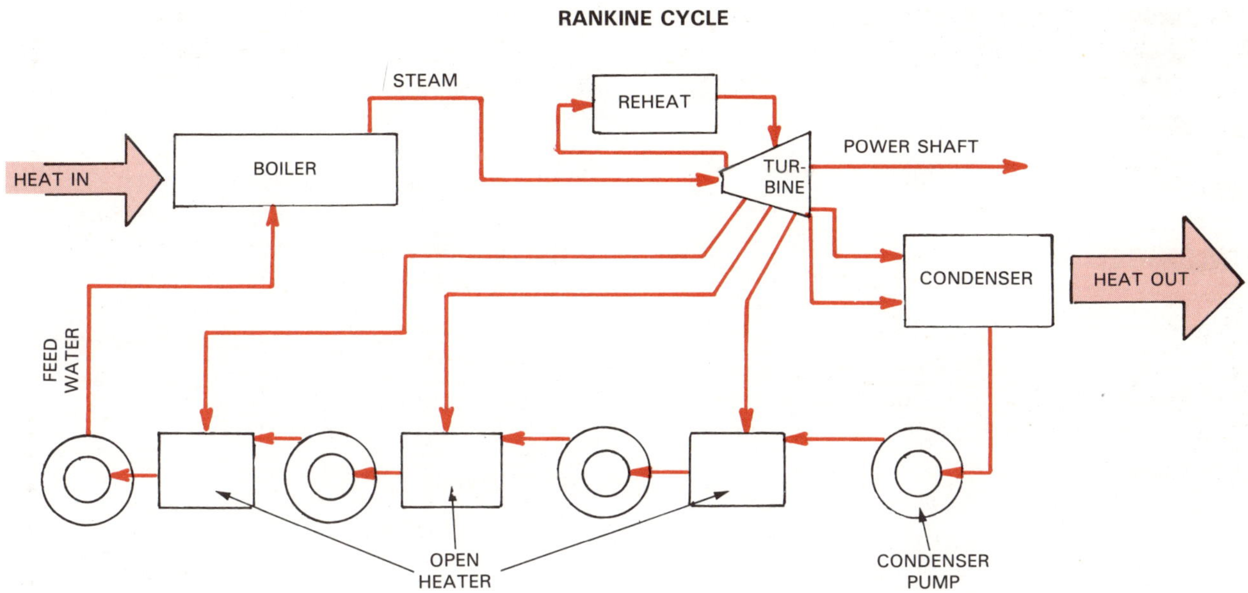

Fig. 4-20. Schematic shows basic arrangement of a Rankine engine. Its most common use is for driving air conditioners.

stored chemicals to produce electricity without any moving parts or any intermediate steps.

Scientists and engineers are working on devices which will convert directly from one form of energy to another. This is desirable for several reasons:

1. For one thing, direct conversions are generally more efficient.
2. Another reason is the belief that a direct conversion device will be more reliable, less prone to breakdown. This factor would make direct converters attractive for use where repairers would not be on hand to fix them. Small units of this type already operate the power systems in satellites and rockets used in the space program.

There are several direct conversion systems. Some of them are in limited production. Others are just being developed.

FUEL CELLS

A fuel cell, Fig. 4-21, is a device somewhat like a car battery. It controls a chemical reaction that produces electricity from a fuel and oxygen.

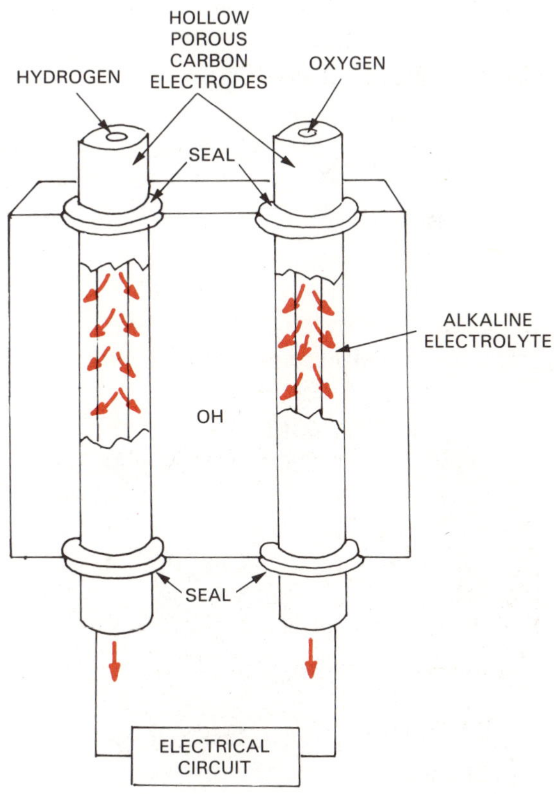

Fig. 4-21. A fuel cell is like a car battery as it produces electrical energy from chemicals. However, unlike a battery, it does not need to be recharged.

The fuel cell differs from the battery in two respects:

1. It needs a continuous supply of fuel to maintain production of electricity.
2. The fuel cell does not need to be recharged. Its chemical composition is always the same.

HYDROX CELL

The principle of the fuel cell has been known for a long time. Little was done with it from its discovery early in the 19th century until the exploration of space. Then engineers needed a different method of supplying electricity to space craft. The hydrox fuel cell was the answer. It uses hydrogen as a fuel and oxygen as an oxidizer. (An oxidizer is a substance which supports combustion or a chemical reaction.)

FUEL

Fuel cells can operate on many different types of fuel. In fact, any hydrocarbon fuel can be used, such as natural gas or biogas from garbage or sewage. A special processor breaks down the fuel into hydrogen.

HOW IT WORKS

A single fuel cell is made up of a container which holds the electrolyte, a solution of phosphoric acid. On either side of the electrolyte are two carbon plates called electrodes. (An electrode is a terminal or conductor of electricity. The electrodes are part of the path through which electricity moves.)

Hydrogen and oxygen are fed into the cell. The fuel flows down the positive electrode called the ANODE. The fuel loses electrons to the anode. The anode becomes negatively charged.

Meanwhile, the CATHODE (positive terminal) is supplied with oxygen, the oxygen collects electrons from the cathode leaving it positively charged. This sets up the fuel cell for the next operation.

When a load, such as a light is connected across the anode and cathode, free electrons on the anode travel through the conducting wire, through the light bulb to the cathode, Fig. 4-22.

While this is happening, positive ions (particles) of hydrogen leave the cathode, travel through the electrolyte, and combine with positive ions of oxygen. This combination forms water.

There are other waste products besides water. Included are nitrogen from the air and carbon dioxide from the fuel as well as heat from the process. Usually there is enough heat to produce steam during the process.

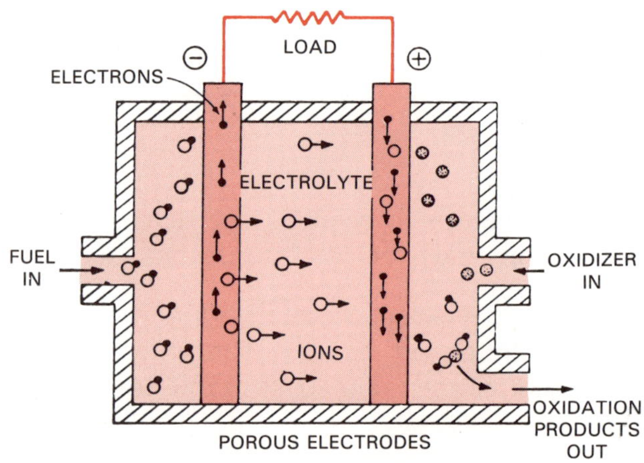

Fig. 4-22. Schematic of a fuel cell. The cell will provide a steady current across its terminals to drive electrical motors or provide lighting.

MHD GENERATOR

In the MHD generator, hot gases (called plasma) at 4000° to 5000°F (2200° to 2800°C) are passed across a magnetic field at right angles to the magnetic lines of force. Electrodes attract the free electrons in the gas. The collected electrons will flow through the conductors connected to the electrodes. Fig. 4-23 shows a simple diagram of an MHD generator.

THERMOELECTRIC CONVERTERS

Thermoelectric devices turn heat directly into an electric current. The oldest such device of this

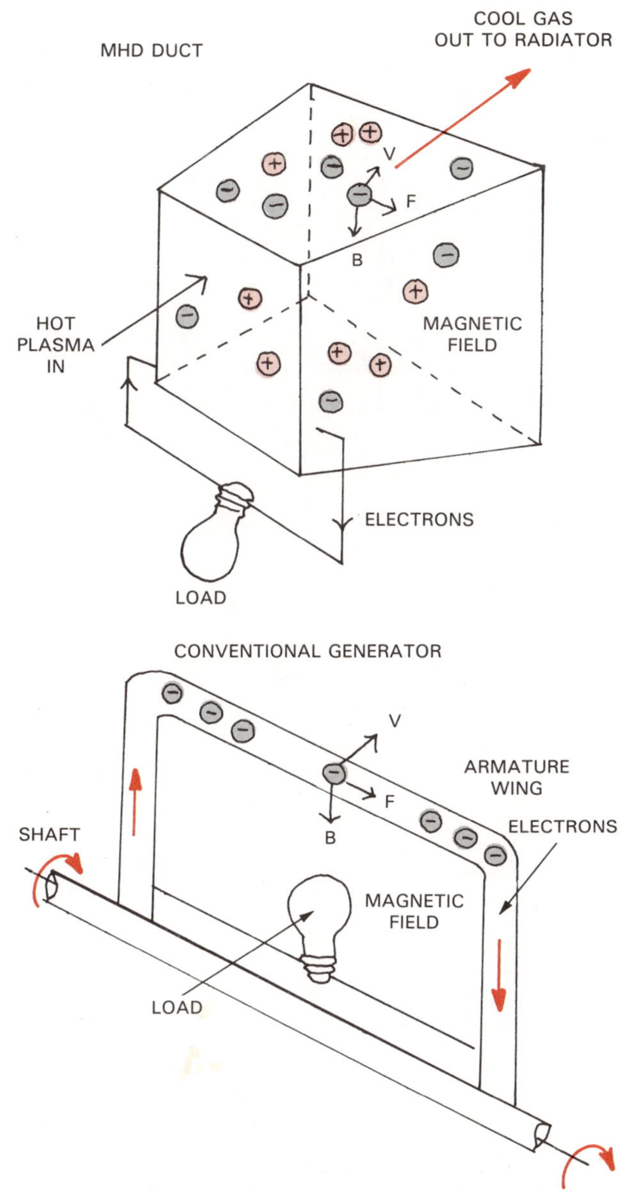

Fig. 4-23. Diagram of a magnetohydrodynamic (MHD) generator. It uses superheated gases to produce electric power.

type is the thermocouple. In 1821 a German physicist discovered that if two dissimilar (unlike) metals were joined and heated, a voltage or electromotive force could be measured across the free ends.

For over 100 years, no practical use was made of this knowledge. Then the semiconductor was invented. It is a better thermoelectric material. Semiconductors have impurities built into them without which they would be insulators.

One type of semiconductor has impurity atoms with some free electrons. The extra electrons are free to move about. This is known as an n-type (negative type) semiconductor.

A second type of semiconductor has impurity atoms with too few electrons. This leaves positive holes in the semiconductor. These holes can move around just like positive charges.

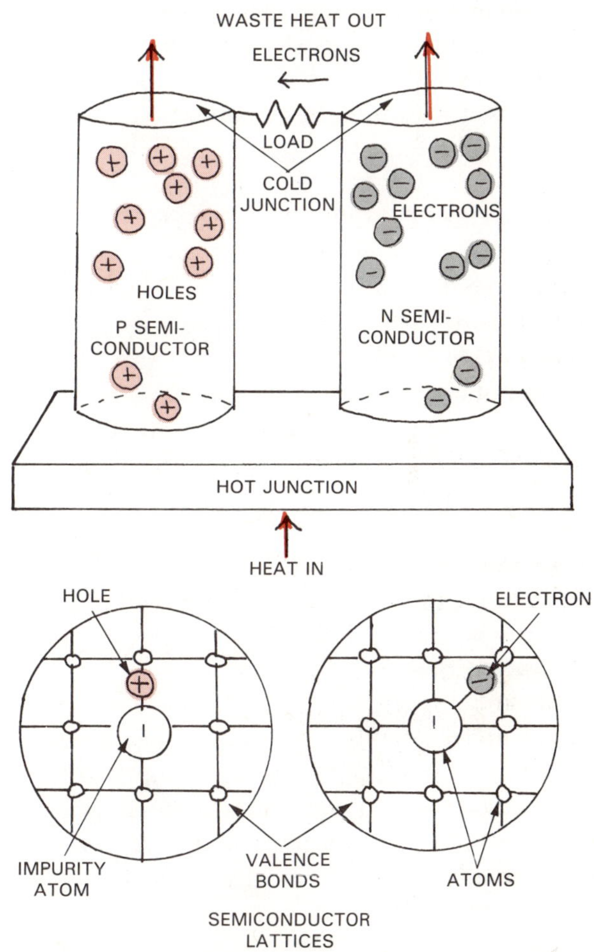

Fig. 4-24. Diagram of thermoelectric device. Top. Thermoelectric couple is made from pair of semiconductors linked by a hot junction and a load. Bottom. Impurity atoms (I) are different in each leg. One has too many, the other too few electrons. Heat drives both holes and electrons toward the terminals.

Suppose that you connected a p-type and an n-type conductor as in Fig. 4-24. Then, if you applied heat to the plate marked "hot junction" this is what would happen:

1. Heat at the hot junction drives the loose electrons and holes toward the cold junction. You can think of them as gases being driven through the semiconductor material by the temperature difference.
2. The holes will naturally travel up one semiconductor and the electrons up the other.
3. The holes create a positive terminal and the electrons a negative terminal.
4. The flow across the load connected to the terminals is driven. This sets up the flow of electrons which causes the voltage.
5. If a load such as a light bulb is connected across the terminals, the voltage will turn on the light.

Fig. 4-25. Small thermoelectric generator. It is powered by a small propane heater. (Global Thermoelectric Power Systems, Ltd.)

THERMOELECTRIC POWER GENERATORS

Thermoelectric generators can be located in remote areas where it is not possible to attend them. They will operate quite satisfactorily without maintenance. Small units operating with nuclear power have been launched into space aboard satellites. They are also used in polar regions and beneath the ocean.

Small propane units are being manufactured for use on campsites, ocean buoys, and in remote places where only a small amount of electricity is needed. See Fig. 4-25.

The efficiency of the thermoelectric generator is about 5 percent. This is low compared to the 35 to 40 percent efficiency of the steam powered power plants. Efficiencies would have to become much higher before thermoelectric power replaces steam powered units.

SOLAR CELL

The solar cell is a converter which captures sunlight and produces electricity from it. It has been found that the junction between p-type and n-type semiconductors will generate electricity if light is shone upon them. The electricity moves through the semiconductors in the same way described for a thermoelectric device. The solar cell and its operation is more fully described in Chapter 8.

Other conversion devices are under study and development. They include:

1. Thermionic converter consisting of two metallic plates. When one of the plates is heated it causes electrons to "boil off" and create an electric current.
2. Ferroelectric conversion. A nonconducting material is made to absorb heat by producing change in molecules of the substance.
3. Thermomagnetic conversion which uses heat to make a change in the magnetic field of a magnet core.

These devices are still in the development stage. However, they may provide important energy conversion capabilities for the future.

DO YOU KNOW THESE TERMS?

Cathode, diesel engines, direct conversion, electrode, energy conversion, ferroelectric conversion, free piston engine, fuel cell, hydrox cell, indirect conversion, MHD generator, rocket, rotary engine, semiconductor, Stirling engine, Thermionic converter, thermoelectric converter, thermomagnetic conversion, turbofan engine, turbojet engine, Wankel engine.

SUGGESTED ACTIVITIES

1. Make a list of power systems in your community, name the type of energy converter each uses and, if you can, list the type of energy it converts.
2. Take the list of energy converters developed for Activity 1 and discuss the types of pollution each converter produces. Take part in a class discussion on possible methods of reducing or eliminating the pollution of the conversion process.
3. Again, considering the list of energy converters in Activity 1, take part in a class discussion on what you would do to replace these converters of energy if their energy supply was exhausted. List alternative fuels and converters.
4. With your instructor's help, build a simple model or toy demonstrating one of the types of energy converters discussed in this chapter.

TEST YOUR KNOWLEDGE

1. An energy converter which changes energy into several different forms before it is used to do work is called a (direct, indirect) converter.
2. A heat engine which produces the burning of fuel inside a cylinder is called:
   a. A reciprocating engine.
   b. An external combustion engine.
   c. A two-stroke cycle engine.
   d. An internal combustion engine.
3. A rotary engine has ___________ power thrusts in every revolution.
4. State Newton's Third Law of Motion and show how it applies to jet propulsion.
5. The Stirling cycle engine has how many power strokes for every revolution?
   a. One.
   b. Two.
   c. Three.
6. List two reasons why direct conversion systems are desirable.
7. A ___________ ___________ controls a chemical reaction that produces electricity from fuel and oxygen.
8. There was little practical use for a thermocouple until the ___________ was invented.

Fig. 5-1. Swampy areas such as this were the beginnings of the great coal beds of the world.

5 FOSSIL FUELS—COAL

The information given in this chapter will enable you to:

- ☐ Describe the process which turned dead plant matter into coal and list the stages in the coal forming process.
- ☐ List and describe the types of coal mining in the United States.
- ☐ Explain methods used to transport coal from mine to where it is used.
- ☐ List the processes for converting coal to gaseous and liquid fuels.
- ☐ Discuss the future of coal as an energy source.
- ☐ Explain the impact of coal on the environment and describe efforts to reduce pollution from burning of coal.

Coal is a fossil fuel as you learned in Chapter 1. The coal is thought to be the remains of plant life which decayed and became buried millions of years ago. It is a rocklike material though it is much softer. Coal is usually black or brown depending on its age.

Heat, pressure, and bacteria acting on the dead matter made new substances that are rich in carbon (82 percent) and hydrogen (12 percent). The substances are called hydrocarbons. The mixture burns well and makes excellent fuel.

Coal began forming in the earth about 500 million years ago. The process took about 85 million years and was completed over 235 million years ago.

FORMS OF FOSSIL FUELS

In addition to coal, we find fossil fuels in two other forms:

1. Liquids or semisolids. This form includes petroleum, tar, and tar sands.
2. Gases. This form of fossil fuel is often found with petroleum. It is called natural gas.

The world's great beds of coal are the remains of great near-tropical forests. The leaves and wood fell into swampy water and decayed, Fig. 5-1. First it was a slimy, watery mass. Later, it lost most of its water and the rotting vegetation went through various stages which are explained later.

Coal was formed as seas advanced and receded in cycles all over the earth. The waters deposited heavy layers of sandstone, shale, and other rock on top of the decomposing plant life. The deposits caused pressure on the decaying matter and it became more and more dense.

STAGES OF COAL

Coal is found in several stages, depending on its age. The first stage is a product that looks like decayed wood. It is called PEAT. As pressure continues, the peat turns slowly into a soft, brown fibrous material known as LIGNITE. This is young coal. It burns well but does not give off much heat. It is not too useful as a commerical fuel.

Older coal becomes more dense and turns black. This type is called BITUMINOUS or soft coal. Its carbon content is much higher and it produces a hotter fire. It is the most plentiful and most used. See Fig. 5-2. ANTHRACITE coal is older and much harder than bituminous. It burns with a hot, blue smokeless flame. However, it is not very plentiful. In the United States, Pennsylvania is the only major producer.

Fig. 5-2. Bituminous coal, black and much denser than lignite, burns well and produces much heat.

COAL SUPPLIES IN THE UNITED STATES

Coal is abundant in the United States. It makes up 90 percent of all our known fossil-fuel reserves. Our coal reserves contain 10 times the energy in all of Saudi Arabia's petroleum.

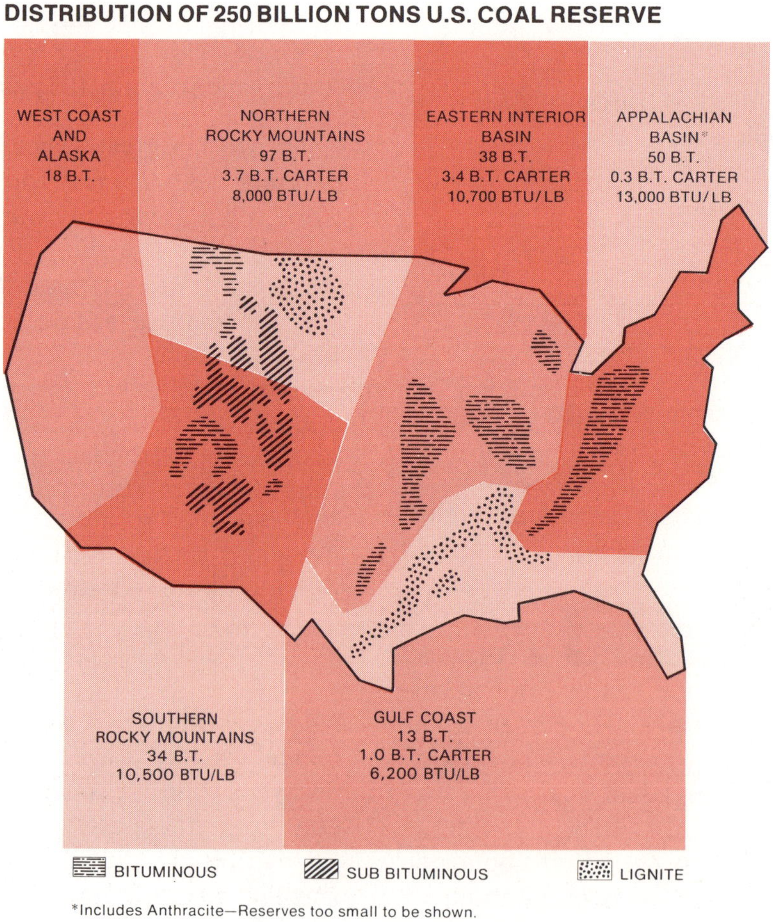

Fig. 5-3. Major coal deposits in the United States. (Exxon Corp.)

Most of the coal is found in two regions:

1. Eastern and midwestern.
2. Western.

Fig. 5-3 shows the location of coal deposits in the United States.

MINING COAL

There are several types of mining.

1. If the coal is near the surface, the coal is recovered through SURFACE or OPEN-PIT MINING, Fig. 5-4.
2. If the coal is located in hilly country or if it lies too deep beneath the surface for surface mining, the mining company will use an UNDERGROUND MINE, Fig. 5-5.

The type of mining to be used depends upon how deep the coal is buried in the ground. Most western and midwestern coal can be strip mined because it is near the surface. Eastern coal, being located deeper underground, must be extracted through underground mining.

SURFACE MINING

Before an open-pit mine is developed, engineers must study the TOPOGRAPHY of the land. (This is the way the land lays.) It might be

Fig. 5-4. In open pit mining, large quantities of earth and rock are moved to get at the coal. Dipper of big shovel in foreground holds 125 cu. yd. of earth and rock. (National Coal Assn.)

Fig. 5-5. An underground mine. This type is necessary in hilly country or where coal is buried deep in the ground.

Fig. 5-6. Developing a surface or open pit mine. Earth and rock, called overburden, are being removed by dragline to spoil area. (Gulf Oil Corp.)

mountainous, rolling hills, or flat land. They also study the thickness of the coal bed and the amount of OVERBURDEN. This is the rock and soil lying on top of the coal.

Obtaining this information may require aerial survey maps. These are photographs over which map lines are laid. The engineers will study the maps to:

1. Locate coal outcroppings.
2. Estimate the coal reserves (amount of coal in the ground).
3. Determine property lines.
4. Select areas where the excavated earth can be dumped. These are called the SPOIL AREAS. See Fig. 5-6.

PROSPECTING THE COAL SEAM

To find the size of the coal bed, the engineers will make test borings. These are the earth samples taken out of deep holes. They examine the materials brought up by the drill. This is called "prospecting."

The engineers will also make a SEISMIC ANALYSIS (study) to determine the hardness, depth, and nature of the overburden. Explosives are set off in holes. Shock waves travel into the ground and "bounce" back. Instruments measure the time lapse. An operator studies a SEISMOGRAPH (chart measuring the shock waves) and determines the type and thickness of underground materials. These data help engineers decide whether mining the coal seam is worthwhile.

If surface mining is being considered, the data collected from drilling and blasting will also help the engineers determine whether the overburden can be easily removed with a STRIPPING MACHINE or whether it must be drilled and blasted first. See Fig. 5-7.

STRIPPING

Removal of the overburden above a coal seam is called STRIPPING. A bulldozer may be used first to level off the surface so that other, larger machines can work there.

Usually, specially designed machines will do the actual stripping to lay bare the coal seam. One type is the WHEEL EXCAVATOR, Fig. 5-8. It

Fig. 5-7. Vertical drill bores 12 in. holes in earth and rock lying above coal seam. Explosives set in the holes break loose the overburden so it can be removed by power shovels or draglines. (National Coal Assn.)

Fig. 5-8. Huge wheel excavator such as this is sometimes used to strip away overburden. Conveyor carries it to spoil area in background. (Consolidation Coal Co.)

Fig. 5-9. In 1969 this dragline was the biggest moving land machine in the world. Its bucket picks up 220 cu. yd. of earth and rock at every bite. Its boom is longer than a city block.

has a cutting wheel at the end of a long boom. The wheel rotates, cutting and scooping up the overburden with bucketlike claws. Another machine used for stripping is the WALKING

Fig. 5-10. Dipper of surface coal mining shovel removes about three carloads of earth and rock at one bite. (National Coal Assn.)

DRAGLINE. Shown in Fig. 5-9, it is basically a large bucket suspended from the end of a boom.

A giant SURFACE MINING COAL SHOVEL may also be used. The dipper of one, shown in Fig. 5-10, will hold 210 tons of earth and rock or coal.

STRIP MINING METHODS

Two popular methods of strip mining are:
1. Scraper haulback method, Fig. 5-11.
2. Truck haulback method, Fig. 5-12.

In either method, overburden is moved to a portion of the pit already mined.

LOADING AND TRANSPORTING

Coal from surface mining operations is often loaded by electric power shovels into huge trucks, Fig. 5-13, and moved to a preparation plant. A conveyor belt may be used for the same purpose.

UNDERGROUND OR DEEP MINING

Coal deep beneath the earth must be extracted through underground or deep mining.

There are three types of underground mines. They are named for the kind of entry shaft used to reach the coal.
1. Drift mine. The shaft is driven horizontally into a hill, Fig. 5-14.
2. Slope mine. The mine shaft slopes gently downward.
3. Shaft mine. Its opening is sunk straight down, Fig. 5-15.

The MINE ENTRY serves two purposes. It takes miners and equipment into the earth to reach the coal. It also provides a way to transport the coal out.

In Fig. 5-16, the shaft on the right is designed for ventilation and has a blower forcing fresh air into the mine. It can also be used for raising and lowering workers and materials.

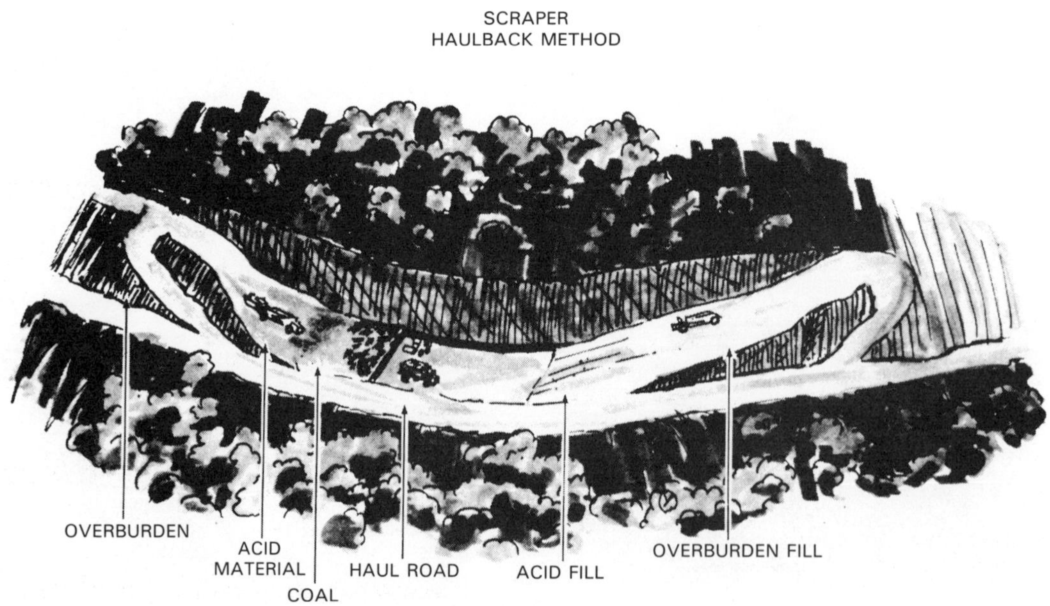

Fig. 5-11. Scraper haulback method of strip mining. Overburden is moved to a mined out area by scrapers. Acid fill is laid down first. Rock and clay are put on top. (Caterpillar Tractor Co.)

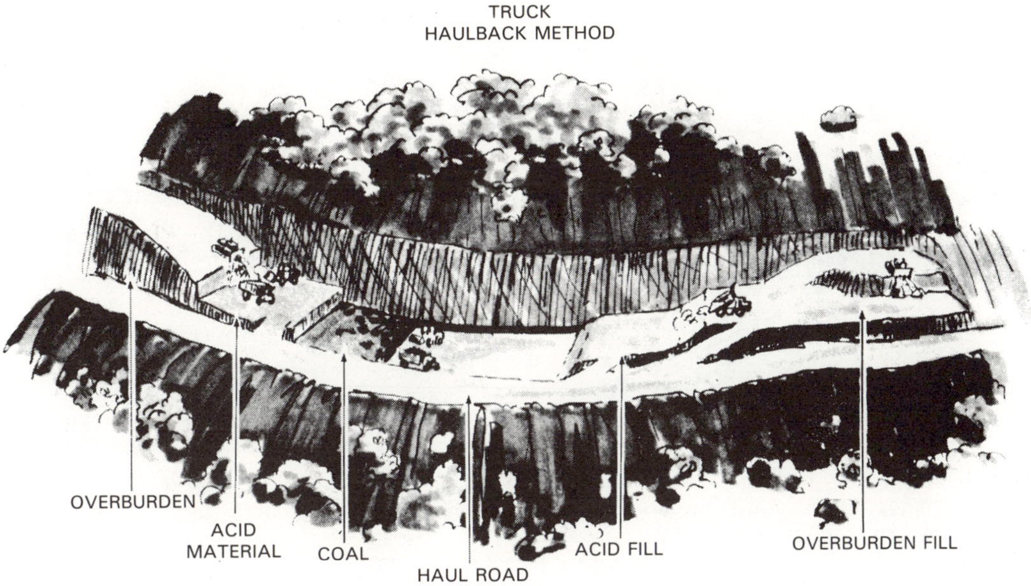

Fig. 5-12. Truck haulback method. Trucks replace scrapers to haul acid fill and overburden. (Caterpillar Tractor Co.)

The mine in Fig. 5-17 has two levels. The lower one is provided for drainage and/or to get at another coal seam.

DEEP MINING METHODS

There are two methods for removing the coal from a deep mine:

Fig. 5-13. Electric shovel loads coal onto truck for transport to coal preparation plant. (National Coal Assn.)

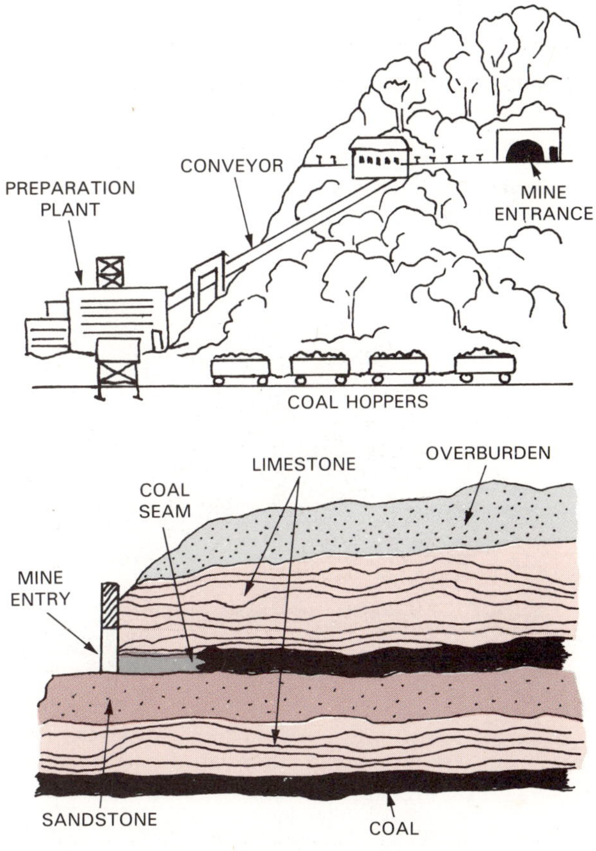

Fig. 5-14. Drift mine. Top. Frontal view. Bottom. Side view.

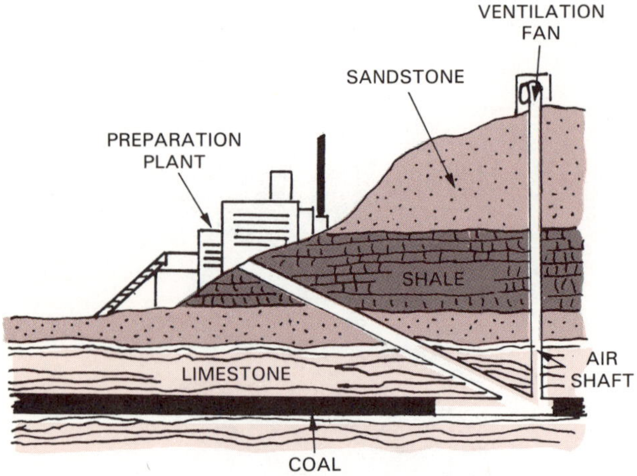

Fig. 5-15. Slope mine. Mine shaft slopes gently downward to coal seam.

1. Conventional mining.
2. Continuous mining.

CONVENTIONAL MINING

Mining by conventional methods involves four steps: cutting, drilling, blasting, and loading. The cutter looks like a giant chain saw. It can make horizontal or vertical cuts. Portable electric drills bore holes for the explosive charges.

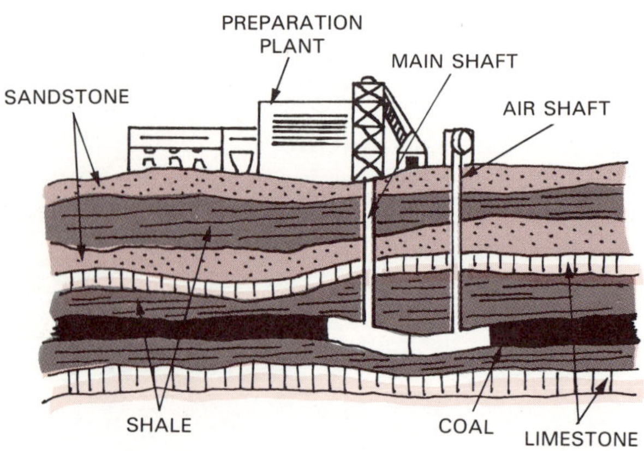

Fig. 5-16. Shaft mine is expensive to dig. There must be two shafts, one provides access. The other is for ventilation or an escape route should the main exit be blocked.

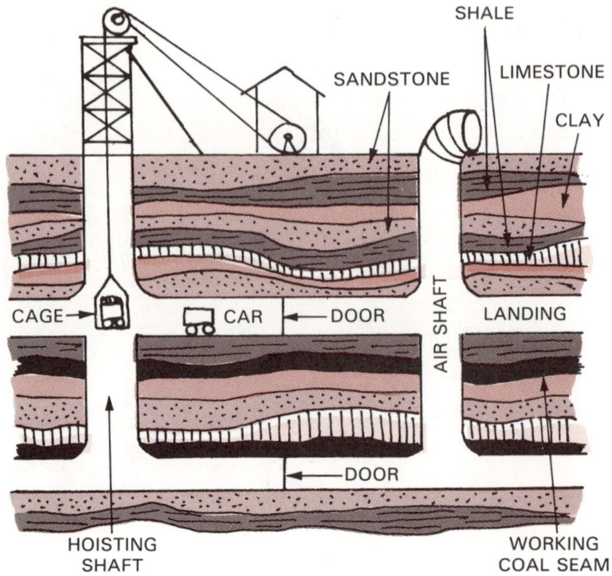

Fig. 5-17. Cross section of vertical shaft coal mine. Smaller shaft at right is used for ventilation.

Loosened coal is scooped up by a LOADER. This is an electrically powered machine with clawlike arms. An attached conveyor carries the coal to a shuttle car, Fig. 5-18, behind the loader. The shuttle car moves the coal to a conveyor or to a spot where it can be reloaded onto rail cars.

Fig. 5-18. Low-slung electric truck, called a shuttle car, moves mined coal from a mining machine to conveyors. Walls and roof of mine are coated with powdered limestone as a safety measure. (National Coal Assn.)

CONTINUOUS MINING

In continuous mining, special mobile (moving) machines mine and load coal in a nonstop operation. No explosives are used.

Machines used in continuous mining include:

1. A continuous miner which tears the coal from the seam with horizontal rotating teeth, Fig. 5-19.
2. A boring miner which tears coal from the seam by boring into the vertical working face.

A

B

Fig. 5-19. Continuous mining machine. Rotating teeth rip coal from seam. Hopper below (arrow) gathers up mined coal and moves it onto a conveyor. A—Unit at work with cutter head only inches off the floor of the mine. (Consolidation Coal Co.) B—Cutter head raised to show gathering head which scoops up loose coal. Note bolted ceiling. (Lee-Norse Co.)

With the addition of special equipment, continuous mining machines can bolt the roof at the same time that coal is being mined, Fig. 5-20.

Bolting is a system for supporting mine roofs. Holes are drilled in the roof to accommodate long bolts. The bolts are anchored at their upper end by an expansion shell. As the bolts are tightened, they bind the rocks together to act like supporting beams, Fig. 5-21.

Fig. 5-20. Roof control drill supports mine roof while drilling holes for roof bolts. (J.H. Fletcher & Co.)

DEEP MINING SYSTEMS

Coal must be mined in such a way that all the coal can be removed without having workers and machines trapped by cave-ins. The most used system in the United States is the ROOM AND PILLAR or BOND AND PILLAR system. In this method, coal is removed by making a series of "rooms" in the coal seam, Fig. 5-22. Large pillars of coal are left at regular intervals to act as roof supports.

Usually, the pillars are removed after the coal seam has been worked out. The roof is allowed to cave in as the operation works back to the entry shaft.

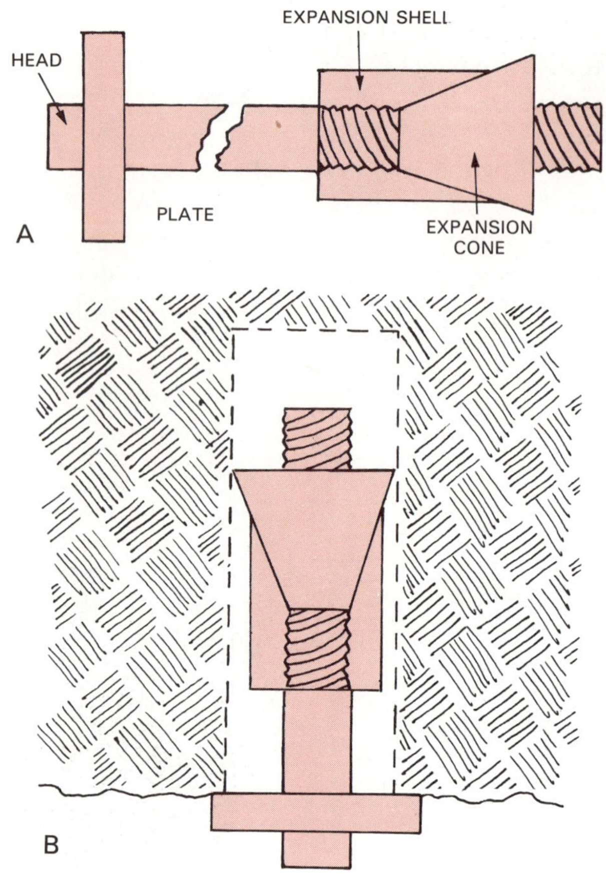

Fig. 5-21. A—Drawing of roof bolt designed to strengthen roof against cave-ins. B—Roof bolt installed. Expansion cone grips sides of hole.

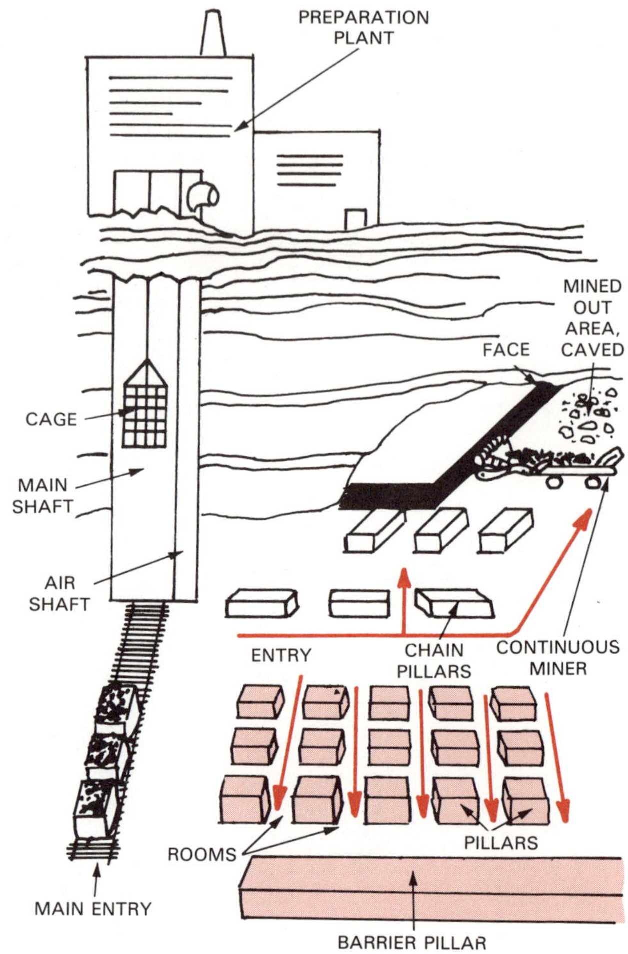

Fig. 5-22. Room and pillar method of underground mining. Pillars of coal are left to support roof during mining operations. They, too, are mined out eventually.

A second system, LONGWALL continuous mining, is used in some coal mines. In this system two parallel ENTRIES (like hallways) are driven through the seam. They may be up to 750 ft. apart. The entries are joined together at the far end by another hallway called a CROSSCUT. The coal face formed by the crosscut is called the longwall. A mining machine moves back and forth along this face removing coal as it goes. Fig. 5-23 shows a machine designed for longwall mining operations.

Fig. 5-23. Longwall miner uses rotating teeth to rip coal from face of coal seam. A—Conveyor. B—Cutters. (Consolidation Coal Co.)

PROCESSING OF COAL

Unprocessed coal is called RUN-OF-THE-MINE coal. It contains rock and other impurities, such as iron sulfide, which must be removed.

Fig. 5-24. View of coal preparation plant at Percy, Illinois. Coal is cleaned in plant in background and stockpiled in foreground or delivered directly to unit train on tracks at right. (National Coal Assn.)

Mined coal is hauled to a coal preparation plant, Fig. 5-24. At the same time, the coal is crushed, sized, and blended, Fig. 5-25.

The plant where bituminous coal is prepared for use is called a TIPPLE. Anthracite is prepared in a building called a BREAKER.

COAL STORAGE

Coal is stockpiled against the time when it will be used. Storage areas are usually located at shipping terminals or at electric power generating stations. Huge traveling stackers, such as the one pictured in Fig. 5-26, unload the coal from unit trains. Then the stackers move along steel rails and store the coal in gigantic piles. Some units are capable of unloading 10,000 tons of coal in less than six hours.

When the coal is reclaimed from the storage piles, a bucketwheel on the stacker transfers the coal at rates up to 6000 tons an hour. It can be moved by conveyor to dockside or to a power plant. The bucketwheel is also shown in Fig. 5-26.

MOVING COAL

Coal can be moved by rail, barge or ship, trucks or by pipeline. Much coal is moved by unit train. The most useful type of rail car is the Gondola car. It has an open top for easy loading.

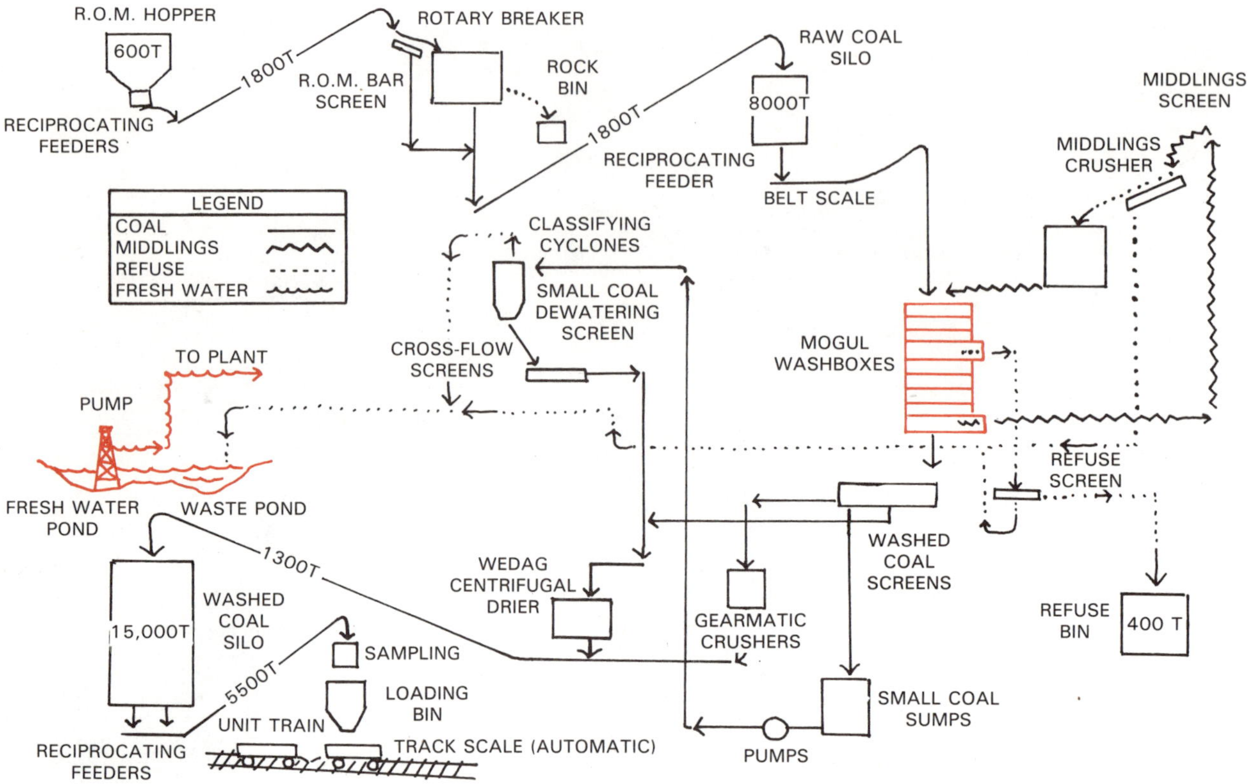

Fig. 5-25. Flow diagram showing how coal is processed in a coal preparation plant.

Fig. 5-26. A traveling stacker and bucketwheel reclaimer. Constructed at Thunder Bay, Ontario, it stockpiles lignite being shipped by boat to eastern Canada. Left. Stacker at work. Right. Bucketwheel retrieves stored lignite. (Allis-Chalmers)

Fig. 5-27. Slurry preparation plant for a coal slurry pipeline. (Black Mesa Pipeline)

Some have hopper bottoms. The floor of the car opens up like double doors to unload the car quickly without need for additional equipment.

PIPELINE TRANSPORT

In 1970, Black Mesa Pipeline, Fig. 5-27, began to operate an 18 in. slurry pipeline. It supplies 4 million tons of bituminous coal a year to a power plant in Nevada. The line is 273 miles long and moves coal from the Black Mesa coal field in northeast Arizona to the Mojave generating station at the southern tip of Nevada.

PIPELINE SLURRY TECHNOLOGY

Moving coal by pipeline requires the use of several pieces of equipment. The process must move through several steps:

1. The coal is ground to a powder with the consistency of granulated sugar. See Fig. 5-28.

Fig. 5-28. Cage mill crushes coal into smaller lumps. (Black Mesa Pipeline)

Fig. 5-29. Rod mill reduces crushed coal to consistency of sugar and mixes it with equal weight of water.

2. A second crushing in a rod mill reduces the coal to even finer particles. Water or some other liquid is added, Fig. 5-29.
3. Pumps, Fig. 5-30, move the slurry through a pipeline at speeds of 5 to 6 ft. per sec. Pumps are located at intervals of 40 to 95 miles.

Fig. 5-31 is a flow diagram of a coal slurry system. It is said that pipeline transport is cheaper. Unit train delivery costs 2.5 center per ton/mile; pipeline delivery, 1.5 cents.

POSITION OF COAL AS ENERGY SOURCE

Even though oil became the most used fuel in the 1960s, the use of coal has continued to in-

Fig. 5-30. Coal slurry pumping equipment. The pumps force the slurry through the pipeline at speeds of 5 to 6 ft. per second. (Black Mesa Pipeline)

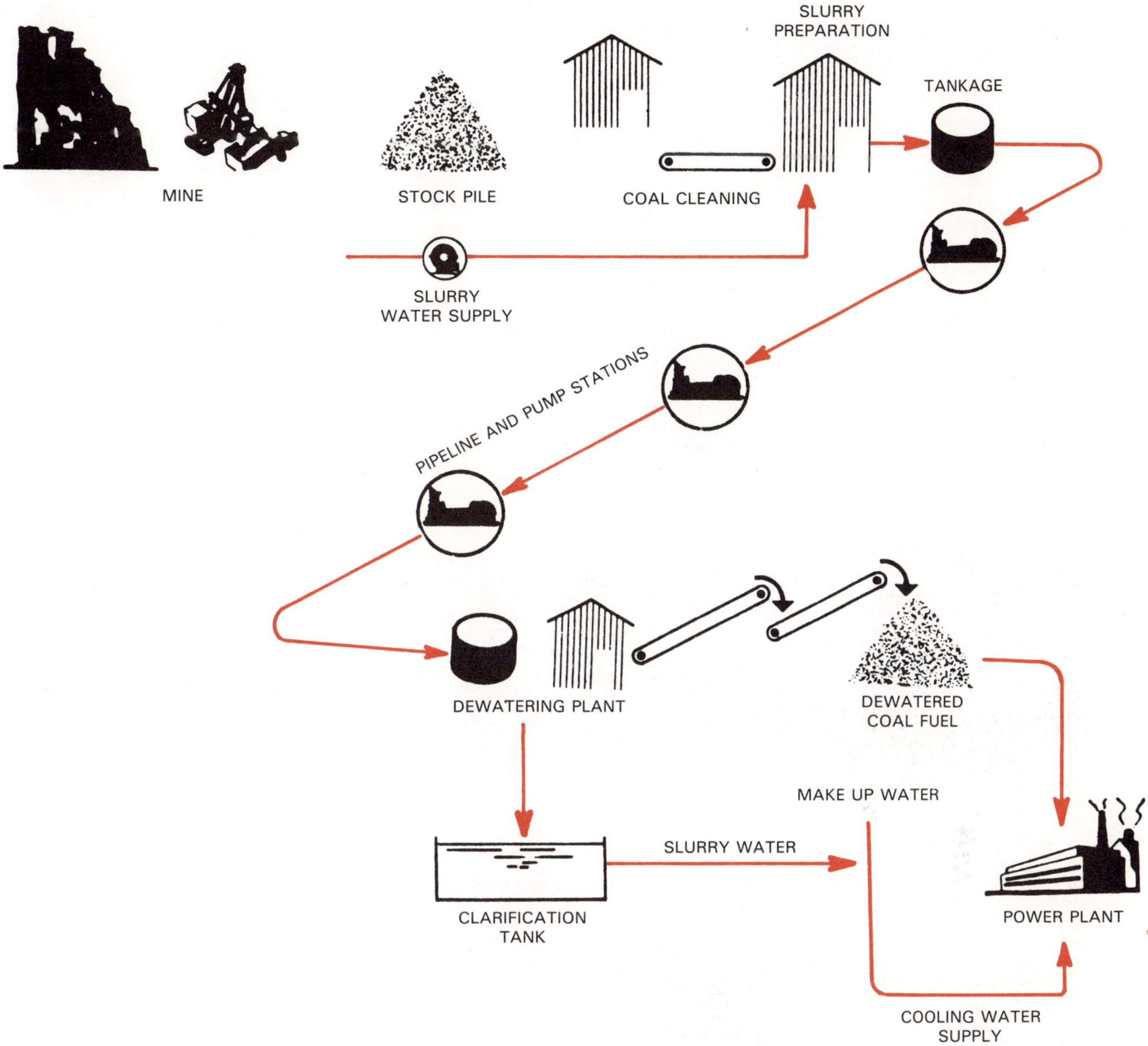

Fig. 5-31. Simple diagram of a complete coal slurry system. (Black Mesa Pipeline)

crease. In 1949, for example, the United States used a little more than 483 millions tons of coal. In 1960, we consumed only 398 million tons. Over the next years, however, consumption had grown so that in the year 1970, 523 million tons were burned. Only 501 million tons were used in 1971. Then an international shortage of oil caused users to become concerned about future supplies. Ever since, use of coal has grown an average of 20 million tons a year. See Fig. 5-32.

HOW MUCH IS THERE?

The United States is second only to Russia in coal reserves. The supplies of unmined coal can be roughly estimated by geologists using seismographic soundings and borings.

The amount of these reserves for the United States in about 1.5 trillion tons. How long this will last is not certain.

INDUSTRY AND MISCELLANEOUS USE IN SHORT TONS							
Year	Electric Utilities	Coke Plants	Other Industry and Miscellaneous	Total	Transportation	Residential and Commercial	Total
1952	107.1	97.8	117.1	214.9	39.8	92.3	454.1
1953	115.9	113.1	117.0	230.1	29.6	79.2	454.8
1954	118.4	85.6	98.2	183.9	18.6	69.1	389.9
1955	143.8	107.7	110.1	217.8	17.0	68.4	447.0
1956	158.3	106.3	114.3	220.6	13.8	64.2	456.9
1957	160.8	108.4	106.5	214.9	9.8	49.0	434.5
1958	155.7	76.8	100.5	177.4	4.7	47.9	385.7
1959	168.4	79.6	92.7	172.3	3.6	40.8	385.1
1960	176.7	81.4	96.0	177.4	3.0	40.9	398.1
1961	182.2	74.2	95.9	170.1	0.8	37.3	390.4
1962	193.3	74.7	97.1	171.7	0.7	36.5	402.3
1963	211.3	78.1	101.9	180.0	0.7	31.5	423.5
1964	225.4	89.2	103.1	192.4	0.7	27.2	445.7
1965	244.8	95.3	105.6	200.8	0.7	25.7	472.0
1966	266.5	96.4	108.7	205.1	0.6	25.6	497.7
1967	274.2	92.8	101.8	194.6	0.5	22.1	491.4
1968	297.8	91.3	100.4	191.6	0.4	20.0	509.8
1969	310.6	93.4	93.1	186.6	0.3	18.9	516.4
1970	320.2	96.5	90.2	186.6	0.3	16.1	523.2
1971	327.3	83.2	75.6	158.9	0.2	15.2	501.6
1972	351.8	87.7	72.9	160.6	0.2	11.7	524.3
1973	389.2	94.1	68.0	162.1	0.1	11.1	562.6
1974	391.8	90.2	64.9	155.1	0.1	11.4	558.4
1975	406.0	83.6	63.6	147.2	—	9.4	562.6
1976	448.4	84.7	61.8	146.5	—	8.9	603.8
1977	477.1	77.7	61.5	139.2	—	9.0	625.3
1978	481.2	71.4	63.1	134.5	—	9.5	625.2
1979	527.1	77.4	67.7	145.1	—	8.4	680.5
1980	569.3	66.7	60.3	127.0	—	6.5	702.7
1981	596.8	61.0	67.4	128.4	—	7.4	732.6
1982	594.1	40.6	65.6	106.2	—	7.1	707.4

Fig. 5-32. Chart of coal consumption. Use of coal declined in late 1960s and early 1970s.

Estimates are that, at 1975 levels of usage the U.S. would have enough coal to last 600 years. However, as other sources dwindle and as technology finds more ways to use coal, usage will increase. Assuming that usage would grow at 2 percent a year, coal reserves might be expected to last for 150 years.

COAL AND THE ENVIRONMENT

Extracting coal and using it as a fuel affects the environment in a number of ways. We need to consider the following and decide how best to mine and utilize coal:

1. Open pit mining gouges deep holes in the ground and leaves unsightly piles of overburden. The overburden pollutes streams and is blown by the wind. Little vegetation can grow on it. About 2 million acres of land in 26 states have been disturbed by surface mining operations.
2. Pit mines can collapse causing the land to sink and sag. This has happened in the Pennsylvania coal region and in southern Illinois.
3. Coal, used as power plant fuel, is a greater pollutant than if natural gas or oil were used. Coal also produces carbon dioxide, a small amount of radiation, and thermal (heat) pollution. Possibly the heat produced warms up the biosphere. (The biosphere is the part of the earth and its atmosphere in which life is normally found.)

Fig. 5-33. Coal companies are demonstrating that mined out land can be returned to production of crops. This mine area in North Dakota produced lignite for a power station and wheat for bread. (Consolidation Coal Co.)

4. Low sulfur coal pollutes less than high sulfur coal when used as power plant fuel.
5. Surface mining creates noise pollution.
6. Pit mining uses very little land and creates very little noise.
7. Open pit or strip mining uses up farmland in some regions such as southern Illinois but not in western regions because of mountainous terrain.
8. Large quantities of water are used to mine, clean, and wash coal.

LAND RECLAMATION

Most of the land disturbed by surface mining can be reclaimed. It can be made productive again and so continue to be a benefit to humankind. Mined sites can be redeveloped for many different purposes:

1. Production of agricultural or horticultural crops.
2. Recreation and wildlife habitat.
3. Industrial and residential building sites.
4. Planted to trees which will yield lumber and paper products.

The forestry service of the U.S. Department of Agriculture is concerned about the proper reclamation of such lands. Much study has gone into the development of proper reclamation methods. Mining companies are also experimenting with reclamation methods. See Fig. 5-33.

NEW COMBUSTION PROCESSES

Different methods of burning and processing coal are lowering the pollution levels which have always made it a dirty fuel. Fluidized combustion, Fig. 5-34, is one process that has proven successful in cutting down sulfur and nitrogen oxide (No_x) emissions and fly ash from burning coal. The coal, reduced to small lumps, is fed into a heated bed of sand or ash while air is bubbled up through the sand.

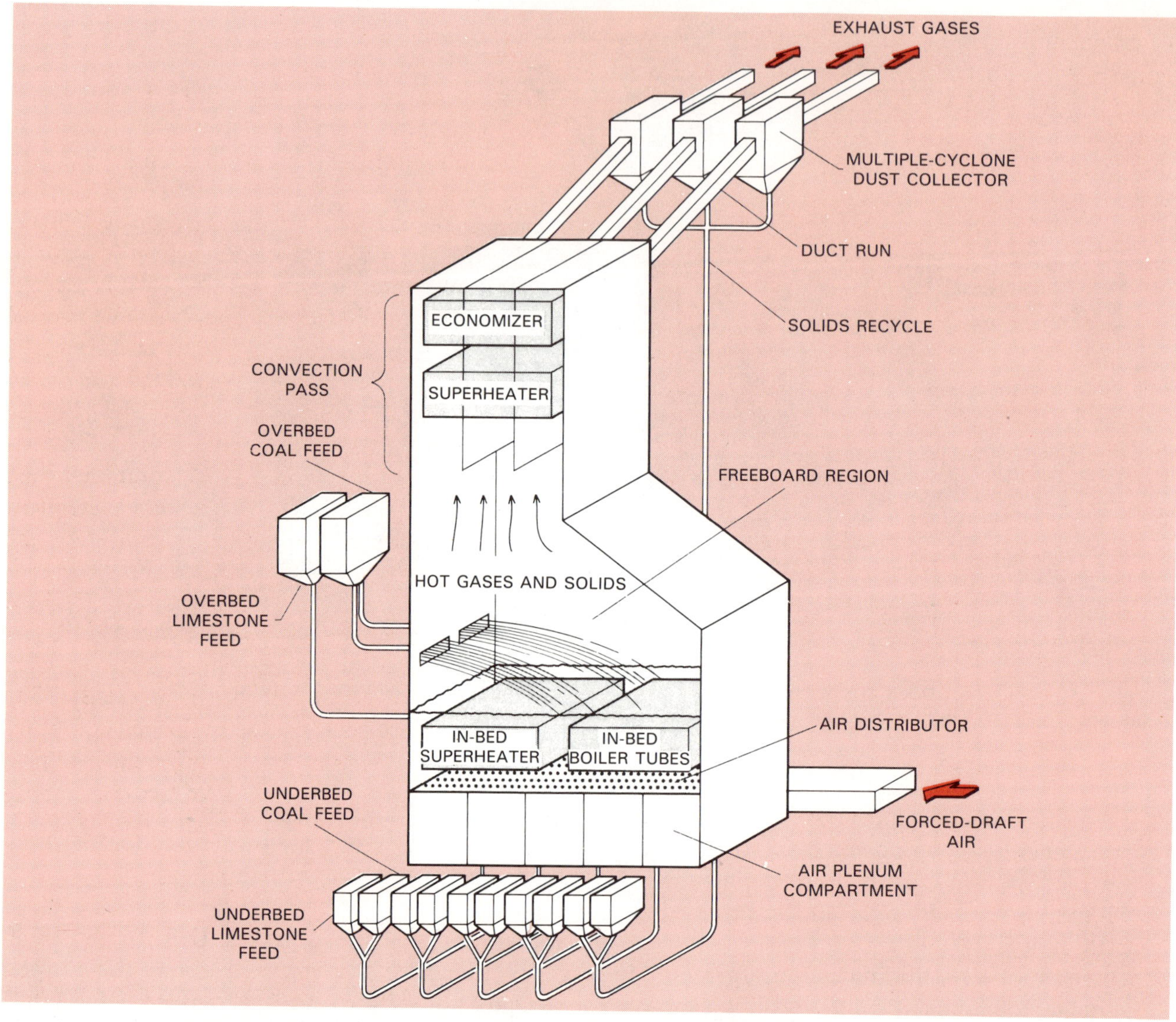

Fig. 5-34. Sketch of an atmospheric fluidized bed combustion boiler. This method of firing coal without the heavy air pollution of standard boilers is being tested in a pilot plant in Pennsylvania. (EPRI Journal)

At the same time, calcined (powdered) limestone is fed into the combustion chamber to absorb the sulphur dioxide (SO_2). Heat exchangers under the fire pot draw off the heat so that combustion temperature stays at about 1550°F (843°C). At this temperature the lime captures the sulphur, and nitrogen oxide gases are kept below required pollution control standards.

It is also possible to break down coal into gaseous and liquid fuels.

NEW FUELS FROM COAL

Producing gaseous and liquid fuels from coal are not recent processes. Very early in the industrial revolution, kerosene was made from coal and is still called coal oil in some rural areas. Another product of coal was the TOWN GAS or WATER GAS used in the cities for cooking and light in the 19th and early 20th century.

It was only the discovery of large quantities

of crude oil and natural gas that caused these coal conversion industries to be abandoned. During World War II, Germany improved the technology of conversion and demonstrated that large quantities of oil and gas could be produced this way.

COAL GASIFICATION

Gasification of coal is the process of turning it into a gas fuel. Coal is almost pure carbon. It has only small amounts of hydrogen, oxygen, sulfur, and nitrogen. Gasification (or any of the processes which convert coal to other fuels) adds hydrogen atoms to the carbon atoms.

In coal there are 16 carbon atoms for every hydrogen atom. Many more hydrogen atoms are found in gas or liquid fossil fuels, Fig. 5-35. Producing any of these fuels from coal requires large amounts of hydrogen.

RATIOS OF CARBON TO HYDROGEN IN SEVERAL FUELS		
Fuel	Number of Carbon Atoms	Number of Hydrogen Atoms
Coal	16	1
Heavy Fuel Oil	6	1
Gasoline	1	2 or 3

Fig. 5-35. Comparison of coal, with liquid fuels. Coal has a great deal of carbon atoms in its molecular structure.

ELEMENTS IN THE PROCESS

There are four primary ingredients in the gasification process:

1. Carbon. This is supplied by the coal.
2. Hydrogen. Some is supplied by the coal. The rest usually comes from breaking down natural gas (methane is mostly hydrogen) or by introducing steam which then reacts with the coal. (Water, from which the steam is made has two hydrogen atoms for every oxygen atom.)
3. Oxygen. Either it is supplied in its pure form or from air.
4. Heat is supplied by burning coal or a gas which is a by-product of the process. Fig. 5-36 illustrates the basic chemical reaction. The gas produced by this reaction, carbon monoxide and hydrogen, will burn but has only a third of the heat value of methane. It can be burned to carry on the process but it is not worth moving long distances by pipeline.

Instead, it is put through another process or step called METHANATION. The hydrogen and carbon monoxide are combined with a catalyst (something that triggers a chemical reaction) to make methane. Several steps are necessary to produce methane. First the carbon monoxide reacts with steam to produce carbon dioxide (CO_2) and hydrogen. The carbon dioxide is then separated along with another gas, hydrogen sulfide (H_2S). The latter is formed when the hydrogen combines with the sulfur impurities in the gas. At present, the process is being refined and developed for possible commercial use, Fig. 5-37.

LIQUID FUELS FROM COAL

As with gasification, liquification of coal is aimed at reducing the high carbon-to-hydrogen ratio (16 to 1) to at least 6 to 1. This is the carbon-hydrogen ratio of fuel oil. There are three different processes which will accomplish this:

1. Hydrogenation. Pure hydrogen is added. It produces a heavy oil that can be used in power plants.
2. Pyrolysis. Coal is heated in the absence of oxygen.
3. Catalytic conversions.

Pyrolysis produces three fuels: a pipeline gas, a synthetic crude oil, and a carbon residue called CHAR. The char can also be burned as a fuel if it does not contain too much sulfur.

Catalytic conversion produces the same reaction as gasification described earlier. The carbon

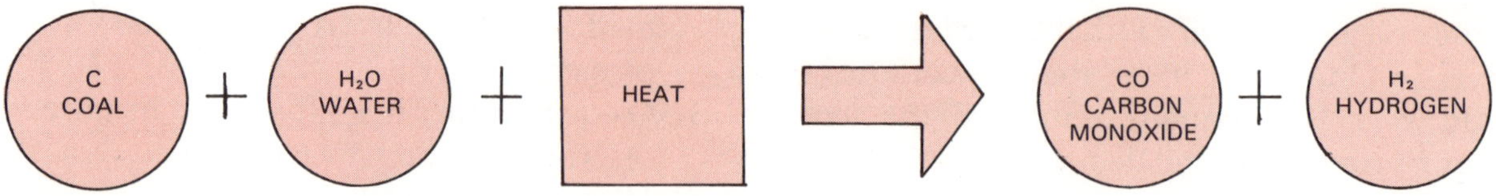

Fig. 5-36. Basic chemical reaction for coal gasification.

monoxide and the hydrogen are then combined with a catalyst to make a liquid fuel.

PRODUCING SOLID FUELS

Making solid fuel from coal is called SOLVENT REFINING. The solid fuel produced by this method is called solvent refined fuel or SRC.

First the coal is crushed and mixed with a SOLVENT such as light oil. This mixture is heated at high temperature and pressure. As the coal dissolves, the ash and the impurities can be separated from it. The fuel resulting is like a solid tar and can be crushed or heated and melted. Its heat value is higher than coal and it is a much cleaner fuel.

COAL CONVERSION AND THE ENVIRONMENT

Gasification and liquification have some serious drawbacks. Among the most serious is the heavy use of water which not only provides

Fig. 5-37. This coal gasification plant is providing research data to support commercial development of the process. It converts coal to synthetic gas for use in steam boilers and gas turbines. (Texaco Inc.)

cooling but becomes a raw material in production of hydrogen. About 1 1/2 to 3 lb. (roughly 2 qt.) are needed for every pound of coal processed. This is double the normal electric power plant usage.

Then too, most of the coal which will be gasified is located in areas where water is scarce. (Heavy coal deposits are found in the southwest, and the north central regions of the Dakotas, Montana, and Wyoming.)

Some pollution problems are also present. Air pollution controls will be needed to collect particulate pollution and SO_X. There will be solid wastes also. Some of these, of course, could be returned to the mined out areas.

An added health concern comes with production of synthetic fuels: presence of carcinogens. These are chemical elements that produce high risk of cancer for those who work with or come in contact with them. The high temperatures used in synthesizing coal produces molecules known as POLYCYCLIC AROMATIC HYDROCARBONS (PAH). There is plenty of evidence that these materials can cause cancer.

OUTLOOK FOR COAL CONVERSION

Experts say that it is difficult to predict how extensively synthetic fuel conversion will be practiced. Developing the huge commercial plants needed will be costly and the process itself is expensive. Gas companies are reluctant to undertake the development of such plants since they would place a serious strain on their assets.

DO YOU KNOW THESE TERMS?

Anthracite, bench, bituminous, carcinogens, catalytic conversions, continuous mining, drift mine, fluidized combustion, gasification, gondola car, hydrocarbons, hydrogenation, liquification, longwall mining, open pit mining, peat, polycyclic aromatic hydrocarbons, pressure cylinder, pryolysis, reclamation, room and pillar mining, seismic, shaft mine, shooting, slope mine, slurry, solvent refining, spoil, stemming material, tipple, topography, town gas, underground mine, walking dragline, wheel excavator.

SUGGESTED ACTIVITIES

1. Have a class discussion on how the U.S. can use coal to conserve its dwindling petroleum supplies without creating widespread pollution.
2. Build a model of any type of coal mine.
3. Draw up a list of the products which use coal as a raw material.
4. Have a coal company representative speak on coal's future.

TEST YOUR KNOWLEDGE

1. List the stages of coal.
2. Coal makes up ________ percent of U.S. fossil fuel reserves.
3. Coal which is found near the surface is mined by the following process (check correct answer):
   a. Open pit mining.
   b. Surface mining.
   c. Shaft mining.
   d. Drift mining.
4. In the ______________ method of underground mining machines rip out the coal along the working surface of the coal seam and haul it away in a single nonstop operation.
5. Why must run-of-the-mine coal be cleaned?
6. List the methods of transporting coal.
7. U.S. coal reserves amount to:
   a. 1 million tons.
   b. 500 million tons.
   c. 1.5 billion tons.
   d. 1.5 trillion tons.
8. Coal is our most plentiful fossil fuel and is expected to last somewhere between ______ and ______ years.
9. Coal is a greater pollutant than gas or oil. True or False?
10. Name three processes by which coal is converted to gas or liquid fuels.

Some of our country's petroleum reserves are found mixed with other elements such as shale and sand. This aerial view shows a tar sands extraction plant. Tar sand pile (arrow) descends by gravity to belts that carry the raw materials to the extraction plant. Steam rises from utilities. (Gulf Oil Corp.)

6 FOSSIL FUELS— PETROLEUM AND NATURAL GAS

The information given in this chapter will enable you to:

- Define petroleum and natural gas and explain briefly their chemical makeup.
- Discuss the historical development of the oil and gas industry.
- Describe methods of extraction and refining.
- List the products which come from petroleum and natural gas.

Like coal, petroleum or CRUDE OIL is a fossil fuel. It is the natural state of liquid hydrocarbons. As it comes from the well, crude oil is a mixture of semisolids, liquids, and gases. The gaseous portion of the petroleum is called NATURAL GAS.

Natural gas is always found wherever oil deposits are discovered, Fig. 6-1, but, often, it

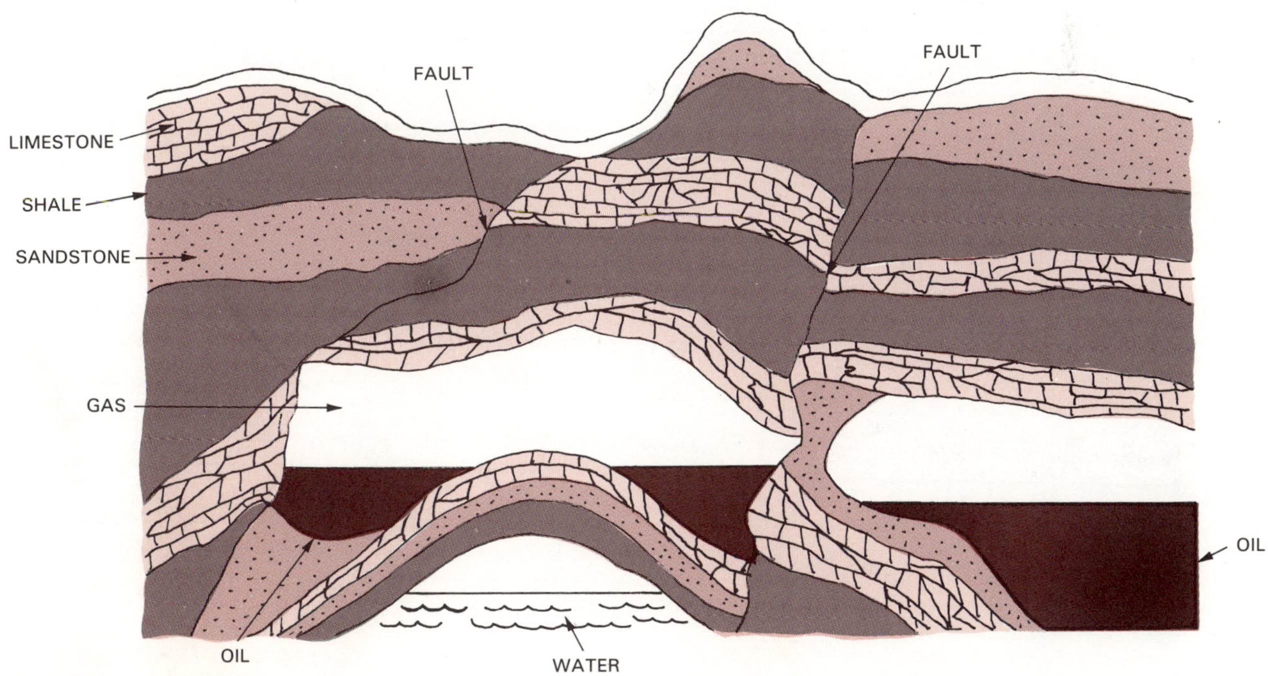

Fig. 6-1. Cross section of oil bearing rock formations.

is found in underground pockets which contain no oil. Natural gas is made up chiefly of methane, ethane, propane, and butane. All are combustible (burnable) gases. Small quantities of inorganic gases such as carbon dioxide, hydrogen, and nitrogen are also mixed with the volatile gases.

HISTORY OF PETROLEUM

Petroleum was first used as a medicine. Semi-solid forms, such as asphalt, were used for waterproofing as early as 3000 B.C. Egyptians and Chinese drank it as medicine and applied it to wounds. It was not used as a fuel until near the end of the 1700s. Then the need for a cheaper fuel for heating homes and factories and for driving machinery caused rapid development of the petroleum industry.

FIRST OIL COMPANY

The modern petroleum industry really began with the Pennsylvania Rock Oil Company. In 1859 the company drilled the first oil well at Titusville, Pennsylvania. See Fig. 6-2. It was accomplished with redesigned salt well boring equipment. Two years later the first gusher (well) was drilled nearby at Oil City. From then on, the oil industry boomed. By 1900, wells in the United States were producing 64 million barrels a year. Demand led to deeper drilling and more scientific methods. The first off-shore drilling was attempted off the California coast in 1896. In 1938 a well was drilled from a platform built off the Louisiana coast.

Fig. 6-2. First oil well was drilled in Pennsylvania in 1859. (Shell Oil Co.)

Need for gasoline to fuel the automobile led to development of the refinery industry. The first successful cracking (separating) process was introduced in the United States by two chemists, William M. Burton and Robert E. Humphreys.

LOCATING PETROLEUM AND GAS DEPOSITS

Geologists (experts on rocks) look carefully for surface rock formations that may give them an idea of the earth structure below. For instance, outcroppings of shale, sandstone, dolomite, or limestone may indicate the presence of oil or gas far below. Aerial photographs are sometimes taken because they show rock formations not noticeable from the ground.

Rock formations may be examined further if they seem promising.

TEST INSTRUMENTS

Tests will be made with sensitive instruments to determine the underground rock structure. The rock structure does not prove the existence of oil, but it tells the geologist if the subsurface rock formations are the type in which oil occurs.

Three instruments are used for these tests:

1. Gravimeter. Dense rock exerts more force of gravity than less dense rock. The gravimeter, Fig. 6-3, measures gravity and will indicate the density of rock deep in the ground.
2. Magnetometer, Fig. 6-4. This device measures the strength of a magnetic field. It is used to

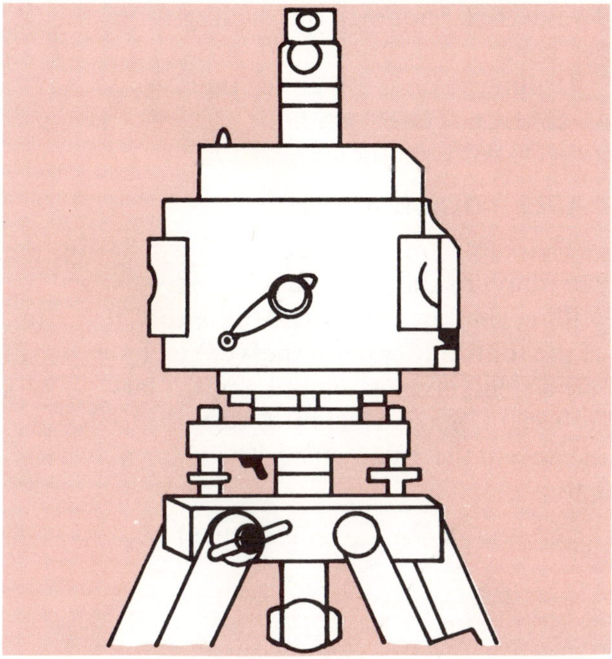

Fig. 6-3. Gravimeter is used by geologists to detect density of rock deep beneath the earth.

locate rocks likely to carry oil. Oil itself has no magnetism, but rocks do. The magnetometer is usually towed behind an airplane, but it can be moved about by truck.

Fig. 6-4. Magnetometer locates rock formations likely to contain petroleum deposits. (Shell Oil Co.)

3. Seismograph. A modification of the instrument used for detecting and measuring the intensity of earthquakes, the seismograph measures shock waves bouncing back from rock formations underground. Dynamite is placed in small holes bored in the ground. When exploded, the dynamite causes shock waves that bounce off the rock formation. These are picked up by the seismograph, Fig. 6-5. Recordings of the shock wave are "read" by the geologist. They tell him or her what kind of rock is below, how large the formation is, and its depth.

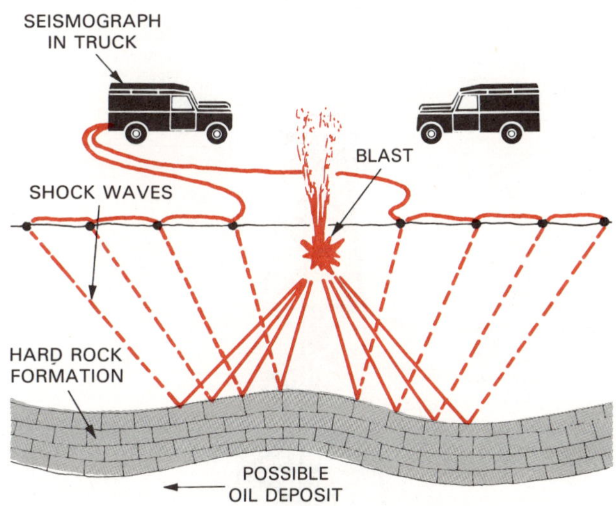

Fig. 6-5. A seismic study. Measuring echo of dynamite blast tells geologist where large rock formations are to be found underground.

DRILLING FOR OIL

Testing can tell geologists where there is likely to be oil. However, the only way to be sure is to drill a well. There are two types of wells:

1. The exploratory well which is drilled to find a new pool, to extend an existing oil field, or to find an entirely new oil field. In other words, it is drilled in search of more oil. Often these are called WILDCAT wells.
2. The DEVELOPMENT well which is drilled in an area that is already producing oil.

Well drilling is a big operation. More than 45,000 are drilled every year for exploration or development.

TEST CORES

During drilling, test cores are examined. These are cylindrical chunks of rock and other subsoil taken from the hole. Geologists study them looking for fossils and the type of rock which might suggest the presence of oil far below.

DRILLING METHODS

Before drilling can begin, workers build an open steel tower called a DERRICK, Fig. 6-6. The derrick supports the drilling tools.

Drilling is done by either the:
1. CABLE TOOL method.
2. ROTARY method.

CABLE TOOL DRILLING

The cable tool method, adopted during the early days of the petroleum industry, is used for drilling shallow wells. The cable tool string (all of the tools that go into the well) includes a drill bit, a stem, connecting links called jars, sinkers for weight, and a cable for getting the tools down the hole to the drilling face. The string may weigh 2 tons.

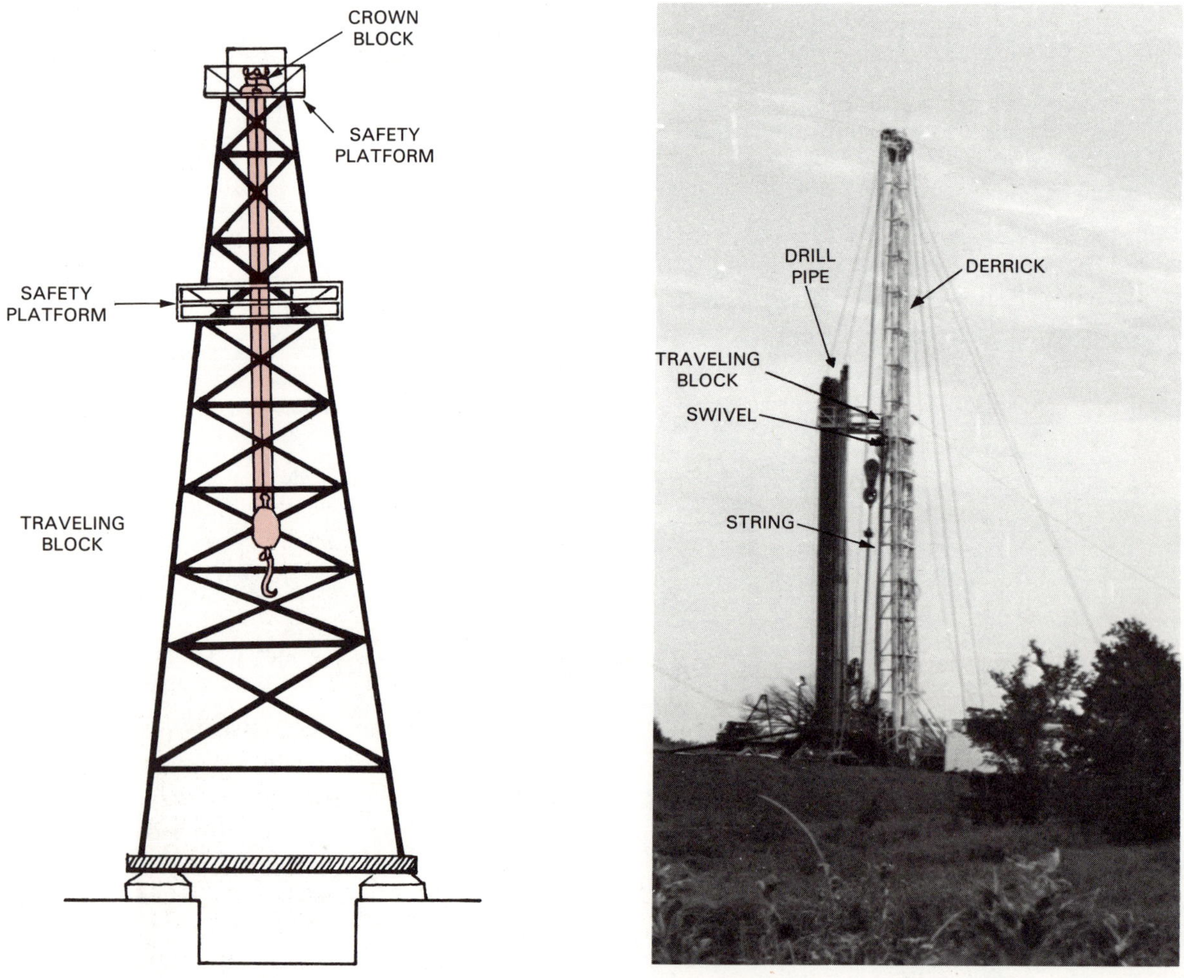

Fig. 6-6. An oil derrick supports tools used in well drilling. Left. Sketch of well derrick. Note block and tackle for hoisting pipe and drilling bits. Right. Derrick on well drilling site in Oklahoma.

The hole is punched into the ground by the alternate raising and lowering of a heavy chisel-like cutting tool. The tool is suspended on a cable.

During the drilling operation small amounts of water are fed into the hole. The water mixes with the rock dust and chips produced by the pounding of the bit. It forms a sludge that can more easily be removed.

When the sludge becomes so thick that it interferes with the drilling, the tool is pulled out and the sludge is removed. More clean water is fed into the well and drilling continues. Fig. 6-7 is a sketch of a rig set up for cable tool drilling.

ROTARY DRILLING

In rotary drilling the cutting tool is a heavy steel bit. This is connected to a length of steel pipe called a DRILL PIPE. A turntable supported on a platform at the base of the derrick turns the drill. The turntable is powered by diesel engines. As the drilling goes deeper into the earth, additional lengths of drill pipe must be added to the bit.

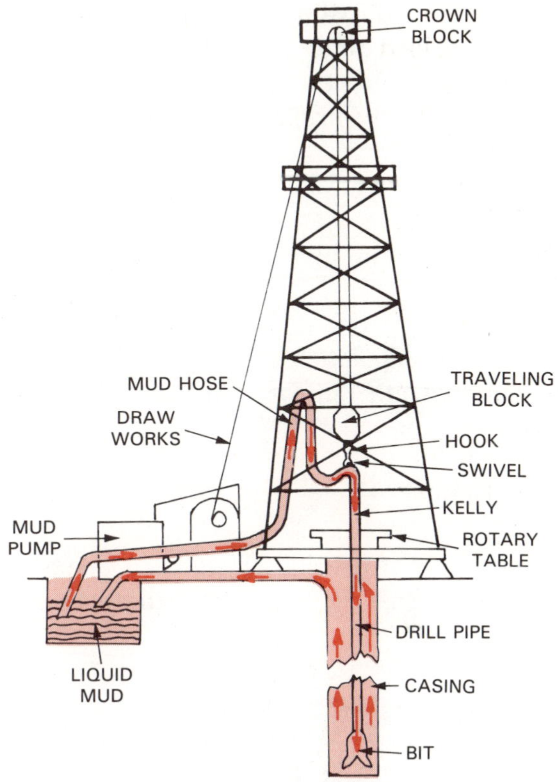

Fig. 6-7. Drilling operation. Mud is pumped down the well casing to cool the drill bit and bring up broken rock loosened by the drill.

Fig. 6-8. Drilling crews work around the clock finding new petroleum reserves. (Texaco, Inc.)

A special mixture of water, clay, and chemicals is continuously fed into the well during drilling. It cools the bit and carries away the rock pieces broken loose by the drilling. Then, this mixture, called MUD is forced back up the hole. On the way up, it lines the walls of the hole to prevent cave-in or seepage. Later, a steel casing will be installed to line the hole. Fig. 6-8 shows a drill crew at work.

When the recirculated mud reaches the top, the rock particles are removed and the mixture is reused.

DRILL BITS

Many kinds of drill bits have been designed, Fig. 6-9. Selecting the right one is important. It has a direct bearing on the speed and efficiency of drilling. For example, a FISHTAIL bit is best for soft clay and sand. CONE AND ROLLER BITS are designed to cut through certain kinds of rock.

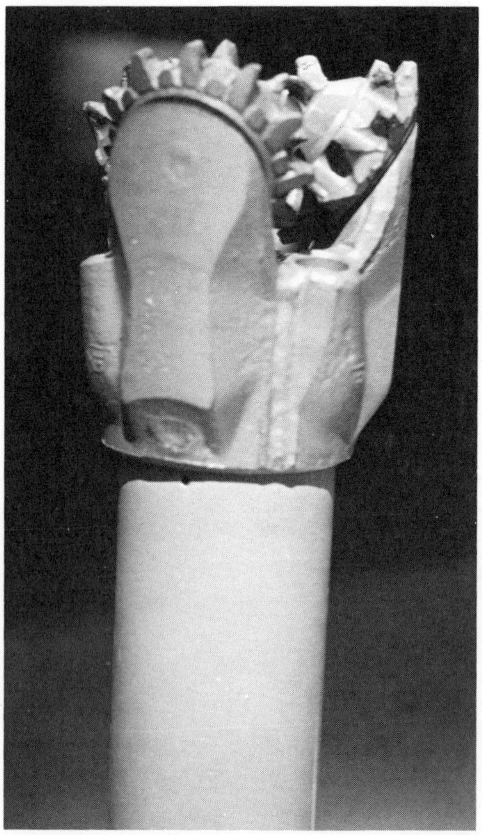

Fig. 6-9. Bits are also called rock drills. Toothed wheels turn against each other and crush rock broken loose at bottom of hole.

Since abrasive rock would wear out an ordinary bit, TUNGSTEN CARBIDE (a hard metal) is used or the die bit may be studded with tiny industrial diamonds.

Even so, drill bits quickly become dull and must be exchanged frequently for sharp ones. This is the reason for the tall derricks. To change bits all of the pipe must be pulled from the hole. If each 30 ft. length had to be disconnected, this method of drilling would be slow and inefficient. The great height of the derrick makes it possible to pull three or four sections from the well and stack them on end inside or near the tower.

OTHER WELL DRILLING METHODS

While cable tool and rotary drilling are the most common, there are other drill methods designed for special needs. These include:

1. Turbodrilling. In this method, the force of the mud being pumped into the well powers a small turbine just above the bit. Its main advantage is the nearness of the power to the bit. There is no need for the heavy lengths of cable called "string."
2. Electrodrilling. This type takes its power from an electric motor located just above the bit.
3. Sonic drilling. The bit vibrates at the same level as a high-frequency sound generated at the surface. This makes the bit cut faster.
4. Flame drilling. Jets of burning liquid rocket fuel burn a hole through rock and soil.

TESTING THE WELL

As drilling progresses, the drill cuttings are examined under ultraviolet light to see if oil is present. When the cuttings indicate that the drill has struck oil, a TESTING TOOL is lowered into the well to check for BOTTOM HOLE PRES-

SURE. This is the pressure of fluids and gases at the bottom of the well. The tool records these pressures as the fluids and gases enter it. Then the tool is sealed and brought back to the surface. Its contents are inspected.

SEALING

After the discovery of oil the well must be sealed. Steel pipe casing is lowered into the well. Sections are telescoped so the points seal out water. Cement is pumped down through the casing where it rises up the outside of the pipe. This completes the seal that separates the oil from other formations. It keeps out sand, mud, and groundwater that might contaminate (mix with) the oil.

PIERCING THE CASING

At this point the oil also is unable to enter the well. Piercing of the well casing opens up the bottom of the sealed well. This job is done with an electric perforating gun. The gun is lowered into the well where it fires "bullets" or shaped charges through the casing into the oil bearing rock. The rock fractures (breaks) and oil flows into the well through the openings created by the perforating gun. See Fig. 6-10.

Tubing, about 2 in. in diameter, is lowered into the well. Oil, gas, and, sometimes, salt water flow up through the tube.

At the top, a set of valves are installed to control the flow from the well. Known as a CHRISTMAS TREE, the valves control distribution of the oil and gas and prevents unwanted flow.

Some of the valves are BLOWOUT PREVENTERS. They prevent the violent GUSHERS which formerly were a common occurance during drilling of new wells. Other valves direct the flow from the wellhead to gathering lines, collecting tanks, or separators. Separators remove the water and gas from the crude oil.

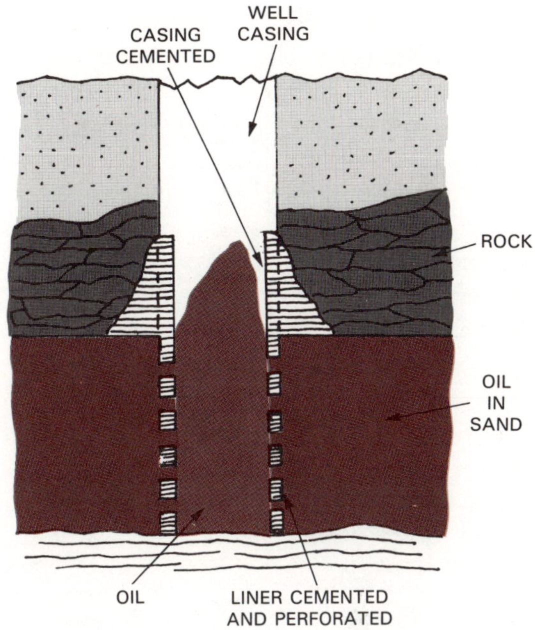

Fig. 6-10. Sketch of the bottom of an oil well. Oil flows into casing through holes and is pumped to the surface.

RAISING THE OIL

Bringing the oil to the surface from several thousand feet down takes a great deal of force. This force is called DRIVE or LIFT. Several methods are used:

1. Natural or reservoir drive.
2. Artificial drive.

NATURAL DRIVE

Natural drive uses the pressure within the oil pool itself. Most new wells have this pressure. It brings the oil to the surface without the aid of pumps or any other type of artificial force. There are three types of natural drive:

1. Water drive, the most efficient. Often, a large pool of salt water has been trapped around the oil pool. As shown in Fig. 6-11, the salt water, being under tremendous pressure, tries to expand. It pushes the oil through the fractures in the rock into the well casing and up the tube.
2. Gas-cap drive. Pockets of natural gas, under

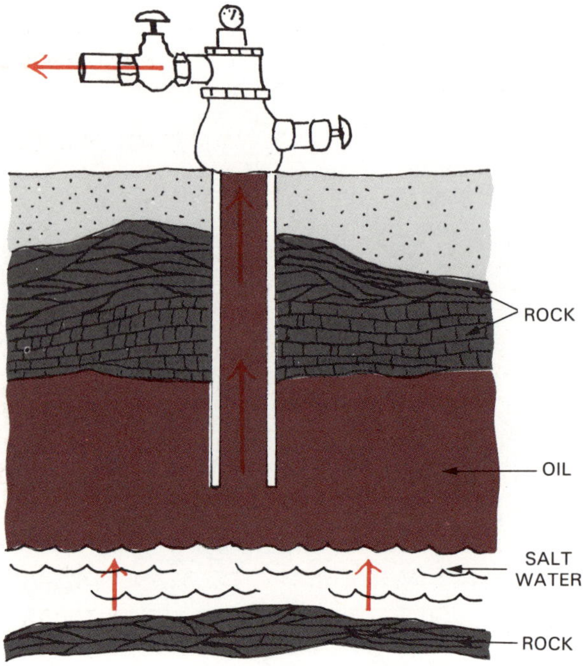

Fig. 6-11. Sketch of water drive. Salt water occurring naturally drives oil into well, up the pipe to the surface.

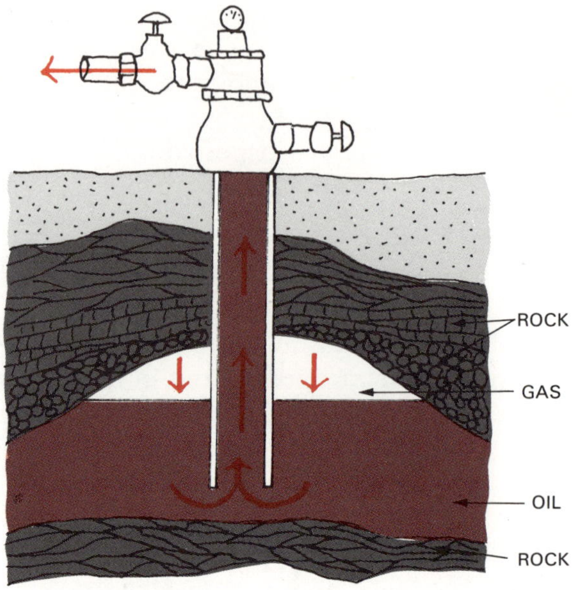

Fig. 6-12. Gas-cap drive. Trapped underground, natural gas in dome exerts pressure on oil to drive it to surface.

tremendous pressure, drive the oil up the tubing. See Fig. 6-12.

3. Dissolved gas drive. There is no gas cap in this type of drive. The gas is dissolved in the oil, Fig. 6-13. The low pressure in the well casing causes the trapped gas to move into the well, dragging oil with it. It is not as efficient as the water or gas-cap drive. Some of the oil gets left behind.

ARTIFICIAL DRIVE

As wells give up their oil, natural pressure drops and eventually the flow stops. Then the remaining oil must be brought up by some other means. One of several artificial methods is used:

1. Pumping. This is the most common. Some wells never flow naturally and must be pumped from the start, Fig. 6-14.
2. Gas-cycling. This method uses gas previously removed from the well to maintain the pressure. The mixture of oil and gas is separated as it comes from the well. The oil goes to storage; later it will be transported to a refinery. The gas goes to a natural gas processing plant. Being a mixture of other

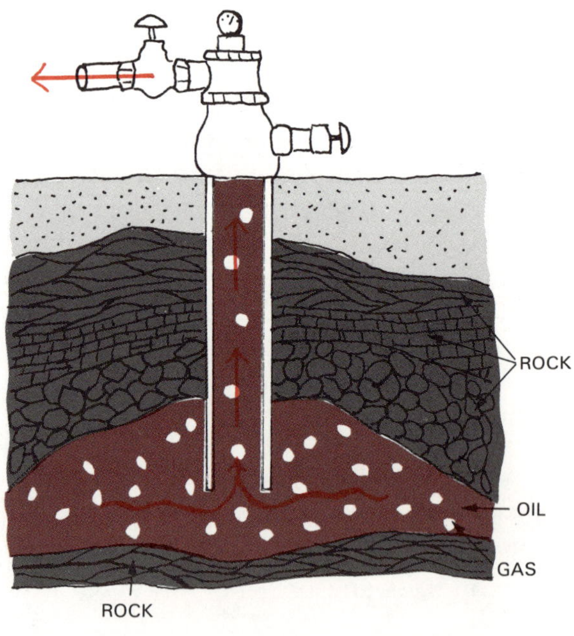

Fig. 6-13. Dissolved gas drive. Trapped gas mixes with oil dragging mixture into well and to surface.

Fig. 6-14. Pumping with a lift pump is the most common method of bringing oil to surface. Top. This rocking beam pump is a familiar sight in oil fields. Bottom. Pumped petroleum is stored temporarily in tanks nearby.

petroleum products such as propane, it is known as WET GAS. These by-products are removed and some of the gas is returned to the oil pool by way of an INPUT WELL, Fig. 6-15. It helps maintain the pressure and keeps the oil moving into the well.

3. Repressuring. In this method the oil reservoir is flooded with water or injected with more gas to raise the pressure.

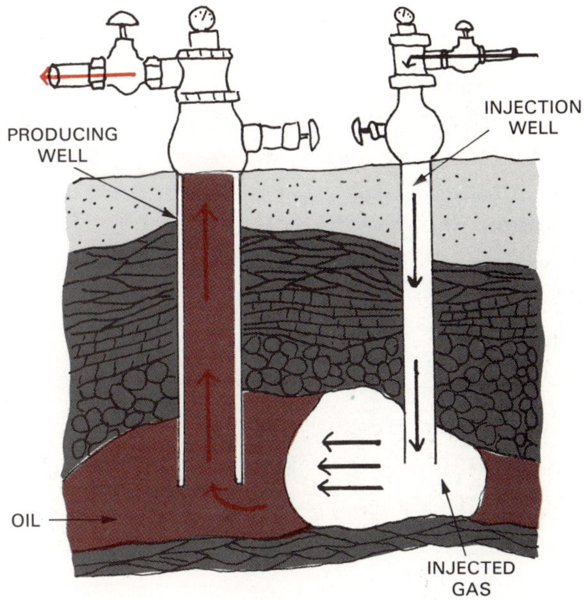

Fig. 6-15. Artificial drive. Gas pumped into an injection well may be used to force oil to the surface.

4. Front burning. This method is used in wells that contain heavy oil. A fire is lighted in the reservoir while air is pumped into it. The pressure of hot expanding gases and the heat make the oil flow better. Another name for this method is "in situ combustion." (In situ means "in place" or "on location.")
5. Steam injection. Steam is generated at the wellhead and fed down the well into the oil-bearing rock. Steam causes the oil to expand and move toward the well. Some of the lighter oils vaporize and these gases provide additional pressure to move the oil.
6. Other methods being considered for oil recovery are: high pressure gases, nuclear explosives, and solar and electrical heat.

INCREASING OIL FLOW

Low pressure and loss of pressure are not the only problems faced in extracting oil. Sometimes low porosity (too dense) of the rock formation keeps oil from the well. Then the porosity of the rock must be improved by:

1. Fracturing treatment.
2. Acid treatment.

FRACTURING

Fracturing actually produces cracks in the rock formations. High-pressure pumps force a mixture of sand and fluid down the hole and into the oil-producing area. The fluids produce cracks in the rock and the grains of sand keep the cracks open as the oil and fluid flow back to the well. See Fig. 6-16.

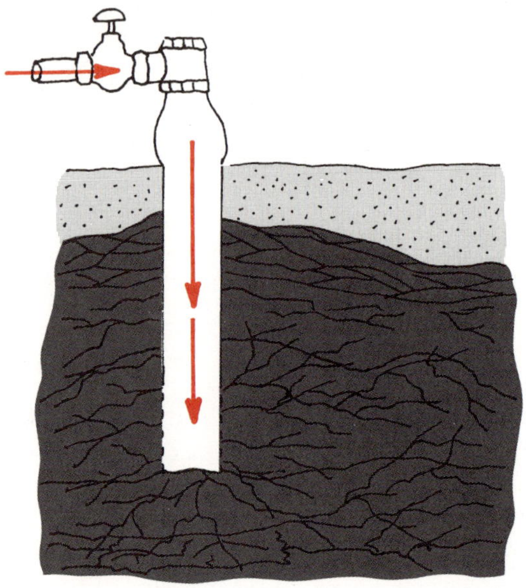

Fig. 6-16. Fracturing of dense rock formation at bottom of an oil well often is done to increase flow of oil in the well.

ACID TREATMENT

Treating the rock formation with acid etches tiny channels that allow the oil to move toward the well. Usually, hydrochloric acid is used. It is pumped into the well along with other chemicals at high pressure. It enters the rock formation at the point where the well has been pierced. Several treatments are usually needed.

OFFSHORE DRILLING FOR OIL

A great deal of oil is buried underwater in coastal areas. Drill rigs are mounted on plat-

A

B

Fig. 6-17. Offshore oil rig. A—Rig is supported by piles sunk deep in the ocean floor. B—Aerial view showing well drilling operation from an offshore platform. (NICOR, Inc.)

forms, Fig. 6-17. The platforms are of three different types:

1. The permanent platform is supported by piles bedded deep in the floor of the ocean. It is large enough to house the drilling crew as well as the machinery and supplies.
2. The small permanent platform, like the larger permanent, is supported by pilings. However, the crew and their supplies are housed in an attached floating tender.
3. A mobile platform is supported by a submersible barge. It is the most economical because it can be used to drill many wells in different locations. In deep water the mobile rig is self-propelling.

UNDERWATER DRILLING RIGS

A newer type of drilling rig can be operated underwater. It is lowered to the ocean floor and anchored. Operation is controlled from the surface.

Another new system is the robot with a mechanical arm. It can perform the same tasks as a deep-diver. The robot is controlled from a boat. Equipped with sonar and television camera, it can locate objects underwater.

CLASSES OF CRUDE OIL

Petroleum is not uniform. There are many complex mixtures of hydrocarbons. Also present in the petroleum are small amounts of sulfur, nitrogen, and oxygen compounds.

The most common petroleum mixtures are:

1. Paraffinic hydrocarbons.
2. Aromatic hydrocarbons.
3. Asphaltic base hydrocarbons.

Also, petroleum may be sweet or sour. Sour crudes contain sulfur and have an unpleasant odor like rotten eggs. Sweet crudes contain very little sulfur and have more pleasant odors.

REFINED PRODUCT	PROCESSES USED TO REFINE	USES OF THE PRODUCT
Light Gases	Pipe still distillation of crude oil	Manufacture of chemicals; Manufacture of gasoline; Gas fuels (liquid petroleum gas)
Gasoline	Distilling of crude in pipe still; Thermal and catalytic cracking; Catalytic and thermal reforming; Hydrocracking and coking of pitch; Polymerization of light olefins; Isomerization of paraffins; Alkylation of olefins with isoparaffins	Fuels for automobiles, trucks, and aircraft; Solvents; Fuels for cooking and lighting
Kerosene	Distilling of crude oil; Cracking of heavy fractions	Jet fuel; Fuel for cooking, illuminating, and heating; Solvents
Gas Oils	Distilling of crude oil; Cracking of heavy fractions; Hydrocracking of residual oils; Vacuum distilling	Residential and light industry fuel; Manufacture of chemicals; Manufacture of gasoline; Solvents; Asphaltic materials for roads
Lubricating Oils	Distilling of crude oil; Hydrocracking of heavy fraction; Solvent refining of residual oils	Many kinds of lubricants and oils; Pharmaceutical white oils; Waxes and petrolatums; Petroleum jelly; Asphalt; Greases
Residual	Vacuum distillation of petroleum; Distilling of synthetic petroleum	Manufacture of gasoline and fuels; Chemicals; Source of coke; Asphalt
Coke	Thermal cracking of residuals and pitches	Fuel; Manufacture of iron and steel; Industrial electrodes

Fig. 6-18. Major products listed above are all derived from crude petroleum.

Fig. 6-18 shows the major products that are made from crude oil. As the crude is received from the well it is not usable. It must be processes. This is called refining.

REFINING OF OIL

Refining is the process of breaking crude oil down into various usable liquids and semisolids. In its original state, crude oil's carbon atoms are arranged in many different ways. Crude with many carbon atoms forms a thick, heavy petroleum such as asphalt. Crude with fewer carbon atoms makes up the lighter elements such as gasoline. The products of crude are called FRACTIONS. They are separated by refining processes which include:

1. Separation.
2. Conversion.
3. Treating.

SEPARATION

The main separating processes are:

1. Distillation.
2. Solvent extraction.
3. Adsorption.
4. Refrigeration.

DISTILLATION

Distillation separates different parts of the petroleum through their different boiling points. Refineries today use a process called fractional distillation. First the petroleum is heated to 700°F (371°C). Heat is provided by a PIPE STILL. This is a type of furnace with many coils of tubing running through it. The oil is heated as it flows through the tubing.

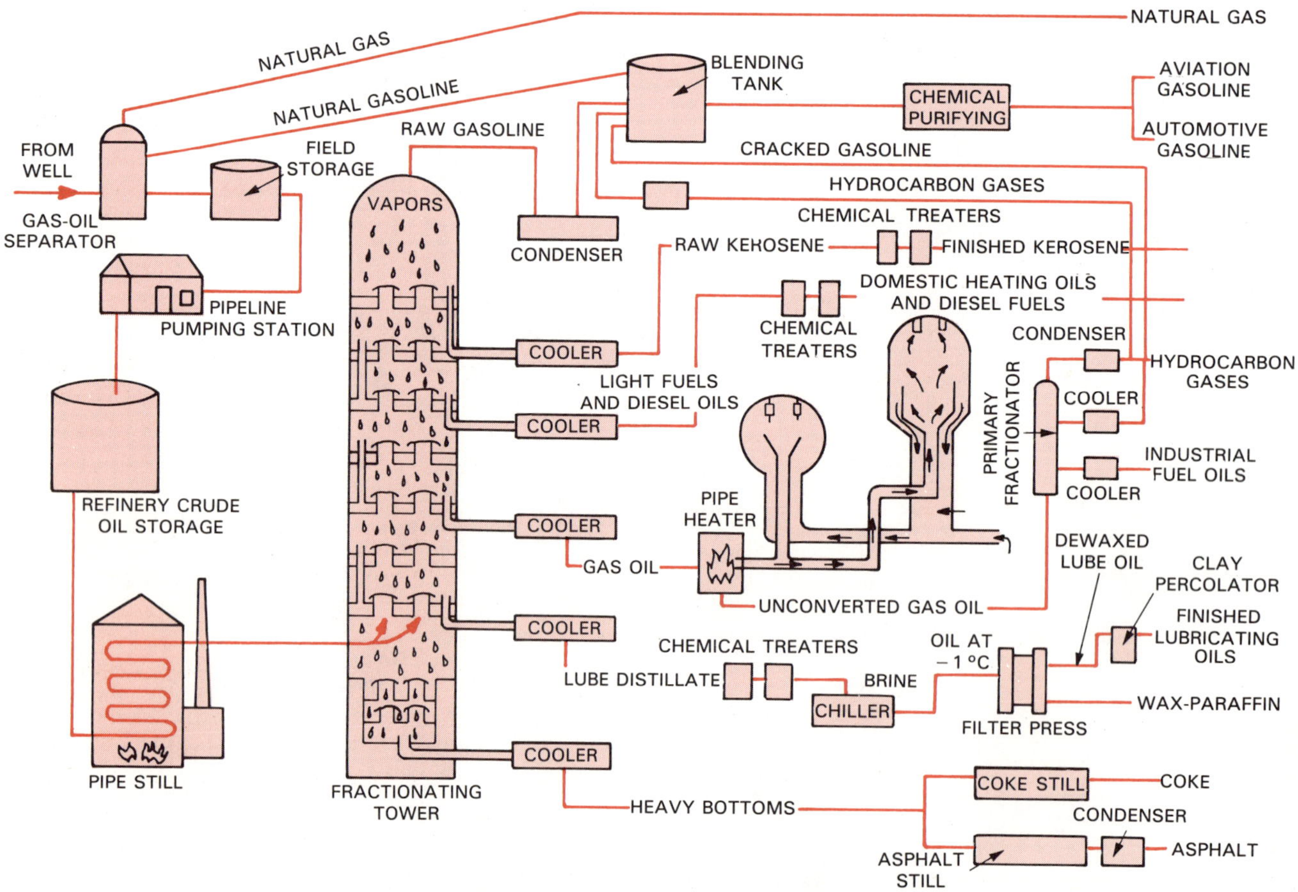

Fig. 6-19. Cutaway of fractionating tower. Vapors rise to different levels and are caught by trays.

Hot liquids and vapors resulting from the heating are released into a FRACTIONATING or BUBBLE TOWER. The tower is about 100 ft. tall and has an arrangement of trays at different levels to collect the products of the distillation. See Fig. 6-19.

The rising vapors pass through holes in the trays and are caught and condensed by BUBBLE CAPS placed above the holes, Fig. 6-20.

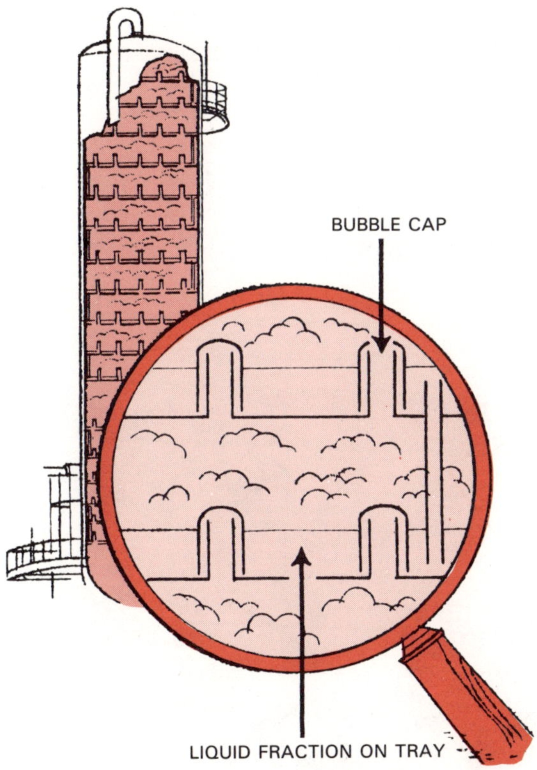

Fig. 6-20. Bubble cap catches rising vapors and deposits droplets into trays.

The various petroleum products condense at different temperature levels in the tower. The heaviest fractions (products) collect at the bottom. They are either processed into asphalt, used as fuel oil, or are reprocessed into lighter products. The residue is sold as coke. This is a pure carbon that is used to make steel.

Gas oils condense at the next level, then kerosene and naptha. Straight-run gasoline condenses near the top of the tower. Some gases do not condense and these are piped off at the top of the tower.

SOLVENT EXTRACTION

Solvent extraction is used mainly for refining different fractions produced in the fractionating tower. The solvents take into solution (mix with) the undesirable substances and can be drawn off with the solvent. One product produced this way is lubricating oil.

ADSORPTION

In the adsorption process a very porous solid material is used to collect the undesirable substances from a petroleum fraction. Silica gel is a popular absorbent for this purpose.

REFRIGERATION

In the refrigeration process, cooling of unwanted substances, such as paraffin, causes them to solidify. In this condition they can easily be removed from the oil.

CONVERTING OIL FRACTIONS

After liquids have been separated by distilling they can be changed into other products. Heavy fuel oils can be made into high octane gasoline. The conversion process is called CRACKING.

Cracking changes the heavy molecules into small, light ones. There are two cracking methods:

1. Thermal. The first to be developed, this method depends upon high temperatures and pressure.
2. Catalytic. The more efficient of the two, this process also required high temperatures. Pressure required is much lower than in thermal cracking. Catalysts, such as silica-alumina, speed up the chemical reactions. Catalytic processes include:

a. Fluid catalytic cracking. The vaporized mixture of gas and oil and a powdered catalyst circulate together at high temperatures through a system of pipes and reactors.
b. Hydrocracking. This process uses hydrogen and catalysts under high pressure. It yields more gasoline and less residue than ordinary methods.

ALKYLATION

Alkylation is another conversion process. It is designed to produce liquids closely related to gasoline from refinery gases. Its importance is not that it is another way to make gasoline. The product of alkylation is highly prized for its antiknock properties. (Knocking is the "pinging" noise sometimes heard in automobile engines upon acceleration. It is caused by the fuel igniting in the cylinder too soon.) Adding the alkylated product to the gasoline causes detonation (burning) of the fuel to take place more slowly.

POLYMERIZATION

Like alkylation, polymerization is a way of making gasoline from refinery gases. However, only the olefinic gases react with the catalyst (material that causes the chemical reaction). Paraffinic gases are not changed by the catalyst which is usually phosphoric acid.

Polymerization was used during World War II to make isooctane for aviation gasoline. The concentrated polymers have antiknock qualities but they are not as good as those obtained from the alkylation process. They are used in the manufacture of lubricating oils and plastics.

ISOMERIZATION

Isomerization converts chemical compounds in refinery gases into its isomers. (An isomer is an ion or molecule of any substance.) The catalyst is usually aluminum chloride, antimony chloride, or platinum.

The product of isomerization is used as an additive in gasoline to increase its octane rating and reduce knocking.

BLENDING GASOLINE

Refineries add antiknock compounds to gasoline produced by fractionating. Usually about a teaspoonful to a gallon is enough. This would amount to about 1 part in 1300.

TRANSPORTING OF OIL

One of the advantages of petroleum and its various products is the ease of transport. The fastest and easiest method is through pipelines, Fig. 6-21. In the United States, pipelines are the largest single oil carrier. A pipeline network of more than 200,000 miles carries crude oil and refined petroleum products.

Fig. 6-21. Pipelines transport petroleum quickly and easily. (Texaco, Inc.)

Pumping stations spaced about 150 miles apart move the petroleum through the pipelines at speeds of two to three mph. Modern pumping stations have pushbuttons or computer controls.

Other modes of transporting oil include tankers, barges, railroad tank cars, and over-the-road tank trucks. See Fig. 6-22. Fig. 6-23 shows an oil unloading facility that has been constructed in the Gulf of Mexico for the use of supertankers. Called LOOP (for Louisiana Offshore Oil Port) it allows three tankers to unload their oil cargo at the same time. The oil is pumped ashore to refineries and storage. Because of a network of pipelines, oil unloaded in New Orleans can be piped as far north as Chicago and Buffalo.

OIL AS FUTURE ENERGY SOURCE

At present, 75 percent of our energy consumption in the United States comes from petroleum.

Fig. 6-22. Tank trucks deliver fuels to point of use. Gasoline delivery to a service station is shown here.

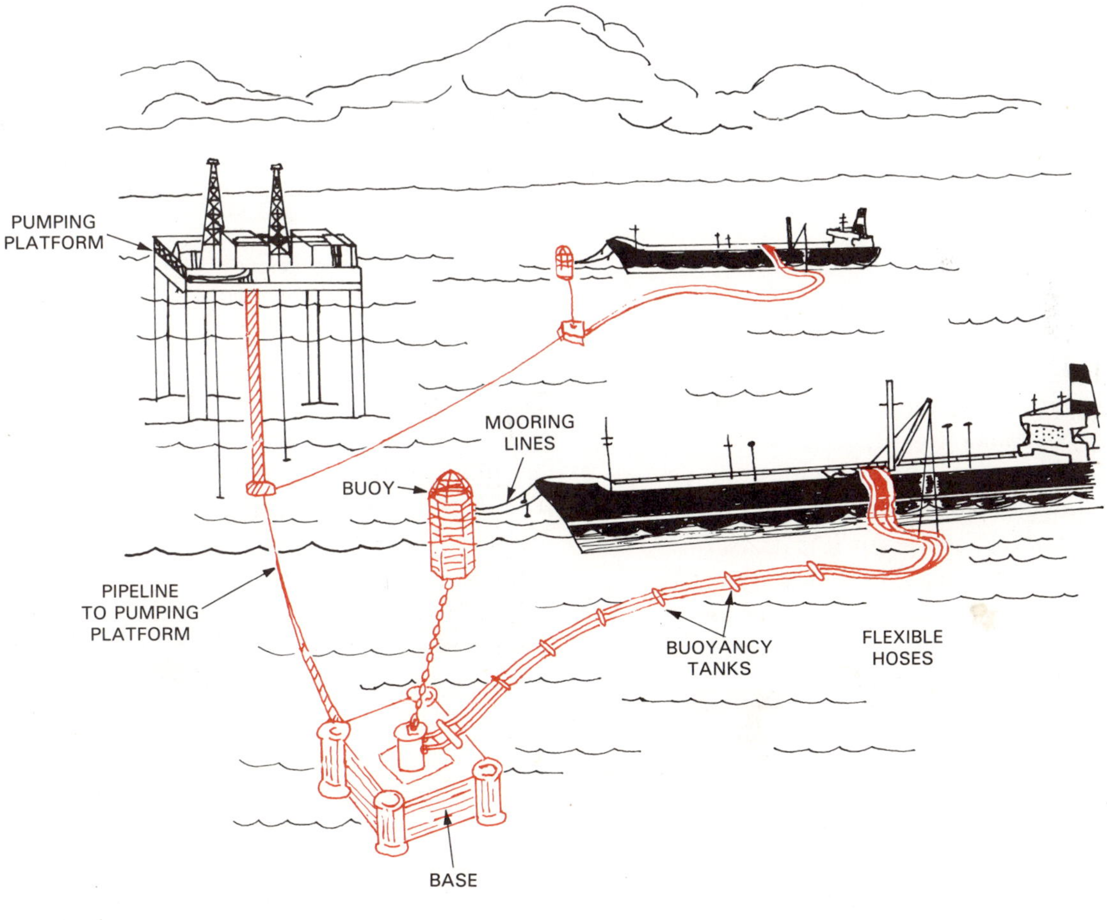

Fig. 6-23. Sketch of offshore oil unloading facility being built in Gulf of Mexico. It unloads oil supertankers which cannot enter ports because of their huge size.

In 1979 about 18 million barrels per day were used. Reserves of petroleum (amount left in the ground) is getting scarce. Future energy supplies must come from other resources. Supplies are discussed in Chapter 7.

NATURAL GAS

As you have learned from a study of petroleum, natural gas is closely related to petroleum. You also know that all oil deposits contain natural gas. However, not all natural gas deposits contain oil.

EXPLORATION AND DRILLING

Gas prospecting is done in much the same way as the search for oil. Sometimes the gas is found seeping up from the underground deposits. Usually geologists look for rock formations that suggest the presence of porous rock layers covered over by solid rock. This is the condition necessary for an oil or gas pool.

GAS DEPOSITS

Gas is often found trapped in natural underground reservoirs. There are three main types of reservoirs:
1. Sandstone. In this type the reservoir is formed by sedimentary rock. This is rock formed of mineral deposits such as quartz.
2. Carbonate. This type is made up on sedimentary rock that came primarily from the shells of sea life. Another name for it is limestone.
3. Other rock formations. These are generally igneous (came from volcanos) or metamorphic rocks. The latter are formed from both igneous and sedimentary rock. Also included in this category are formations of shale. This is a rock made up of mud, clay, or silt.

MAKEUP OF NATURAL GAS

Natural gas is a mixture of hydrocarbon compounds and small quantities of several nonhydrocarbons. It can be found as a pure gas or it may be mixed into a solution with oil in natural underground reservoirs. The hydrocarbon compounds include: methane, ethane, propane, butanes, and pentanes. The nonhydrocarbons include: carbon dioxide, helium, hydrogen sulfide, and nitrogen. The fuel sold under the name "natural gas" is mostly methane. It may also contain small amounts of ethane. Free natural gas which is found alone in its natural reservoir is called NONASSOCIATED GAS. When it is found in the same reservoir as crude oil it is called ASSOCIATED GAS whether it is separated from the oil or in solution with it.

GAS RESERVES

In 1974 the total natural gas reserves in the United States were thought to be about 237 quarillion cu. ft. The largest reserves are found in Texas, Louisiana, and Alaska. These states account for about three-fourths of all the U.S. reserves.

TRANSPORTATION OF NATURAL GAS

Natural gas is usually transported by one of three modes:
1. Pipeline.
2. Railroad tankers, Fig. 6-24.
3. Truck tanker.

The mode being used depends upon:
1. The distance the natural gas must be moved.
2. Whether the terrain is mountainous or suitable for truck or rail transport.
3. How complicated the distribution system is.
4. The kind of gas that is to be moved.
5. The cost of building and operating the transportation system.

PIPELINE TRANSMISSION

Gas must be moved through pipelines by pressurizing it with pumps. These operate somewhat like the compressors used to pump up tires or provide air pressure for paint spraying.

Fig. 6-24. Railroad tank cars are also designed to transport natural gas long distances. (Evans Products)

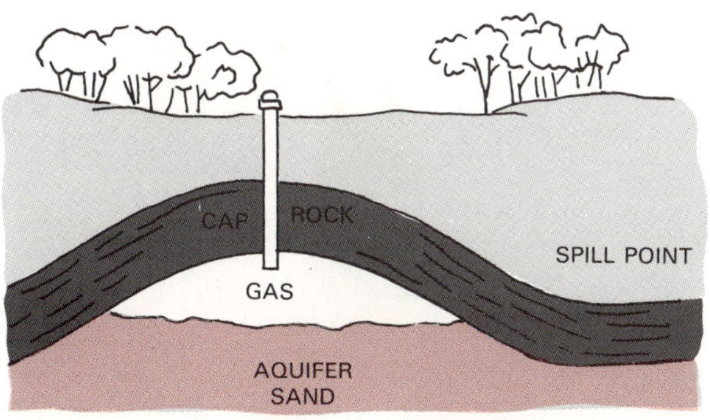

Fig. 6-25. Some natural gas is stored in natural underground aquifers such as this until it is used.

GAS PRESSURE DURING TRANSPORT

Pressure in long distance pipelines can be anywhere from 500 to 1000 psi (3445 to 6890 kPa). Compressors are located about every 50 miles to restore the pressure lost by friction. The natural gas may travel at a speed of about 15 mph (24 kmh). Valves and regulators are also located along the way. These devices are needed to control pressure or to stop the flow entirely in case of a rupture (break) in the pipeline.

NATURAL GAS STORAGE

At certain times of the year, demand for natural gas as a fuel is greater than the ability to extract and transport it. To meet the demand, natural gas is pumped and stored for later use.

Usually, storage is underground. Natural reservoirs receive the gas from pipelines during the warm months when demand is less.

TYPES OF STORAGE

Most gas storage is in old gas or oil fields. Supplies in these reservoirs are at such a low pressure that they no longer are in production. The gas is pumped into them through the wells originally drilled to extract the gas and oil. The gas is forced into the old wells with compressors.

Some gas is stored in AQUIFERS, Fig. 6-25. An aquifier is a rock formation underground which holds large quantities of water that has filtered down from the surface. The rock is porous so that water soaks into the rock. The pores are full of water.

Gas pumped into the aquifier under pressure, displaces the water, forcing it to move deeper into the underlying rock.

The United States has 44 such aquifiers. Most are found in the midwest. Illinois has 19, Indiana 11, and Iowa 6. Several storage reservoirs have been established in abandoned coal mines. Others are in natural salt domes.

MANUFACTURED STORAGE

City gas companies use huge cylindrical tanks to store gas before it is distributed to users. The tanks are closed at the top and open at the bottom. The tanks float in pools of water and can rise or fall according to the supply of gas in them. The weight of the heavy tank keeps the gas under pressure at all times and forces it out into the gas mains for distribution to gas customers.

LIQUIFYING NATURAL GAS

Natural gas may be converted to a liquid for use in transport vehicles. Liquification reduces its volume and makes the gas more practical to move overseas and to store.

DO YOU KNOW THESE TERMS?

Alkylation, aquifer, asphalt, associated gas, blowout preventers, bottom hole pressure, bubble caps, butane, cable tool method, cable tool string, carbonate, combustible, cone and roller bits, cracking, cracking process, Christmas tree, crude oil, derrick, development well, dolomite, drill pipe, drive, electrodrilling, ethane, fishtail bit, fossils, flame drilling, fractionating tower, fractions, geologists, gravimeter, gusher, input well, in situ, isomerization, lift, limestone, magnetometer, methane, mud, natural gas, nonassociated gas, petroleum, pipe still, polymerization, propane, refining, rotary method, sandstone, seismograph, shale, solvent extraction, test core, tungsten carbide, turbodrilling, wildcat well.

SUGGESTED ACTIVITIES

1. Using light wood pieces build a small model of an oil well derrick and explain to the class how it works.
2. Secure literature from an oil company on the history of petroleum. Take part in a class discussion on why it became an important energy source.
3. Draw up a list of energy converters which depend on petroleum for their source of energy.
4. If your lab is equipped with different types of engines, discuss with your instructor methods of testing the suitability of different fuels for each of the engines.
5. Referring to Fig. 6-1 in the text, construct a cutaway model of oil bearing rock formation.

TEST YOUR KNOWLEDGE

1. Geologists looking for oil consider the following geological formations as an indication that oil might be present. Check the correct answers:
   a. Dolomite.
   b. Limestone.
   c. Igneous rock.
   d. Sandstone.
   e. Shale.
   f. All of the above.
2. When drilling for oil a ________________ ________________ bit or one studded with tiny ______________ ______________ is best when drilling through abrasive rock.
3. Oil raised to the surface by its own pressure is

called artificial drive. True or False?

4. Which of the following are products derived from crude oil?
   a. Natural gas.
   b. Gasoline.
   c. Kerosene.
   d. Diesel oils.
   e. Fuel oils.
   f. Lubricating oils.
   g. Petroleum chemicals.
   h. All of the above.
   i. None of the above.
5. What is a pipe still?
6. Where is natural gas found?
7. List the three methods for transporting natural gas.

Some oil reserves are trapped in rock. This oil shale mine is located in the mountains a mile or more above sea level. Private corporations are experimenting with methods of extraction. (Gulf Oil Corp.)

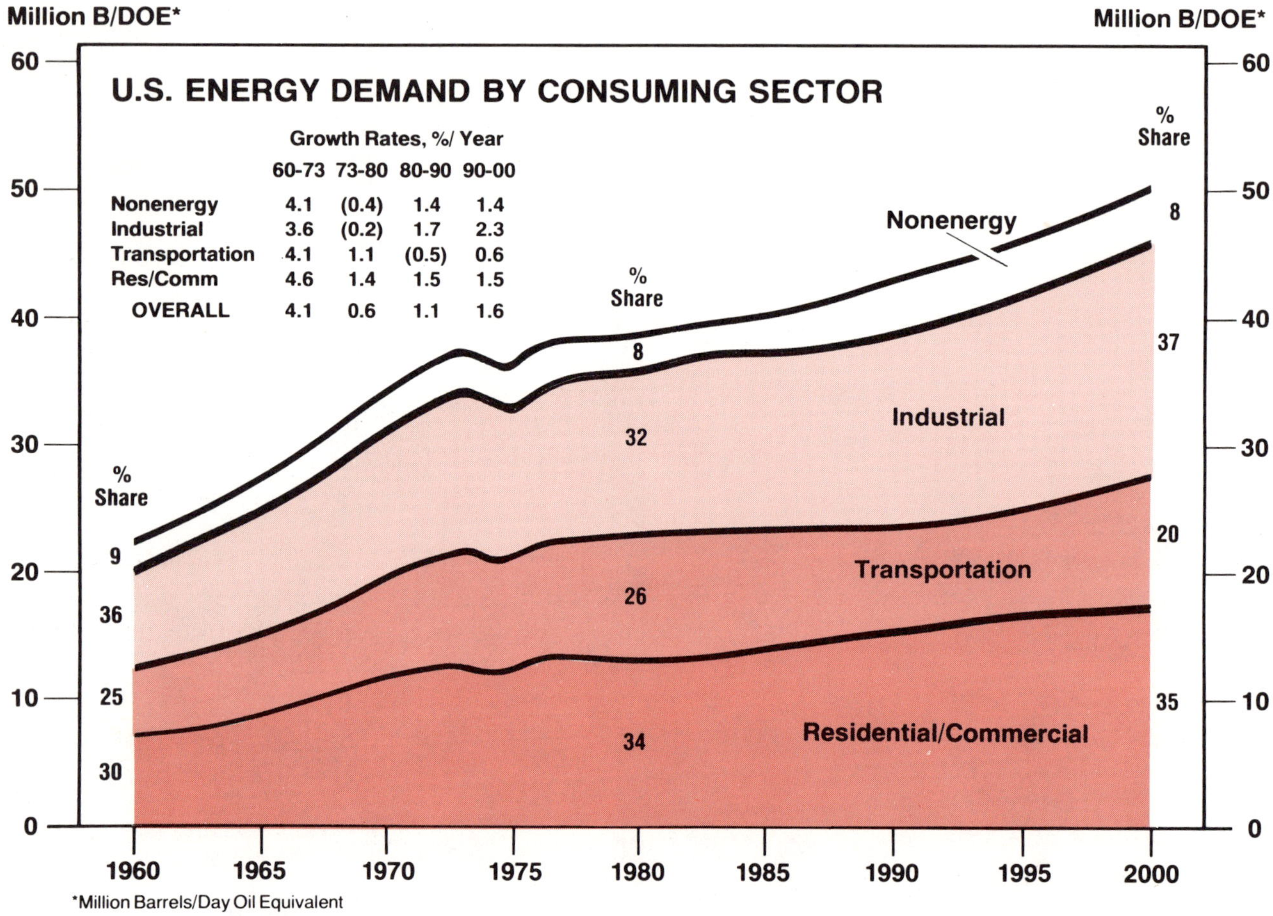

Fig. 7-1. Energy usage in the United States grows year by year. In the year 2000 it is expected to equal 106.5 quad. A quad is 1 quadrillion (1,000,000,000,000,000) Btu. (Exxon Company, U.S.A.)

7 OUR ENERGY CONSUMPTION

The information given in this chapter will enable you to:

- ☐ Discuss the usage of energy in the United States in terms of barrels per day and Btus.
- ☐ List the four major energy-using sectors in our society.
- ☐ Discuss the extent of fossil fuel energy supplies still remaining in the U.S.
- ☐ Define the terms, proven reserves, and future reserves.

Our American society consumes large quantities of energy. In 1980 we used the equivalent of 38.7 million barrels of oil daily to provide enough energy for our needs. By 1990 it is expected that the United States will use 43 million barrels daily and by the year 2000, 50.3 million barrels. See Fig. 7-1.

Our energy consumption is also figured in terms of the amount of heat the energy can produce. The conventional unit of measure for heat is the British thermal unit or Btu. (The metric unit for energy measurement is the joule (J). A Btu equals 1055.6 J or 1.055 kJ.) A barrel of oil is equal to 5.8 million (5,800,000) Btu. In 1980 the United States consumed 38.7 million barrels of oil daily. If we multiply that number by 5.8 million and the result by 365 days we will know the amount of energy consumed that year:

$$5,800,000 \times 38,700,000 \times 365 = 81,927,900,000,000,000 \text{ Btu or } 81.9279 \text{ Quad.}$$

(It is easier to measure such large quantities in quads. A quad is equal to 1 quadrillion [10^{15}].)

A Btu is hard to envision, but it is roughly the amount of heat given off by one kitchen match. Can you imagine almost 82 quadrillion matches?

ENERGY'S IMPORTANCE TO US

Energy is important to our well-being. Without it our industries would not function; our homes and cities would be dark and cold; our planes, automobiles, and trains would be useless. Our means of earning a livlihood would all but disappear. We would be reduced to a level of existence where most work would be done by hand or with the help of animals.

ENERGY-USING SECTORS

Who uses this energy? There are four major groups of energy users in the United States:

1. The NONENERGY sector. This group, Fig. 7-2, includes all those who use fossil fuels as a raw material to produce other products or goods. The energy material—gas, oil, or coal—is used for the manufacture of products such as chemicals, medicines, asphalt, grease and oil, plastics, synthetic fabrics, and steel. In 1980, this sector used about 6.8 quads of energy.
2. INDUSTRIAL sector. In this category, Fig. 7-3, are all types of mining, manufacturing,

Fig. 7-2. Some enterprises such as this asphalt plant use energy as a raw material for products.

and farming businesses. They use energy in all its forms: gas, oil, coal, and electricity. In the year 1980, this sector consumed 26.67 quads of energy.

3. TRANSPORTATION sector. All methods of moving people and materials are part of this group. Included are automobiles, trucks, trains, ships, planes, and pipelines. See Fig. 7-4. Since they can use no other energy source, transport vehicles account for half of the petroleum used by all sectors. Their share of the energy used in 1980 was 20.96 quads.
4. RESIDENTIAL/COMMERCIAL sector. See Fig. 7-5. Included in this group are all private and public buildings such as homes, stores, office buildings, hospitals, and schools. They consumed 27.5 quads of energy in 1980.

ENERGY SUPPLIES

Meeting the needs of all the energy-using sectors just discussed takes a constant, reliable source of supply. For as long as most of us can remember there have been plentiful supplies of petroleum, natural gas, and coal. Where it came from or how long it would last was of little immediate concern to the average person. Then came the Arab oil embargo of 1973. It served to call national attention to the threat of energy shortages.

At the time of the embargo, the United States was already importing large amounts of fuel from foreign countries. Imports included 40 percent of our crude oil, 40 percent of our fuel oil, and smaller percentages of gasoline and jet fuel (kerosene).

The immediate result of the embargo was that imports of fuel fell 3 1/2 million barrels per day below the projected (expected future) demand. The United States had no oil in storage and could not increase production from its own wells. Available supplies had to be spread among all

Fig. 7-3. Energy users from the industrial sector. Top. Farming. Middle. Mining. Bottom. Paper manufacturing.

Fig. 7-4. Transportation sector involves industry that moves people and materials. Top. Trucks and cars. Middle. Train. Bottom. Planes.

Fig. 7-5. Residential sector. Homes use oil, gas, electricity, and sometimes coal or wood to provide heat and light and to operate appliances.

Fig. 7-6. Higher prices for crude oil are reflected in gasoline prices which, by 1979, had advanced to well over $1 per gallon.

energy-using sectors. In 1973, 46 percent of all energy consumed in the U.S. was oil. To keep industry going, the government allowed it almost all the energy it needed. The burden of cutting back was left mainly to the residential/commercial sector. Gas station allotments were cut back. Long lines formed at the gas pumps. Commuters began forming car pools to get to their jobs. States began imposing 55 mph (89 k/ph) speed limits on highways.

Along with the scarcity of fuel came huge increases in the price of crude oil. In 1973, just before the crisis, the price of a barrel of crude ranged from $3.89 for domestic to $4.08 for imported. By 1979, although the embargo was gone and all but forgotten, the price of crude oil had moved up to an average of $21 a barrel. This represented an increase of almost 700 percent in six years. Prices at the gas station pump had moved up to more than $1.00 per gallon. See Fig. 7-6.

LOOKING AHEAD

The crisis of 1973 did two things. It made everyone realize the importance of energy to all sectors of our society. It also drew our attention to the disturbing fact that the fuels we depend so heavily upon will one day be depleted, some of them in our own lifetimes. At the same time, the more scarce these energy sources become, the more expensive their cost.

HOW MUCH IS LEFT?

We can only speculate how long our supplies of petroleum, coal, and natural gas will last. We are able to estimate with some degree of certainty the amount of our reserves. We can also project with somewhat less certainty our annual consumption for some years ahead.

RESERVES

Proven reserves are those fuel supplies which have been located and estimated with reasonable certainty and which can be recovered with present technology. These proven reserves in the United States (including uranium) are 6388 quad Btu. See Fig. 7-7.

There is another classification of fuel supplies known as resources or future reserves. These are stocks that have been estimated to exist with but a small degree of certainty. There is a possibility that they can be recovered sometime in the future. See Fig. 7-8.

U.S. FOSSIL FUEL PROVEN RESERVES

FUEL	QUANTITY	BTU EQUIVALENT (IN QUADS)
Coal	260 billion short tons	5,460
Oil	34 billion barrels	714
Gas	209 trillion cu. ft.	214.225
Liquid natural gas	6 billion barrels	.25
Total		6,388.475

Fig. 7-7. Proven reserves of fossil fuels in the U.S. are a little more than 6388 quad Btu. Uranium adds another 267 quad.

Fig. 7-7 and Fig. 7-8 show that our most plentiful fuel resource is coal. It accounts for 90 percent of all our reserves. At current levels of consumption, supplies could last for about 300 years.

FUTURE RESERVES

ENERGY SOURCE	AMOUNTS AVAILABLE	BTU EQUIVALENT (IN QUADS)
Coal	1.8 trillion tons	37,863
Oil	139 billion barrels	806.2
Natural gas liquids	28 billion barrels	.115
Natural gas	895 trillion cu. ft.	917.375
Total		39,586.49

Fig. 7-8. Future reserves or resources are stocks of fossil energy of which geologists and energy companies are less certain. They may be recoverable in the future. To the fossil fuel future reserves we can add 3.8 billion short tons of uranium supplies.

Oil reserves, estimated at 34 billion barrels, may not last out the turn of the century according to some oil companies and government officials. Others believe that reserves will last to about the year 2020. Oil production in the lower 48 states will, in the words of experts, begin to decline during the 1980s.

Natural gas has been in short supply in some areas for some time. Demand has been outstripping production since 1973. Production from existing reserves will continue to decline into the 21st century.

ENERGY ALTERNATIVES

The history of energy shows that through the centuries, societies have shifted their dependence from one source to another several times. Ancient Rome experienced a shortage of firewood and had begun to turn to other sources. Ruins of ancient buildings show evidence of passive solar construction. Our own country once depended upon wood for a large part of its energy. Later, the shift was to coal. Still later, the emphasis shifted to petroleum which was more desirable because its by-products were less polluting.

Now the need is urgent to change our energy appetites once again. On a short term basis, conservation of energy will help stretch out or save the fossil fuel resources remaining. A long-term approach will include the development of alternate energy sources including solar, nuclear, kinetic, and thermal. Each of these sources will be discussed in the following chapters. Our role in conservation will be the subject of still another chapter.

DO YOU KNOW THESE TERMS?

Future reserves, industrial sector, nonenergy sector, proven reserves, quad, residential/commercial sector, transportation sector.

SUGGESTED ACTIVITIES

1. Prepare a report on how the method of heating homes in the central and northern

regions of North America differs from practices in other cold climates. Ask the school librarian to help you find reference materials.

2. Interview homeowners in your neighborhood on the effect of rising energy costs on home heating bills. Try to estimate average rise (by percentage) from 1973 to the present. Report to the class.
3. Secure an EPA report on estimated fuel consumption for various models of automobiles manufactured. Prepare a graph showing difference in fuel costs between a 4-cylinder compact auto and an 8-cylinder full-size family auto based on 12,000 miles of driving and an average cost of $1.28 for unleaded gasoline.
4. Draw up a list of the products and services you personally require and indicate which energy-using sector contributes the service or product.

TEST YOUR KNOWLEDGE

1. Name the energy consuming sectors in your society and define each one.
2. In 1973, ________ percent of all energy used in the United States was derived from oil.
3. __________ reserves are those fuel supplies which have been located with reasonable certainty and which can be recovered with __________ __________.
4. The proven reserves of fossil fuels and uranium in the U.S. amount to:
   a. 61.63 quad.
   b. Over 6000 billion Btu.
   c. 6163 quadrillion barrels.
   d. 6388 quad Btu.
   e. 6655 quad Btu.
   f. 9055 quad Btu.
   g. none of the above.

8 SOLAR ENERGY

The information given in this chapter will enable you to:

- ☐ State the current theory about the origin of the sun and explain the source of its energy.
- ☐ Define electromagnetic radiation and explain how it travels through space.
- ☐ List the types of radiation.
- ☐ Explain the terms: solar constant, insolation, entropy, Langley, and calorie.
- ☐ Give examples of how solar energy is stored.
- ☐ Cite examples of how solar energy is used as a direct source of energy.

As you learned from Chapter 1, most of our energy comes either directly or indirectly from the sun. "Sol" is a word for sun. Therefore, sun energy is usually called solar energy. However, very little of it is put to use directly. Explaining why is not easy. To put it simply for now, it is very hard to collect widely scattered sunshine. Further, it is expensive to build collector equipment.

To understand the problems of collecting solar rays it is necessary to learn more about the nature of the sun. It is also helpful to understand radiation of energy. (This is the way solar energy travels.)

THE SUN

The sun's diameter is 110 times that of the earth. Being made up of hydrogen, it has half the density of the earth and twice the mass (weight). See Fig. 8-1.

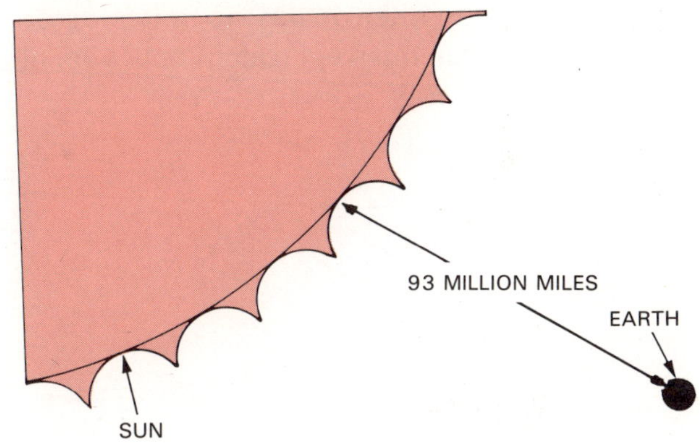

Fig. 8-1. The sun is 110 times the size of the earth. If the sun were 18 inches in diameter the earth would be 1/16 inch across.

Light and heat radiating from the sun are caused by thermonuclear reaction taking place at its core. The temperature at the sun's center is 10,000,000°C or about 18,000,000°F. At the sun's surface, the temperature is 5500°C (10,000°F).

From almost 93 million miles (150 million km) in space, the sun bathes the earth in huge amounts of energy. It is hard to imagine how much heat and light it provides. If it could be collected, two days of sunlight on earth could provide more power than all of the country's fossil fuel reserves. Considering that the sun is expected to shine at least for the next 5 billion years, the supply is certainly renewable.

Scientists estimate that the sun is over 5 billion years old. In that amount of time it has converted four million tons of solar matter every second into radiant energy.

Only one of every 2 billion parts of the sun's rays reaches the earth. This is easier to understand if you consider how much smaller earth is than the sun. See Fig. 8-2. Radiation moves outward from the sun in all directions just like light from a light bulb when it has no reflector or globe. The earth can intercept (catch) only a small amount of the energy.

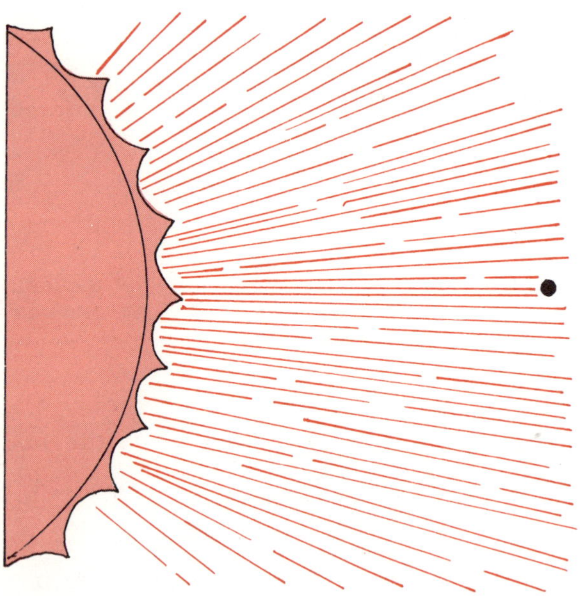

Fig. 8-2. So immense in size is the sun that only 1/2,000,000,000 (one-two-billionth) of its rays even strike the earth.

THE SUN'S ORIGIN

How the sun came to be is not known for certain. One currently held theory is based upon a prediction of Albert Einstein. He reasoned that energy can become matter and that matter can be changed into energy. His prediction is the basis of nuclear physics today. It also led scientists to theorize that the sun (and, thus, the universe) started with a big "bang."

The bang, they theorize, was the creation of a ball of energy. For about the first second of its life the ball had hardly any substance, just energy. It radiated (expanded) with the speed of light and began to cool. The wavelength of the radiating energy became longer, changing to X rays, ultraviolet, visible, and infrared rays. This took billions of years.

As cooling continued, particles of energy paired off and became matter. This process took millions of years. Eventually hydrogen and helium atoms were formed creating the sun.

NUCLEAR FUSION

A joining of the nuclei (centers) of atoms creates the heat and light given off by the sun. The process is called nuclear fusion. It starts up when gravity creates great pressure on the hydrogen atoms. They begin to heat up. The molecules of hydrogen move faster and faster due to the heat buildup. They collide with each other with such force that their nuclei fuse. As this occurs, a small fraction of their mass is released as energy. This energy pushes to the surface of the sun and radiates outward.

HOW SOLAR ENERGY TRAVELS

How is it that solar energy can move 93 million miles through space and still have enough "punch" left to produce heat and light? The answer lies in the way heat and light waves are thought to travel through empty space. This theory is known as ELECTROMAGNETIC RADIATION. You read about it in Chapter 1.

HOW ELECTROMAGNETIC RADIATION IS CREATED

As you know from Chapter 1, an atom is made up of electrically charged particles. A cluster of positively charged particles make up the center or NUCLEI. The protons attract negatively charged particles called ELECTRONS which travel around the nuclei in elliptical or circular paths as pictured in Fig. 8-3. (Elliptical means oval shaped.)

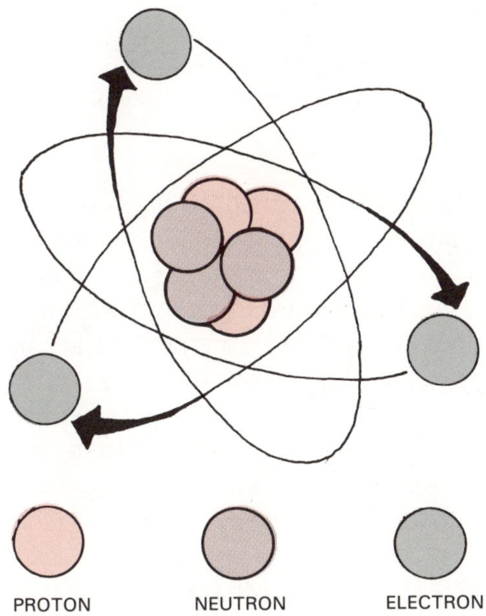

Fig. 8-3. A single atom has a nucleus (core) of particles called neutrons and protons. Around this core are several orbiting electrons that are always in motion.

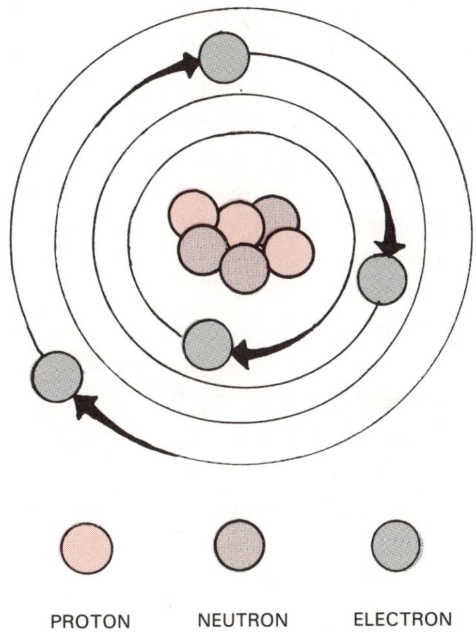

Fig. 8-4. The orbits of electrons around the nucleus of the atom are not along the same paths or at the same distance from the nucleus.

Not every electron travels the same distance away from the nucleus. The paths are at different levels, Fig. 8-4.

A burst of energy occurs whenever one of the electrons moves to another level. This starts a chain reaction as other electrons move to different levels and even into different atoms, Fig. 8-5.

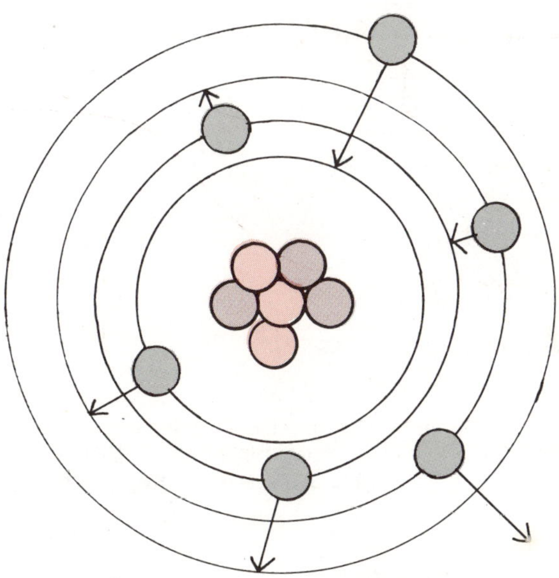

Fig. 8-5. When orbiting atoms leap from one level to another, energy is released. This is the nature of solar energy. Changing of levels creates a chain reaction. Some electrons may even leap out of one atom to join another one.

HOW RADIANT ENERGY TRAVELS

The electromagnetic energy just described produces both electricity and magnetism. The electrical and magnetic energy moves outward in waves. The waves are not all the same length. Short waves have a higher frequency than long ones. Frequency means the time lapse between waves reaching a receiver. You can think of an electromagnetic wave as part of a wiggly line—a series of little hills and valleys. What makes the waves different from each other is the wavelength. This is the distance from one crest to the next. Fig. 8-6 shows a short and long wavelength.

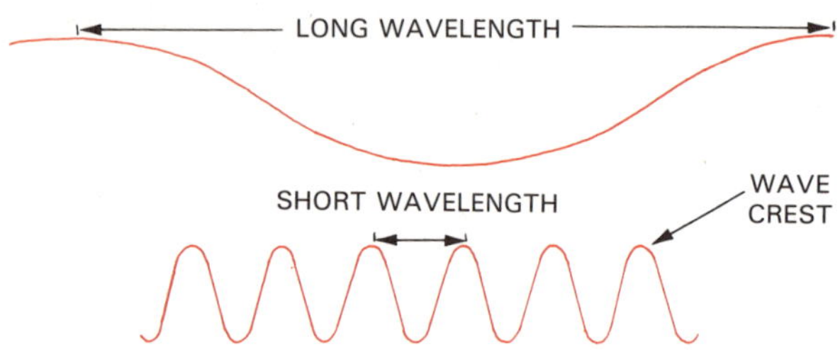

Fig. 8-6. A wavelength is the distance from the top of one wave to the top of the next.

Most electromagnetic waves are invisible. Wavelength determines whether or not you can see them. The different waves from shortest to longest wavelength are:

1. Gamma rays. These are only about one-250 trillionths of an inch long.
2. X rays.
3. Ultraviolet rays.
4. Light rays.
5. Infrared rays.
6. Microwaves.
7. Radio waves.
8. Electric waves.

As Fig. 8-7 indicates, some wavelengths are very long. Others are very short. You can also think of the travel of electromagnetic waves being like the small waves created when a pebble is dropped into water. See Fig. 8-8. Ripples or waves of water move out from the point where the pebble was dropped in ever-widening circles. Eventually the waves reach the shore.

Energy of rays is also related to wavelength.

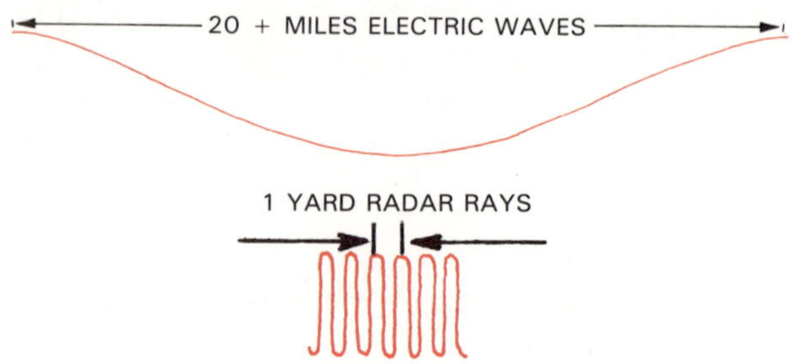

Fig. 8-7. The longest wavelength belongs to electric waves which are 20 or more miles long. Radar wavelengths are only 3 ft. long.

The shorter the wavelength the more energetic its radiation. Most solar energy is made up of short wavelengths. Forty six percent are visible light. This is the wavelength to which the human eye is sensitive. Forty nine percent are infrared radiation. We experience it as heat. The rest of the sun's radiation is in the ultraviolet wavelengths. These waves are shorter than the visible violet wavelength. All wavelengths travel through space at the same speed.

METRIC UNITS OF MEASURE

Three metric units are used for measure of solar radiation.

Fig. 8-8. Electromagnetic waves of energy travel outward from the source somewhat like the ripples travel outward when a rock is dropped into water.

1. Langley. The langley equals 1 calorie per square centimeter. It is also equal to 0.001163 Whr/cm^2 or 4.8 J/cm^2.
2. Calorie. A unit of heat measure. One calorie per cm^2 is equal to 0.0697 W/cm^2.
3. Watts. A unit of power based on energy/area/time. A watt per square meter is equal to 3.152 Btu/sq. ft./hr.

THE SOLAR RESOURCES

Only one-two billionths of the radiation released by the sun reaches the earth. Even so, this is a vast amount of energy, as stated earlier.

We are able to measure the amount of solar radiation we receive every year. These figures are important in a study of solar energy.

SOLAR CONSTANT

The rate at which solar radiation falls on the outer limits of our atmosphere (the power input) has been set at 1.36 kilowatts per square meter (kW/m^2). This is about 130 watts per sq. ft. Actually, the solar radiation varies a little because of the elliptical orbit of the sun. At times, it is as high as 1.4 kW/m^2; at other times as low as 1.3 kW/m^2. See Fig. 8-9.

INSOLATION

Insolation is a shortened form of *in*cident *sol*ar radi*ation*. It is a term for the amount of radiation falling on the ground at any given time. Insolation is usually expressed in Btu per square foot per hour. It is written as: Btu/sq. ft./hr.

There is no constant for insolation as there is for the amount of energy reaching the outer atmosphere. A number of things happen to sunlight as it enters earth's atmosphere. While the radiation is able to pass through airless space quite easily, it is split up when entering the atmosphere.

Some of the rays are scattered by atmospheric conditions and are called DIFFUSED RADIATION. Dry air molecules, dust, water vapor, and thin layers of clouds absorb some of it. Whenever there is heavy cloud cover, all but the scattered radiation is absorbed or reflected by the clouds. See Fig. 8-10.

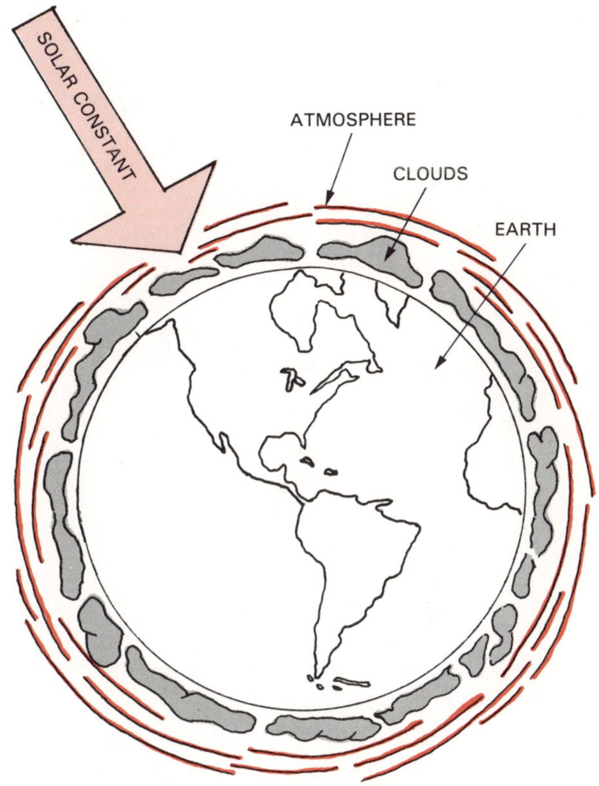

Fig. 8-9. Solar constant is the total amount of solar radiation reaching the earth's outer atmosphere. How much of it reaches the ground depends on clouds, diffusion, latitude, and other conditions.

About 15 percent of the sunlight which strikes the earth is reflected back into the atmosphere. Another 5.3 percent is absorbed by the ground. About 2 percent is absorbed by marine vegetation. Land vegetation absorbs .2 percent. Huge quantities are used to evaporate water and lift it into the atmosphere.

The amount of solar energy reaching the ground depends on latitude, the season, and local weather conditions. However, in the U.S., the average over 24 hours is 177 watts per square

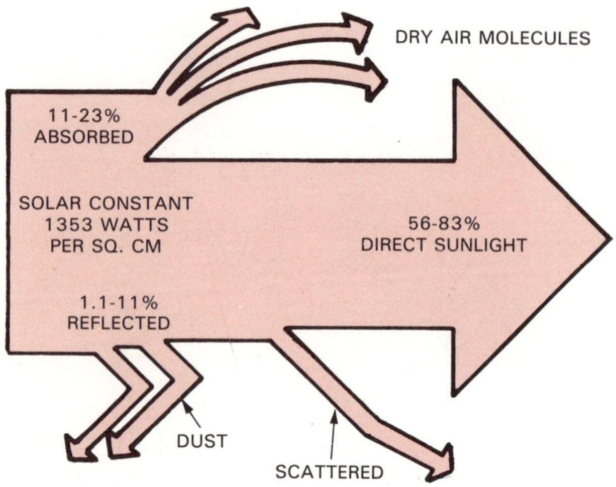

Fig. 8-10. Earth's atmosphere breaks up sunlight so that 17 to 44 percent of it never reaches ground level at all.

meter (W/m²). A collector 1 meter square would provide enough energy for a 200 W bulb. This does not seem like much energy but it adds up. Over 24 hours this amounts to a little more than 3770 calories per square meter. There are 65 calories in a British thermal unit, thus:

$$\text{Btu} = \frac{\text{number of calories}}{65} = \frac{3770}{65} = 58$$

The solar energy falling on each acre of the lower 48 states every day is equal to the energy of 10 barrels of oil. Multiplying this by the 2310 million acres in the contiguous (touching) 48 states yields the equivalent of a little more than 23 billion barrels of oil. This is nearly four times the present consumption of oil. Fig. 8-11 shows the insolation for the United States in the months of January and July.

ENTROPY AND SOLAR ENERGY

You may recall that Chapter 1 discussed the losses of energy due to entropy. Whenever energy is converted from one form to another it loses some of its ability to do work. This reduces the efficiency of the power system. Direct solar energy has an advantage. It goes through but one

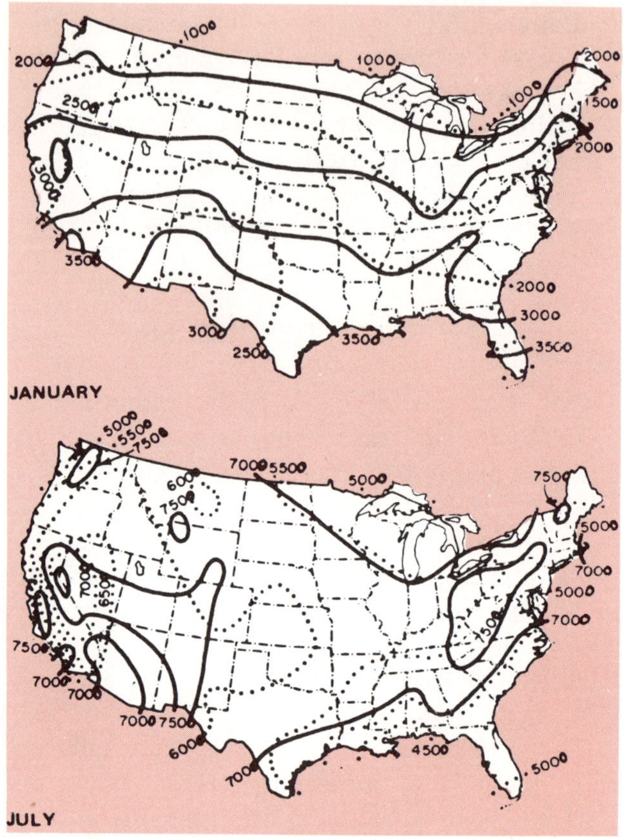

Fig. 8-11. Insolation in the United States for January and July. Areas of equal daily solar energy (on cloudless days) are shown as solid lines. Dotted lines represent equal energy totals for cloudy days. Totals are in kilocalories. (U.S. Dept. of Energy)

conversion from radiant energy to useful heat. Fossil fuels must go through several different conversions before they become a usable fuel. Several additional conversions are necessary before they can produce useful energy.

Consider the light from an electric light bulb. The energy to produce the light was originally solar energy. The solar energy was transformed into plant life. When the plants died they decayed. The decayed vegetation became coal. The coal gives up its energy as heat when burned. The heat creates steam for running a turbine. The mechanical energy of the turbine drives a generator which creates electrical power. The electric power is sent to the customer where it will be converted to light, heat, or mechanical energy. In all, six conversions took place from sunlight to fossil fuel to electric light, Fig. 8-12. The efficiency of the system, having gone through so many conversions, is very low.

NATURE'S SOLAR STORAGE

Large quantities of solar energy are absorbed by the earth through natural systems. Some is stored for only short periods of time. Examples of this are the radiation stored in the form of heat in earth mass such as soil and rock. Large bodies of water also act as storage. The water traps and holds the energy as heat.

Another type of solar storage takes place when plant life interacts with sunlight. Solar rays are stored in plant fiber. Some plant life provides food and energy for animal and human consumption. Some plants provide wood as a fuel. As you learned in your study of fossil fuels, a great deal of energy has been stored underground in the form of solid and liquid carbon.

NATURE OF SOLAR RADIATION

Radiant energy passes through millions of miles of space with little effect on temperature of the vacuum around it. When is strikes matter three things can happen:

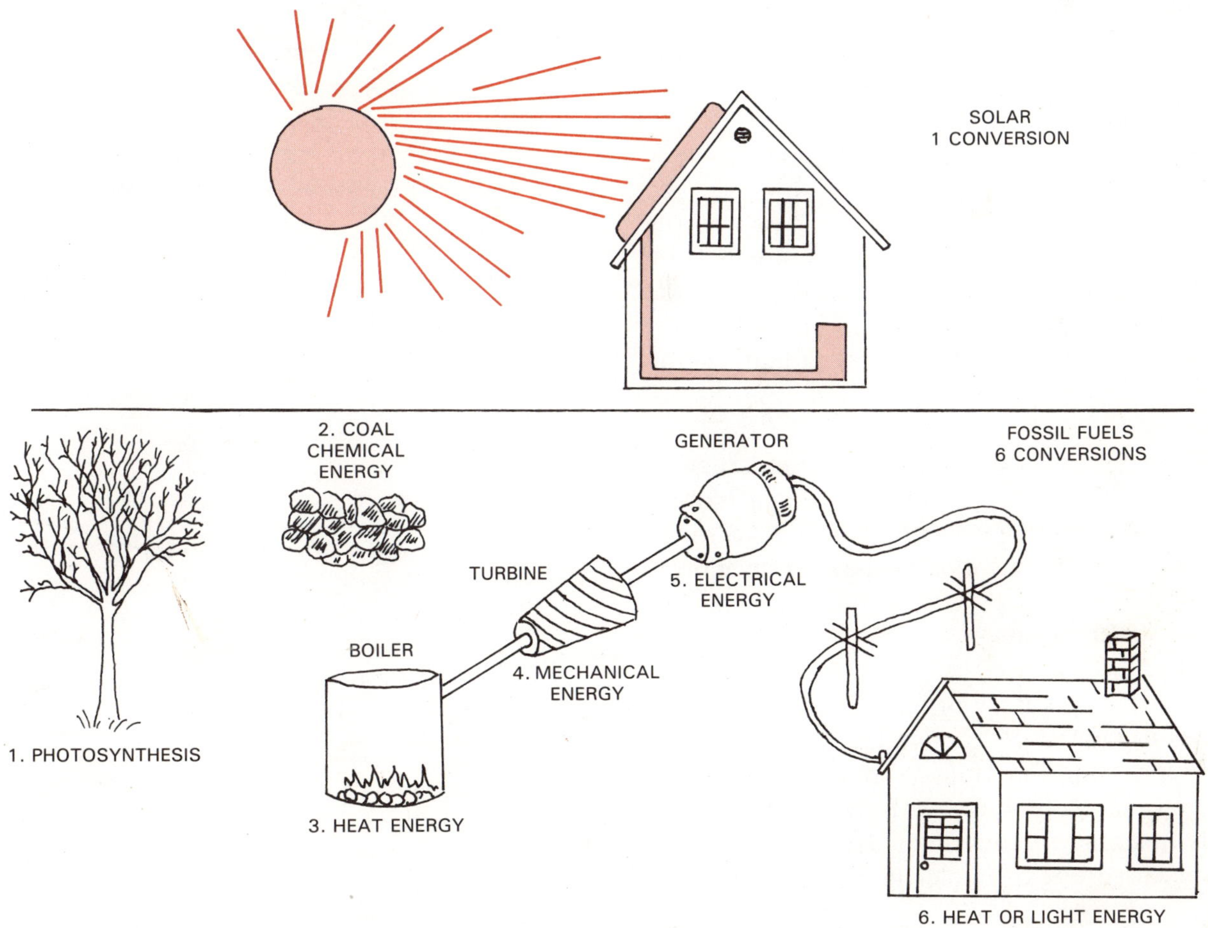

Fig. 8-12. Solar energy is very efficient because it goes from radiant energy directly to heat. Fossil fuels may go through as many as six conversions to produce heat or light.

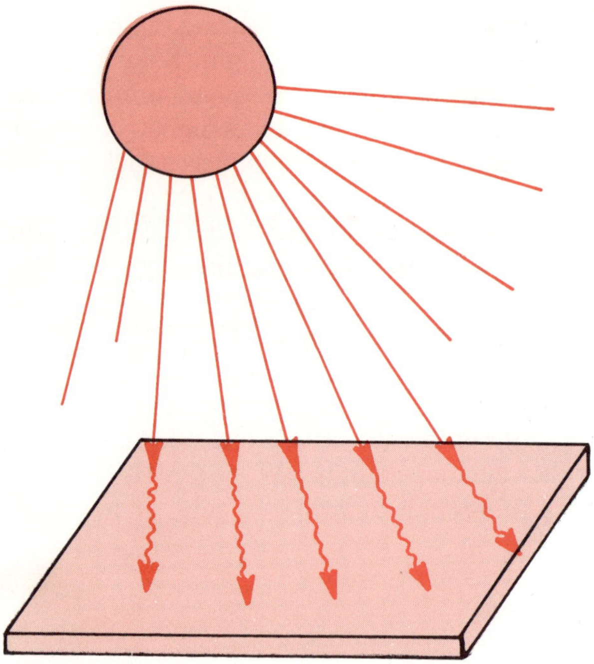

Fig. 8-13. Dark surfaces that are rough will absorb more solar radiation.

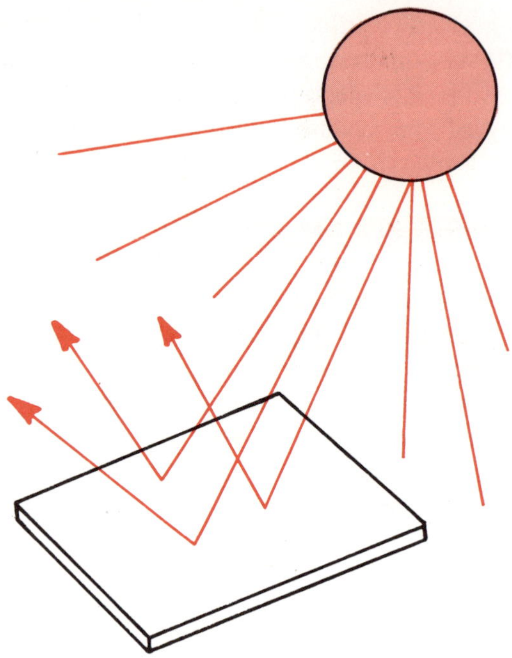

Fig. 8-14. Smooth, shiny surfaces will reflect sunlight.

1. The material may intercept (absorb) the energy and convert it to heat energy, Fig. 8-13.
2. The surface of the material may reflect the radiation back into the air, Fig. 8-14.
3. The material may transmit the radiant energy to surrounding space or matter, Fig. 8-15.

When a substance captures the solar radiation striking it, the light energy becomes heat energy. The molecules of the substance begin to move faster and the temperature of the substance rises.

HEAT TRANSFER

As the temperature of a substance increases, it tries to pass along some of its heat to its surroundings. It is important to remember what you learned earlier. Heat always moves from a warm body to cooler surroundings. Heat will not normally move out of cooler bodies into warmer ones.

The movement of energy to cooler surrounding materials will continue until the temperature of the substance is the same as its surroundings. At this point the substance is said to have reached equilibrium (sameness) with its surroundings.

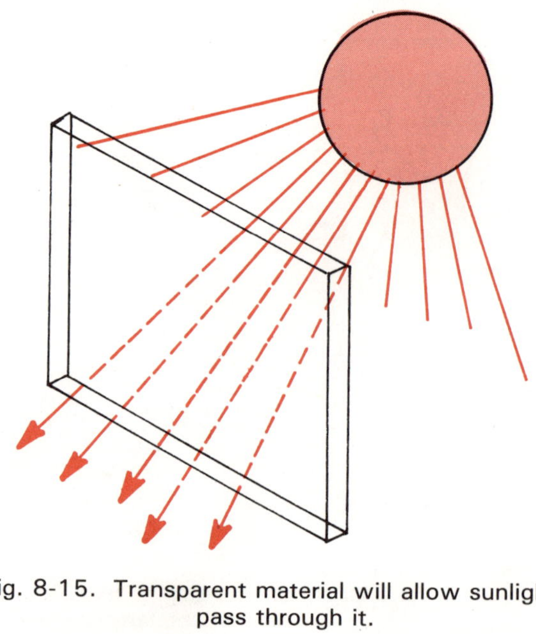

Fig. 8-15. Transparent material will allow sunlight to pass through it.

This transfer of heat is possible because of the three natural processes discussed in Chapter 1: conduction, convection, and radiation. These processes are important in the operation of systems which use the energy of sunlight.

GREENHOUSE EFFECT

Substances which pass along all, or nearly all, of the visible rays they receive are said to be transparent. Glass is one example. Materials which transmit light readily do not necessarily transmit heat waves. Glass will absorb most thermal radiation striking it. Yet, when sunlight travels through glass and is absorbed by materials in enclosed space, the heat given off will not readily pass out through the glass, Fig. 8-16. This heat trapping process is known as the GREENHOUSE effect. You have probably experienced the results of the greenhouse effect when an automobile is left standing in the sun for a few hours. The interior becomes much warmer than the air outside, Fig. 8-17.

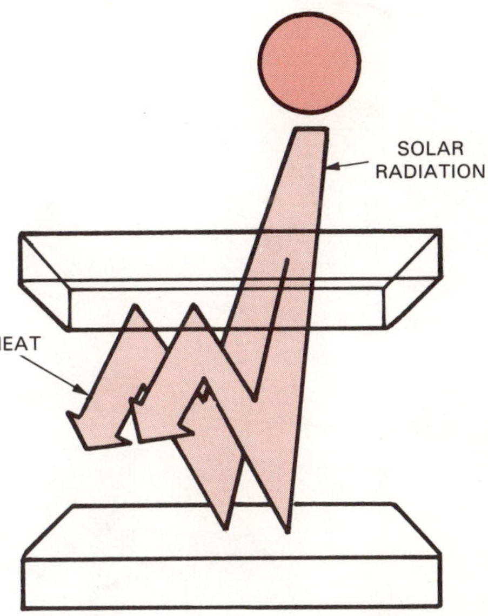

Fig. 8-16. Solar radiation will pass readily through glass. However, heat energy has a different wavelength and will not pass back out through glass as readily.

Polished or shiny surfaces reflect solar radiation to a great extent. Light-colored surfaces also tend to reflect some radiation. A light-colored roof on a house, for example, will not absorb as much heat as a black one. It tends to reflect solar radiation back into the atmosphere.

HISTORY OF SOLAR POWER SYSTEMS

Attempts to use sunshine directly to heat space and power machines are not new. Nearly 2200

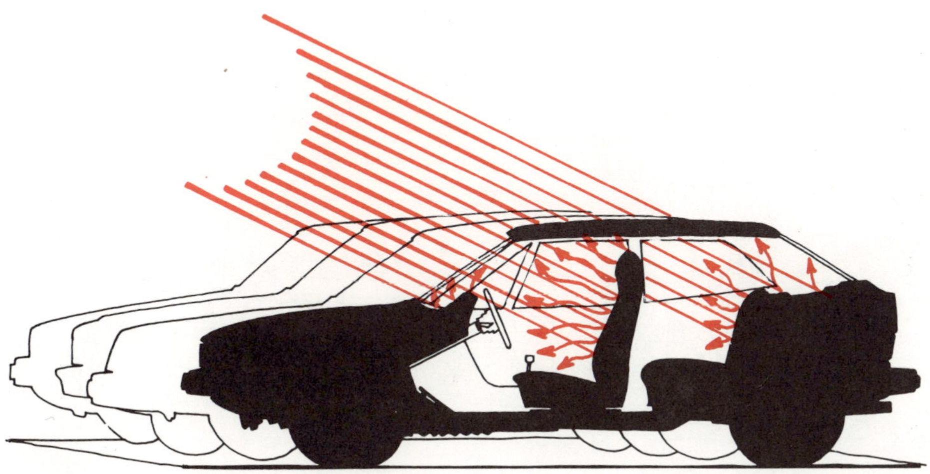

Fig. 8-17. The greenhouse effect (trapped heat) causes autos and homes to heat up naturally in sunlight. (Solar Energy Research Institute)

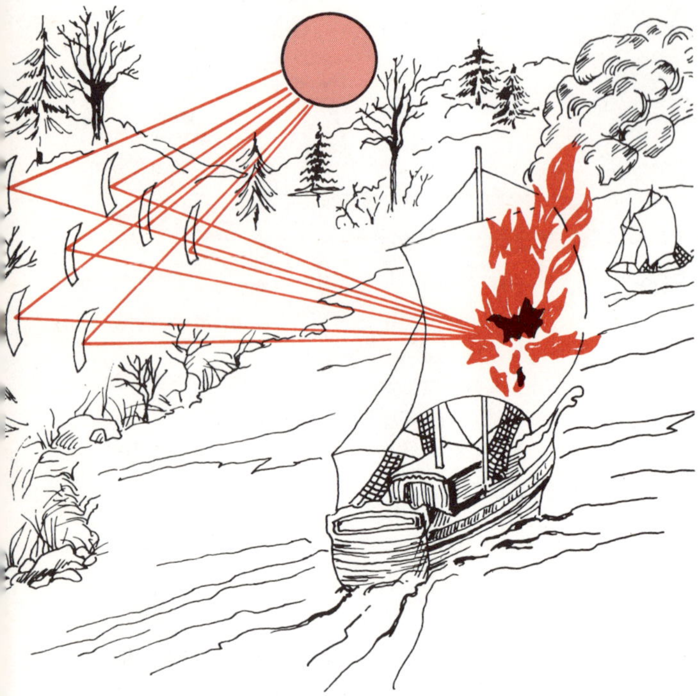

Fig. 8-18. Ancient Greeks were said to have used the sun to set fire to the ships of Romans invading Syracuse nearly 2200 years ago.

years ago, the ancient Greek mathematician, Archimedes, supposedly burned the ships of attacking Romans at Syracuse. A version of the story suggests that he did it with a large hexagonal (six-sided) mirror, concentrating the sun's rays on their sails. In another version, he lined up soldiers carrying polished shields. Solar experts doubt that it could have been done with shields which are convex. There would have been no way of concentrating the solar rays so they could create the heat needed.

One expert believes he could have done it with a number of handheld mirrors, highly polished, all trained on a single target, Fig. 8-18.

SOLAR ENERGY FOR STEAM

Other early inventors hoped to get more than warmth directly from the sun. They tried to use solar radiation for running steam engines. Among the first experimenters with solar steam was John Ericsson, the designer of the warship Monitor. In 1868 he began with solar heated

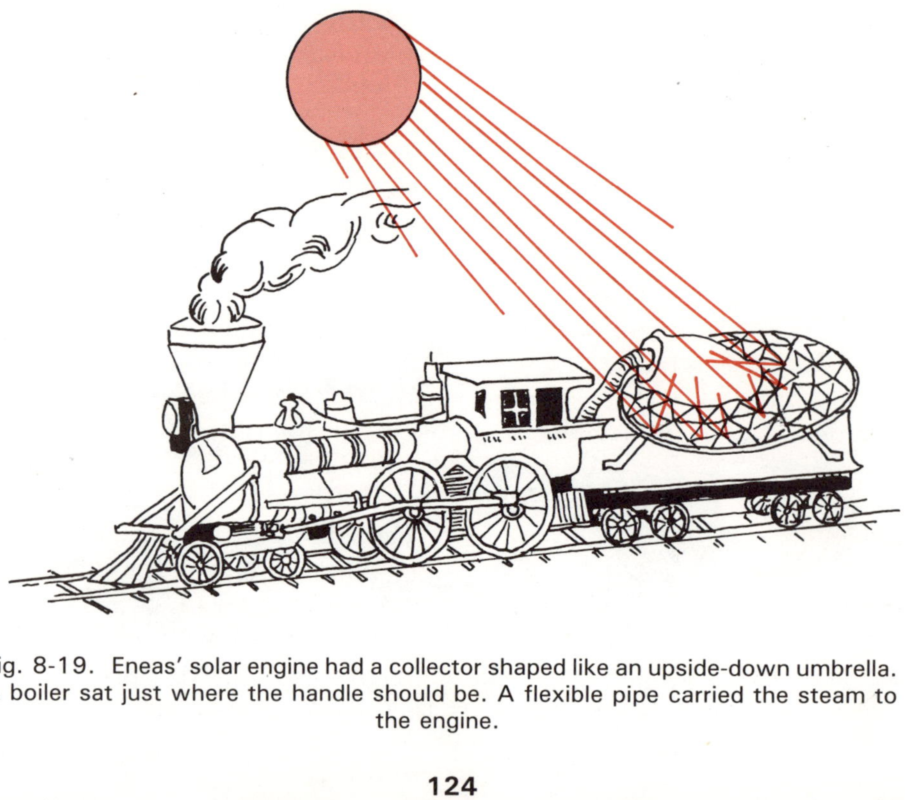

Fig. 8-19. Eneas' solar engine had a collector shaped like an upside-down umbrella. A boiler sat just where the handle should be. A flexible pipe carried the steam to the engine.

water, then switched to heating air because hot air was easier to contain and less dangerous. He built the largest solar engine known up to that time. It had 200 sq. ft. of collectors and developed nearly 2 horsepower.

Meanwhile, similar solar engines were built in India and France. In 1901, A.G. Eneas designed a parabolic collector, Fig. 8-19, which directed reflected sun onto a boiler. According to a newspaper report at the time it resembled ". . . a huge umbrella, opened and inverted at such an angle as to receive the full effect of the sun's rays on little mirrors lining its inside surface. The boiler (of the engine) which is thirteen feet and six inches long, is just where the handle of the umbrella ought to be . . . From the boiler, a flexible metal pipe runs to the engine house near at hand."

During the early 20th century there was another burst of solar steam activity. H.E. Wilsie and John Boyle, Jr. experimented with several designs. Ammonia, ether, and sulfur dioxide were used to raise steam at a lower boiling point. The two inventors designed and built engines of 15 to 20 horsepower and worked out methods of storing heat from the collector to be used as needed. Their work came to nothing. The solar engines, at $164 per horsepower, were not as cheap as conventional steam engines.

In 1911, Frank Shuman tested a 100 hp (74.6 kW) engine on low-pressure steam heated in 10,000 sq. ft. (92,000 dm^2) of troughs lined with reflectors. Later, with the help of Professor E.C. Boys, he moved the engine to Egypt intending to use it for irrigation. He constructed seven more troughs each with more than 13,000 sq. ft. of collector surface. The system worked, developing 50 to 60 horsepower (37 to 48 kW). It was abandonded during World War I, however, because of cheaper human labor.

For the next 60 years, little effort was made to find ways of using solar energy. Gas and oil were so cheap that few bothered. There were some notable efforts, however. During World War II the U.S. military used a small solar still. It was developed to desalinate (take the salt out of) small quantities of sea water and to purify contaminated water.

Simple solar water heaters were common in the southern U.S. until around 1940. Again, cheap and plentiful fossil fuels caused a decline in their use.

STORAGE OF SUNLIGHT

Sunlight can be stored in different ways for later use. All solar energy systems depend on trapping and holding the energy in matter for some period of time.

One way of doing this is to place a bulky material in sunlight where it can be heated. Another is to carry heated air to the material. A third way is to convert the sunlight into electricity. All of these systems will be discussed later.

Some materials can hold more heat than others. Such materials have high specific heat. They are desirable for storing heat.

SPECIFIC HEAT

The capacity of a material to hold heat is called its specific heat. The term is used as a measure of ability to store heat. *Specific heat is the amount of heat, in Btus, a substance can hold when its temperature is raised one degree Fahrenheit.*

In the building trades, however, the amount of a substance is usually given in cubic feet, not pounds. Heat capacity is then given in volume. The heat capacity is called the VOLUMETRIC HEAT CAPACITY.

The volumetric heat capacity of one cubic foot of a substance is found by multiplying its specific heat by its density (number of pounds per cubic foot). Fig. 8-20 compares the specific heat, density, and heat capacity of water with some common building materials.

Material	Specific Heat (Btu/lb/°F)	Density (lb/ft^3)	Heat Capacity ($Btu/ft^3/°F$)
Water	1.00	62.5	62.5
Concrete	0.27	140	38
Brick	0.20	140	28
Gypsum	0.26	78	20.3
Marble	0.21	180	38
Asphalt	0.22	132	29.0
Sand	0.191	94.6	18.1
White Pine	0.67	27	18.1
Air (75 °F)	0.24	0.075	0.018

Fig. 8-20. Specific heat and heat capacity of some building materials are compared to water. Water is one of the best materials in which to store heat. (HUD)

How sunlight is stored for use in heating buildings depends on what type of system is used for collecting the solar energy. Materials commonly used are concrete, concrete block, water, ceramic tiles, wood, rocks, and air. Another material showing promise are salts which can be made to change from a solid state to a liquid state. They are called CHANGE OF STATE or PHASE CHANGE MATERIALS.

When salts such as sodium sulfate decahydrate are heated, they change to a liquid at temperatures around 90°F (32°C). When the heat is drawn off, it becomes a solid once more. In the process of liquifying and solidifying, the salts can collect and then release huge quantities of heat. Thus, a very small amount of the salts can store great quantities of heat energy.

SOLAR HEATING OF BUILDINGS

On a clear day, sunshine delivers about 250 Btu (264 kJ) of radiant energy per square foot to the surface of the earth. This radiation produces enough energy to heat buildings.

Two basic systems are used:

1. Passive solar systems which have no moving parts.
2. Active solar systems.

PASSIVE SOLAR HEATING

Passive systems collect and use heat inside a building without the aid of any mechanical means

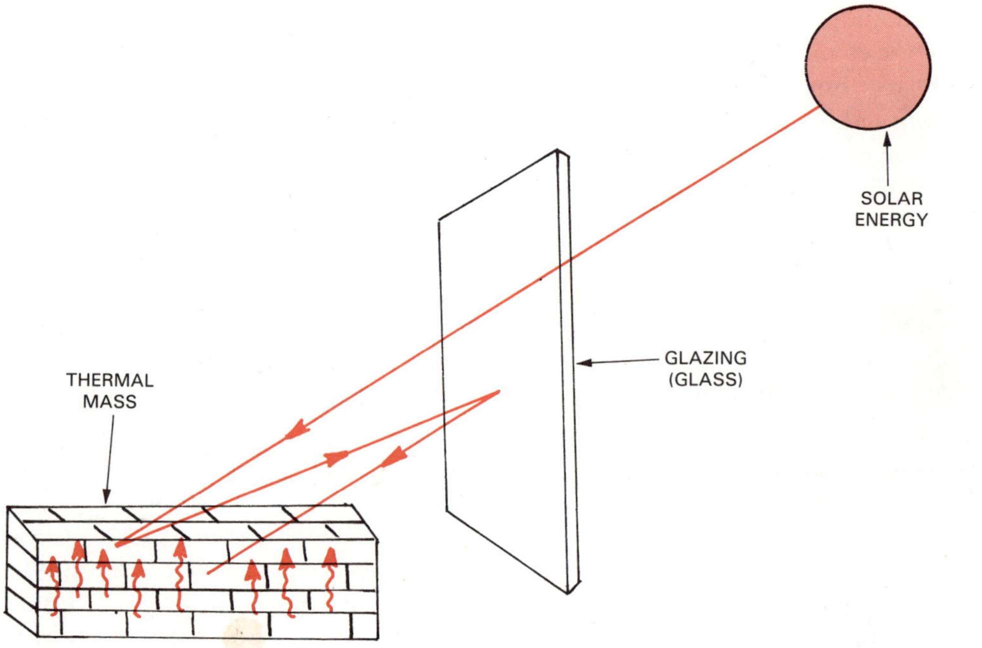

Fig. 8-21. Passive solar energy operates wherever the sun is shining through glazing and the sunlight is stored in thermal mass such as a masonry floor.

such as fans, motors, or pumps. The heat flows by natural radiation, conduction, and convection.

Two things are necessary for a passive solar system.

1. Glass through which the solar radiation passes. The glass traps the heat by preventing it from being reflected back to the atmosphere.
2. A thermal mass to collect the heat, store it, and then pass it on to the air space in the building. Fig. 8-21 shows the basic elements of the passive system.

There are several types of passive systems. Most of them have one thing in common. The elements of the system are part of the structure of the building.

The simplest and least expensive is the DIRECT GAIN SOLAR HEATING SYSTEM. The space to be heated receives direct sunlight, Fig. 8-22. Sunlight, streaming through large south-facing windows falls on floors and/or walls. These parts of the house must be constructed of materials which can do a good job of storing the heat.

One of the best building materials for heat storage is concrete masonry and units such as brick, block, stone, and adobe. Solar radiation strikes the dark surface of the thick masonry. It is absorbed as heat. The heat will remain in the masonry until the room cools. You will recall that it is the nature of warm bodies to pass off heat to cooler surroundings. This principle goes into action as the heated masonry materials pass their heat to the cooler room air. It is normal for such storage walls and floors to be from 8 to 12 in. (203 to 304 mm) thick. The masonry surface must be exposed so it can better receive and radiate heat.

INDIRECT GAIN

Indirect gain solar heating is used to describe a second type of passive solar heating. Sunlight strikes some kind of thermal mass first. The heat is not stored in the living space. It is first absorbed by the mass and then transferred to the living

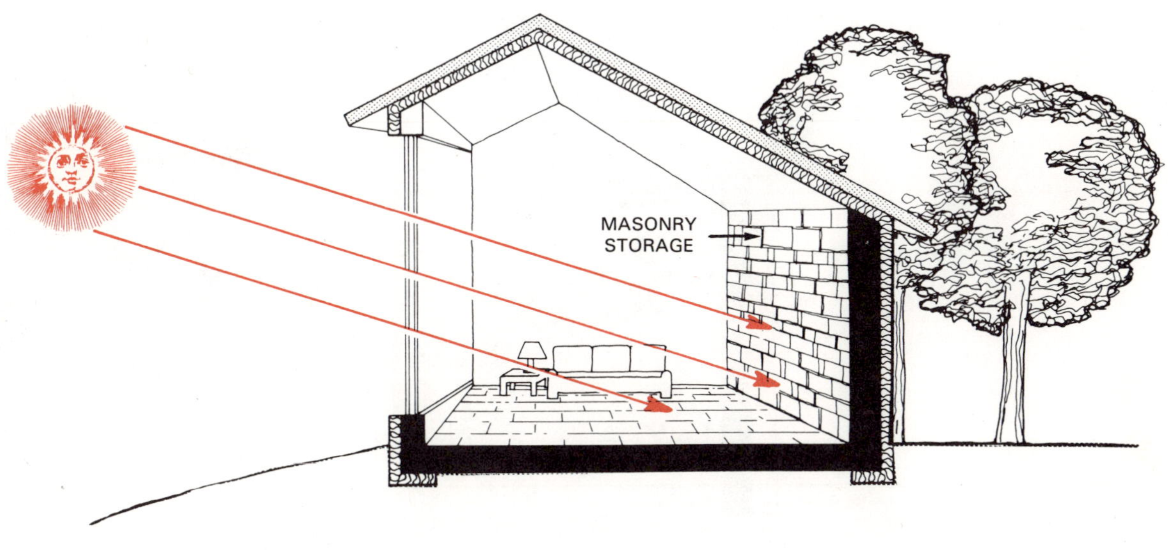

Fig. 8-22. In a direct gain passive solar system, solar heat is collected and stored right in the living space. (HUD)

space as it is needed.

Some indirect gain systems use water for storage. Some use masonry. One way to get thermal mass is to put it into a wall.

Another way is to store water on the roof of the building. This is called a POND. Sunlight heats the water during the day. At night the heated water will keep the space beneath it warmer.

Fig. 8-23 shows a masonry storage wall for an indirect gain system. It is called a Trombe wall because a French scientist, Felix Trombe, produced the original design.

Trombe walls are usually constructed of brick, concrete block, or solid concrete. Wood could be used in some cases. The side of the wall facing the sun is painted black or a dark color so it will absorb solar radiation better.

A second necessary element of the Trombe wall is an area of glazing (clear glass or plastic) in front of it. The glazing allows the sunlight to pass through but prevents most of the heat from escaping back to the outdoors. The glazing is usually about 4 in. (102 mm) from the wall.

As the sun streams through the glazing, it heats both the wall and the air between the glazing and the wall. Some Trombe walls take advantage of the heated air to give the living room immediate heat. Openings called vents are cut in the wall at ceiling level and at floor level. Air, heated in the space between the glazing and the wall expands and becomes lighter than cold air. It rises, passes through the vents at the ceiling, and provides heat to the living space. Meanwhile, cooler air is pulled into the air space between the glazing and the wall through the vents near the floor. This air movement is called THERMOSIPHONING. Fig. 8-24 shows this flow of air.

The Trombe wall heats living space in another way as well. Solar radiation heats up the thick wall during daylight hours. First the side to the sun heats up. Then the heat conducts through the wall until the entire wall is the same temperature. When the side facing the living space is heated up it will pass off heat to the cooler living space.

If the wall is thick enough, this will happen after the sun has gone down. The wall will continue to give off heat as long as it is warmer than

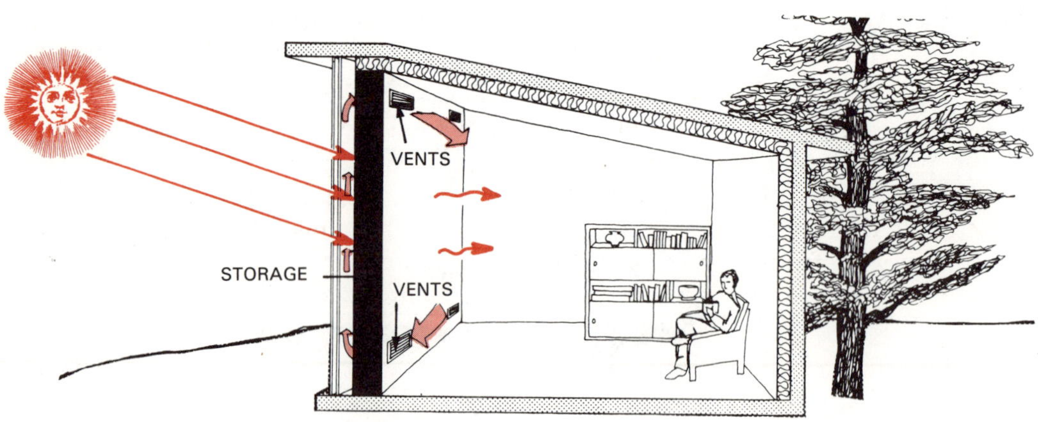

Fig. 8-23. In an indirect gain solar system, heat is stored in thermal mass before it enters the living space. Here a Trombe (masonry) wall has been placed directly behind glass.

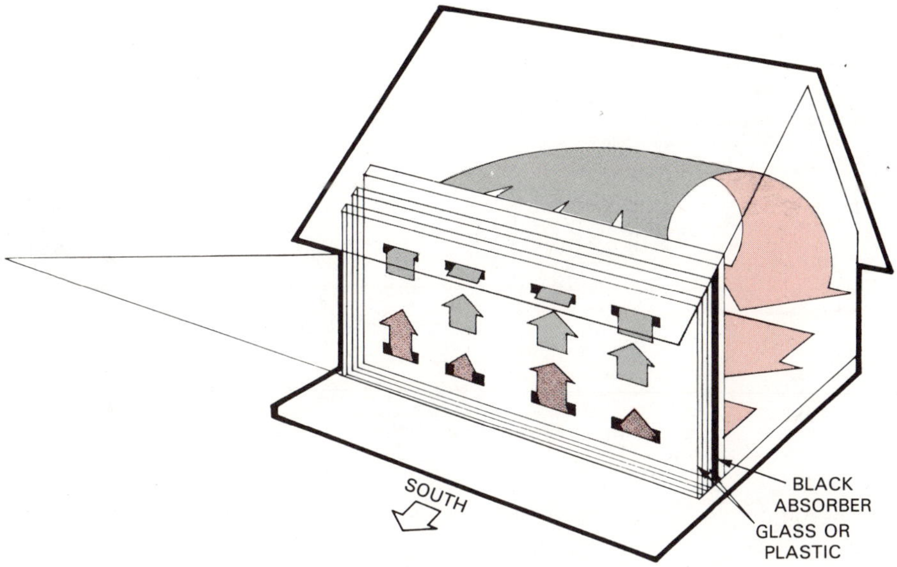

Fig. 8-24. Vents, small rectangular openings in Trombe wall, provide circulation of sun-warmed air into living space. The vents can be closed at night or when heated air is not wanted. (U.S. Dept. of Energy)

its surroundings. The thicker the wall, the longer it will take to heat it up and the longer it will provide heat to its surroundings.

You can see, then, that having the proper thickness is important. If the wall is too thin it will start to heat the living space too soon. The room may become uncomfortable while the sun is shining. It may also pass off its heat too soon and the living space will become cold before the night is over. A wall that is too thick will not be able to pass along its stored heat until well into the night. It will still be passing along heat during the day when sun has begun to heat the room air through the circulation system. Thickness of a Trombe wall may vary from 8 to 16 in. (203 to 406 mm).

As we know from the chart in Fig. 8-20, water is a better storage medium than masonry. It can be stored in barrels, Fig. 8-25, jugs, tanks, or free-standing vertical tubes.

While water will store more energy per pound than masonry, it is hard to contain. Leaking containers not only cause loss of water; they might damage furniture, furnishings, or the structural members of the house.

We mentioned before that water can also be stored in roof "ponds." Here it is contained in tough plastic, Fig. 8-26. During winter nights the pond is covered by a movable insulation. During the day the pond is open to the sunlight so the water can absorb heat. The heat is conducted to the living area later.

ISOLATED GAIN

A third method of passive solar heating separates the collecting and storing of heat from the living space. This type is called isolated gain. Heat can be contained in the system and is drawn from storage only when it is needed. A system of this type is shown in Fig. 8-25, bottom view. The greenhouse attached to the living space collects the solar heat. Storage mass includes the wall separating the greenhouse from the main dwelling and containers of water. A rock bin is another storage area sometimes used. See Fig. 8-27.

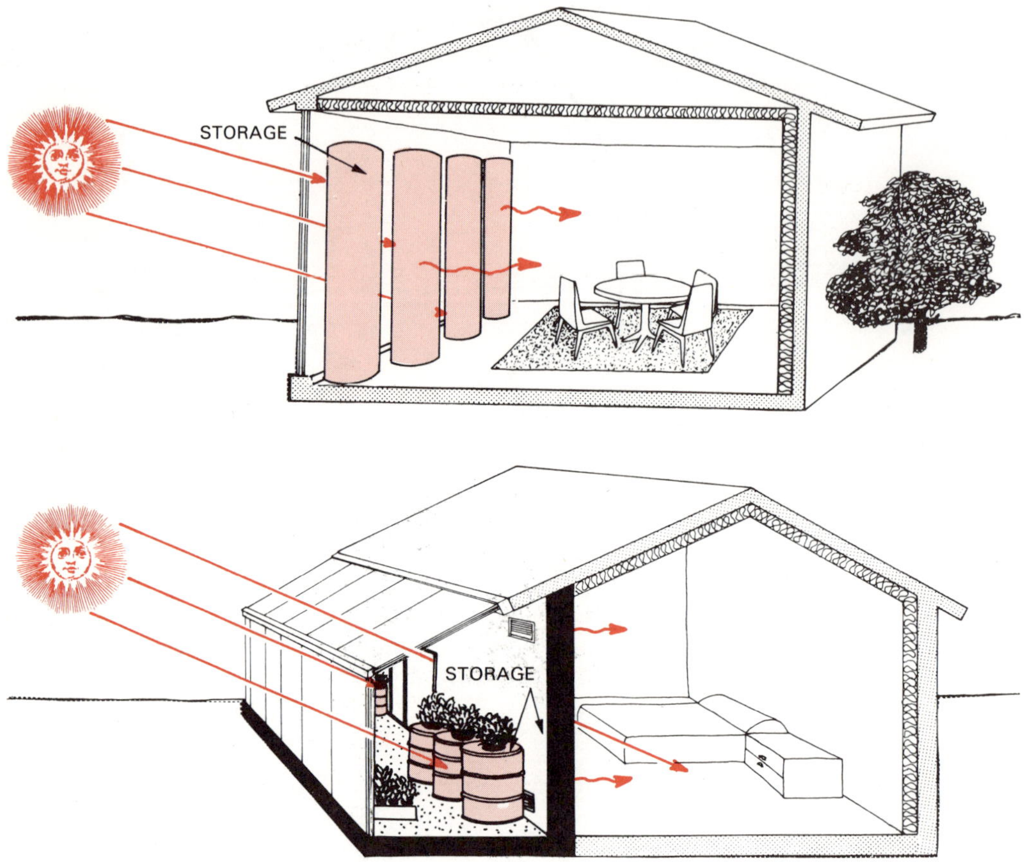

Fig. 8-25. Water wall stores heat and passes it along to living space by convection. Water has twice the heat storage capacity of masonry. (U.S. Dept. of Energy) Top. Freestanding water wall made up of tubes. Bottom. Water storage with an isolated gain system. Solar heat is collected and stored in an area set apart from the living space. (HUD)

Other common methods of getting solar heat by isolated gain are:

1. The convective loop.
2. The window solar collector.

Both of these depend on thermosiphoning to make them work. (Thermosiphoning is the action of rising warm air or water pulling in cooler air or water after them.)

The convective loop, Fig. 8-28, has a flat plate collector connected to a storage tank or rock storage. As water or air in the collector heats up, it rises and enters the storage area. At the same time, cooler water or air from the bottom of the storage container is pulled into the collector. This natural convection (movement) takes place as long as the sun is shining. The flat plate collector is described later in this chapter.

The window collector works on the same principle. It is similar in operation to the convective loop except that:

1. It does not use ductwork. Instead radiation pouring through a pane of glass falls on black absorbent material and heats the air. It rises and pulls cool air in behind it.
2. The warm air is moved directly into the living space rather than into storage. Fig. 8-29 illustrates a window collector.

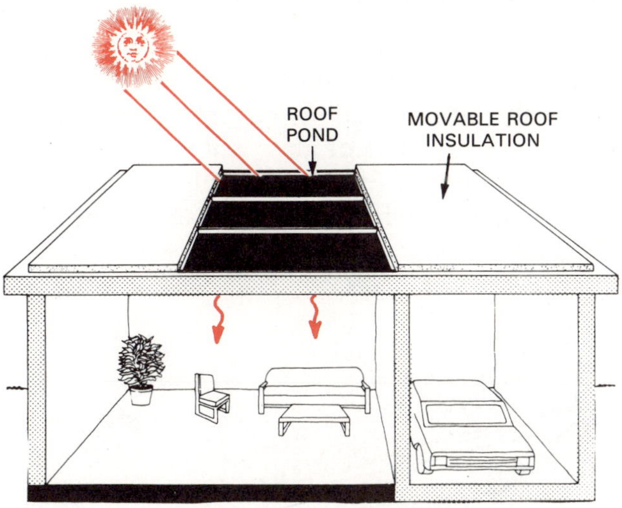

Fig. 8-26. Roof level heat storage in a "pond." Movable insulation covers the roof during a winter night.

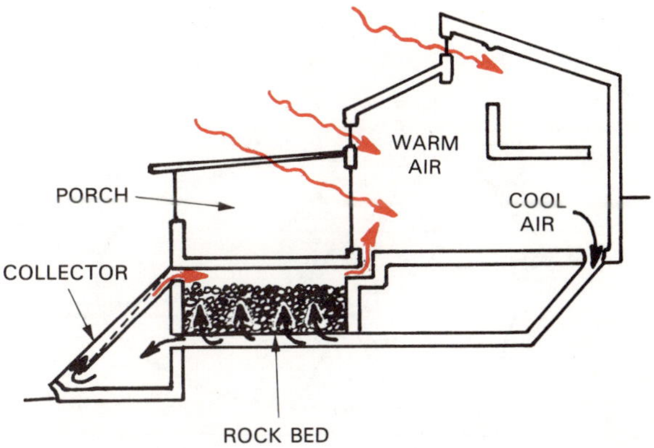

Fig. 8-28. Convective loop system depends on solar heat to drive warmed air or water to storage or living area. (U.S. Dept. of Energy)

CONTROLLING PASSIVE HEATING

Buildings with passive solar heating systems need some methods of control to block out sunlight when it is not wanted. Also, at night some protection is needed to prevent heat losses to the cool night air. Fig. 8-30 shows how the path of the sun varies in summer and in winter. A large roof overhang will cut off direct rays of the sun in the summertime but a low winter sun can easily shine in. See Fig. 8-31.

Greenhouses can be vented to release unwanted heat to the atmosphere. Movable vents can be opened or closed as necessary. At the same time, vents in greenhouse walls leading to living space can be closed in warm months.

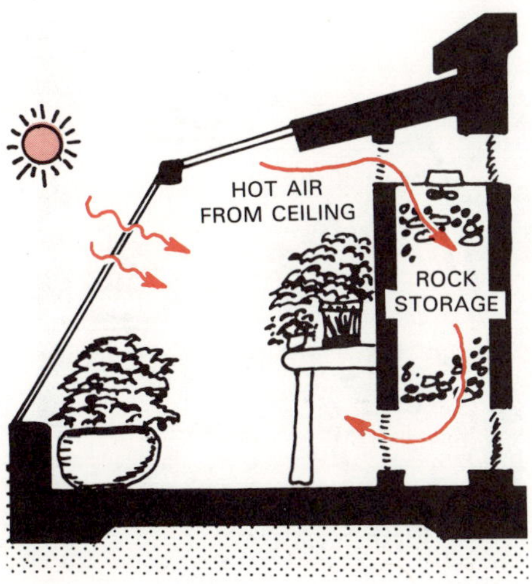

Fig. 8-27. Isolated gain solar heating may store heat in a rock bin.

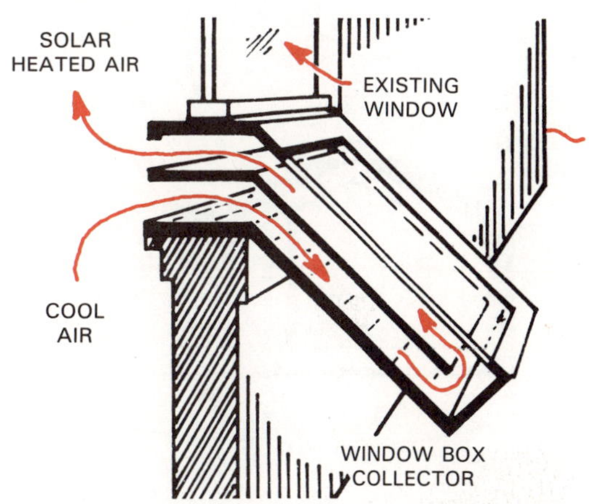

Fig. 8-29. Window collector uses thermosiphon principle. Heating of air in upper section of panel causes the air to rise, drawing in cool air after it.

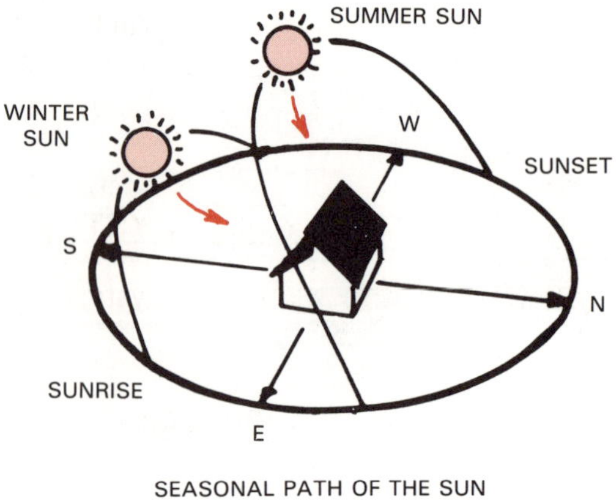

Fig. 8-30. Passive solar construction must take into account the seasonal path of the sun so that the dwelling takes advantage of the low winter sun but shades living space against high summer sun. (Illinois Institute of Natural Resources)

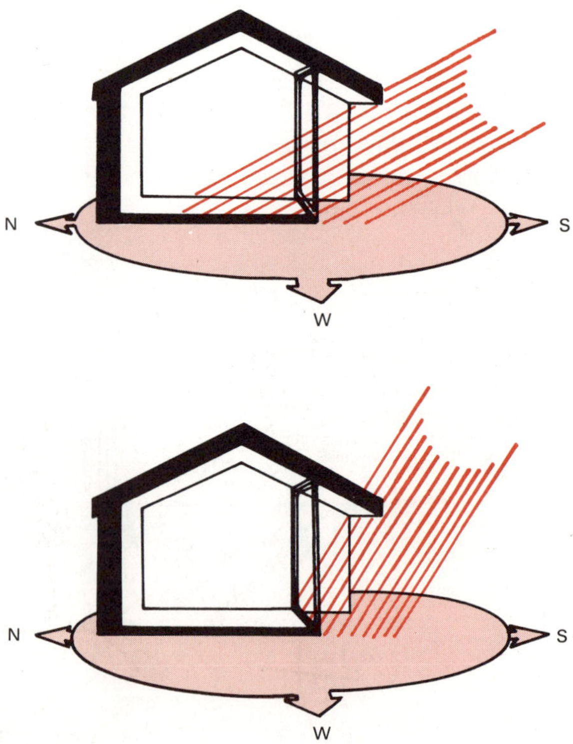

Fig. 8-31. Properly designed overhang on roof blocks summer sun and allows winter sun to shine in and heat living space. (Solar Energy Research Institute)

INSULATION AND SHUTTERS

Both direct gain and isolated gain designs require movable insulation and/or shutters. Spaces with large expanses of glass will lose heat to the cooler night air, Fig. 8-32. Insulating blankets and closures will help control this loss. See Figs. 8-33 and 8-34.

Fig. 8-32. Glass readily conducts heat to the cool night air. (Illinois Institute of Natural Resources)

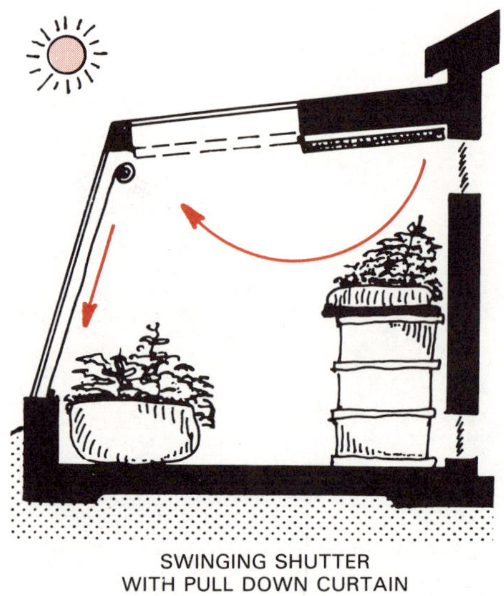

Fig. 8-33. Shutters and pull down curtains will keep out unwanted solar heat.

Fig. 8-34. Insulated shade. (Illinois Institute of Natural Resources)

ACTIVE SOLAR HEATING SYSTEMS

Active solar heating systems are very different from passive systems. Passive systems depend upon natural radiation and convection to distribute the solar heat. Active systems depend on several mechanisms or parts to move and distribute the heat. And, whereas passive structures are usually part of the architecture of the building, active systems are separate from the supporting structure of the building.

PARTS OF AN ACTIVE SOLAR SYSTEM

The active solar energy heating system, Fig. 8-35, is designed to:

1. Collect and trap solar rays.
2. Pick up heat from the collector and move it to where it can be used.
3. Store heat for later use when the sun is not shining.

The various mechanisms which provide all these functions include solar collectors, piping or ductwork, fans, pumps, motors, heat exchangers, and storage bins or storage tanks. See Fig. 8-36.

Fig. 8-35. South facing panels of an active solar heating system. Collectors are used to heat air which is moved away from the panels with blowers.

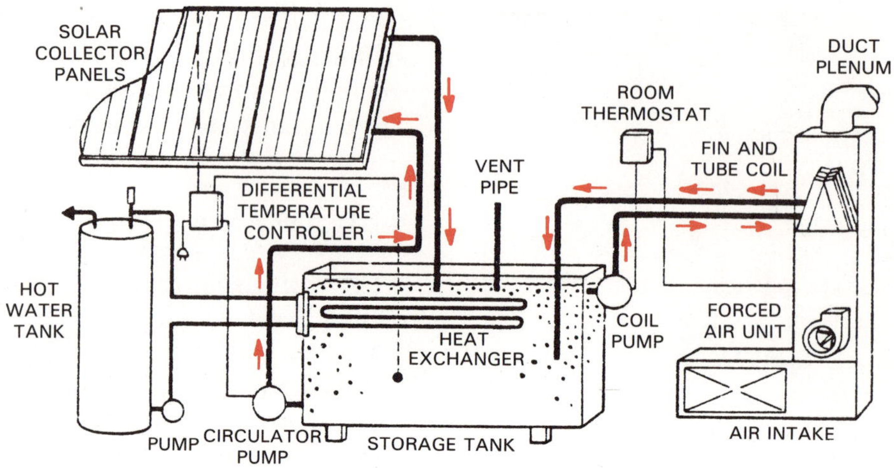

Fig. 8-36. Schematic of an active space and water heating system. (Department of Energy)

SOLAR COLLECTORS

The portions of active solar heating systems which absorb the sunlight are called collectors. There are three basic types:

1. Flat plate collectors. These are most suited to residential use.
2. Evacuated tube collectors. These are more expensive and produce higher temperature heat.
3. Parabolic concentrators. This type is capable of producing very high temperatures.

Basically, a flat plate collector is a closed insulated box with a clear covering that the sun can shine into. The sun heats the insulated black interior, Fig. 8-37.

The surface of the solar collector which absorbs the heat of the solar radiation is called the ABSORBER. It should be black or very dark. The texture should be slightly rougher than an eggshell.

In systems using fluids to absorb the collected heat, the collector has several fluid tubes running lengthwise through it as shown in Fig. 8-38. The sun heats either a plate that conducts heat to the tubes or the tubes are heated. Liquid flows through the tubes picking up the heat.

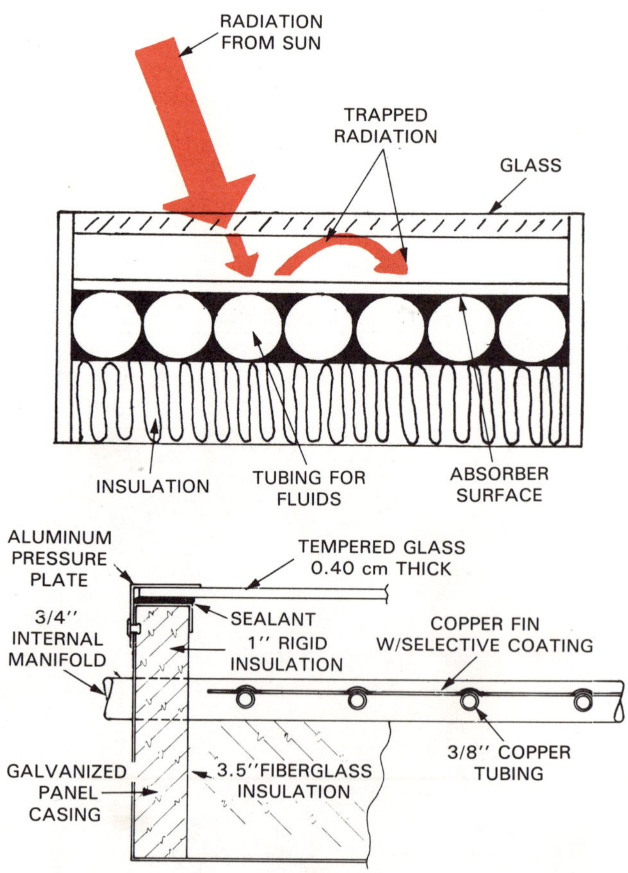

Fig. 8-37. Two cross-section views of collector panels designed for a liquid system. Top. How heat is trapped and absorbed. Bottom. Construction details. (W.L. Jackson Mfg. Co., Inc.)

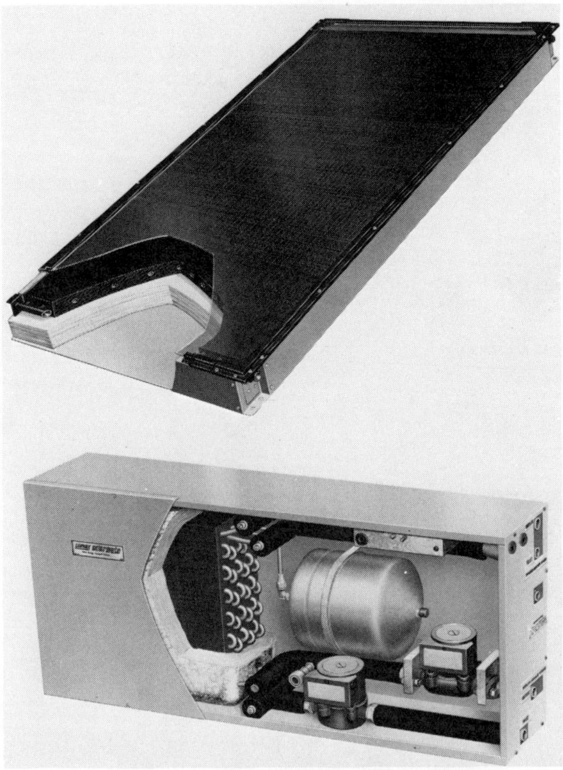

Fig. 8-38. Two components of a domestic hot water system. Top. Solar collector. Bottom. Transfer module. It contains the heat exchanger and the pumps for moving the collected heat to a hot water tank which is the third element of the system. (Lennox Industries, Inc.)

Coatings of lampblack or fine carbon powder produce a dark collector surface which absorbs infrared, ultraviolet, and visible rays of solar energy. Dark surfaces placed in the sun will always get much hotter than white or shiny surfaces. Dark colored objects will heat up to about 253°F (123°C).

A transparent cover serves as a trap for heat. Being clear, it allows the solar radiation to pass through but traps heat.

The cover material is usually glass. It is preferred because it permits more of the radiation to pass through it than other materials. Further, it remains clear indefinitely. This is important since a clear material will transmit more radiation. If collectors are used in cold climates, double glazing may be used. The flat plate is the most commonly used collector. Fig. 8-39 shows an experimental collector design made of plastic.

Fig. 8-39. Low cost black plastic solar collector is being tested in West Germany. (Volkswagen of America)

EVACUATED TUBE COLLECTORS

In the evacuated tube collector, the absorber is a tube within a tube. The two are separated by a vacuum. There is almost no air to carry heat from the inner glazing to the outer glazing. Sunlight can enter the vacuum easily but heat does not move easily through a vacuum. This makes the evacuated tube collector very efficient. It is especially good in cold climates where the temperature difference between the hot absorber and the cold outside air is great.

The absorber tube is coated with a dark absorbing coat. Inside is an absorber plate of copper and fluid tubes to carry away the heat. Fig. 8-40 shows a simple drawing of the evacuated tube collector. It works best where temperatures of 200°F are needed.

CONCENTRATING COLLECTORS

The concentrating collector gathers in solar rays and reflects them onto a very small absorber

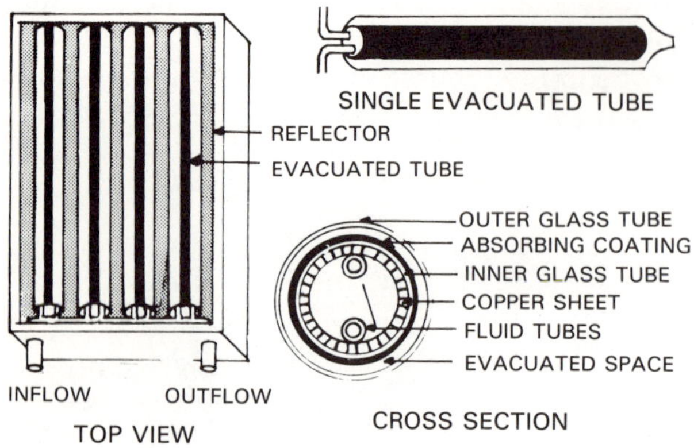

Fig. 8-40. Evacuated tube collector. Tube within a tube surrounded by vacuum makes this design very efficient. Solar radiation can travel through the vacuum to the inner tube, but heat cannot escape to the atmosphere through the vacuum. (HUD)

area. The absorber heats up faster and becomes much hotter. There is much less heat loss because the absorber has less surface area.

Concentrating collectors are easy to recognize. One type is shaped like a dish, Fig. 8-41. Fig. 8-42 shows two types of parabolic troughs. Both dish and trough are lined with a highly reflective metal coating. In the deep trough the light is refracted (bent) through a lens that works roughly like a magnifying glass. The amount of energy directed onto the absorber can be as high as 50 times that of a flat plate collector. Fig. 8-43 shows a trough collector used for domestic water heating.

There are several drawbacks to concentrating collectors:

1. They will work only in direct sunlight.
2. The focusing must be very exact. This requires an expensive tracking system that needs maintenance.
3. Concentrators need heavy frames to control vibration. The weight of the frame and the dish or trough make it unsuited for many applications.

Both concentrator and the evacuated tube are more expensive than the flat plate collector. Few

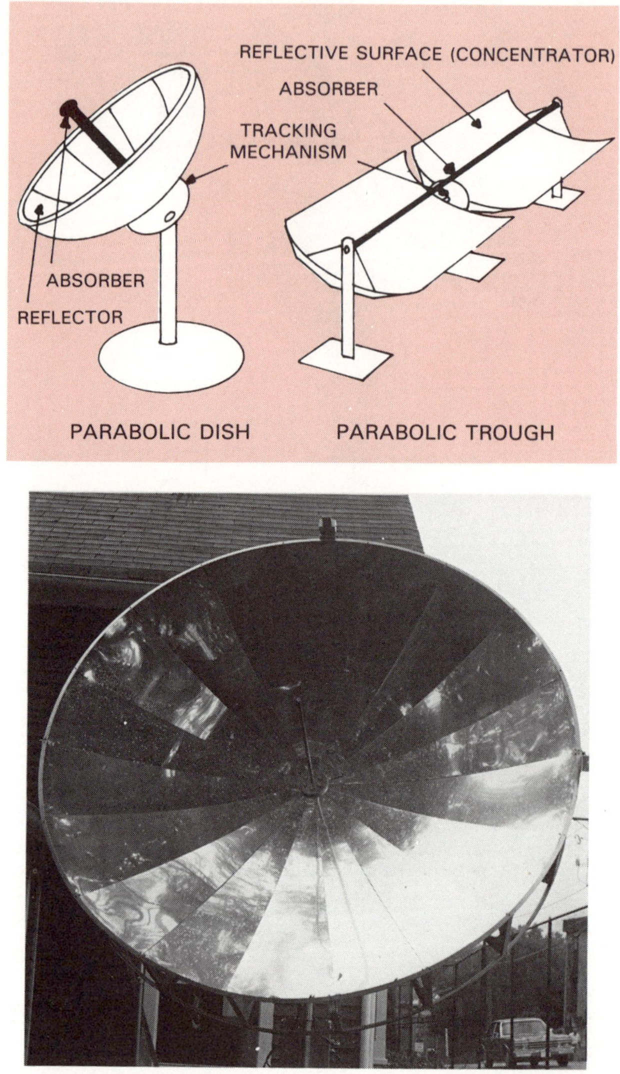

Fig. 8-41. This type of concentrating collector is called a parabolic dish. Top. Basic outline of collector and name of its parts. (HUD) Bottom. View of a parabolic dish unit shows construction of the reflector and absorber.

of them are being manufactured. Fig. 8-44 gives suitable applications for each type of collector.

TRANSFER SYSTEMS

Active solar collectors need devices to carry away the collected heat. Two systems are used:

1. Air system.
2. Liquid or hydronic system.

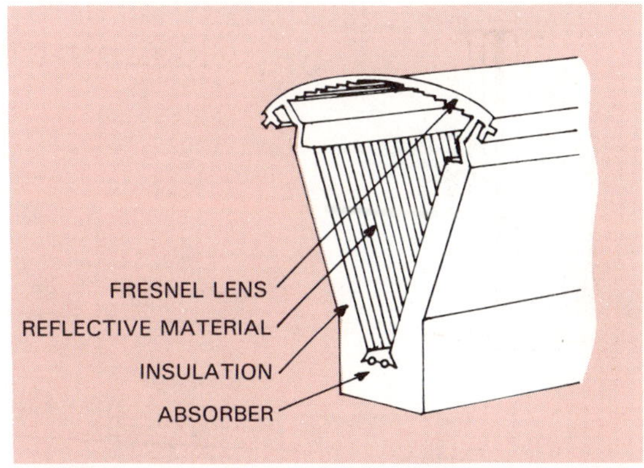

Fig. 8-42. Two types of parabolic trough collector. Left. Shallow trough. Right. Deep trough collector. Note absorber is in the floor of the trough. (HUD)

Fig. 8-43. These parabolic trough collectors track the sun. Attached to the south exposure of a roof they supply domestic hot water for the home.

APPLICATION	COLLECTOR
Agriculture and Pool Heating	Unglazed or Single-glazed* Flat-Plate (often plastic)
Space and Hot Water Heating	Flat-Plate (metallic)*, Evacuated-Tube, Non-Tracking Concentrator
Solar Air Conditioning	High Performance Flat-Plate*, Evacuated-Tube, Concentrator (tracking or non-tracking)
Industrial Process Heat (200-400 °F)	Evacuated-Tube*, Concentrator (tracking or non-tracking)
Industrial Process Heat (over 400 °F)	Tracking Concentrator
Solar Thermal Electric Power	Heliostats*, tracking Concentrator

*Most common choice

Note: All diagrams are descriptive illustrations, not working drawings.

Fig. 8-44. Suggested applications for different designs of collectors. (HUD)

AIR SYSTEMS

Air systems use blowers and ducts to circulate air heated by the flat plate collector. The collector can be located on the roof, on a wall, or on the ground. Ducts carry the warmed air away from the collector and bring in cool air to replace it. A basic air system is shown in Fig. 8-45.

Heated air which is not needed is stored. The storage medium is usually solid material such as a bin of rocks. The air is blown into the bin where it flows over and around the rocks. The hot air gives up its heat to the rock. At the top and bottom of the bin are empty sections. These are called PLENUMS. Warmed air enters through the top. Cooler air is drawn off the bottom and returns to the collector. During cloudy days and at night the blower draws heat from the rock bin.

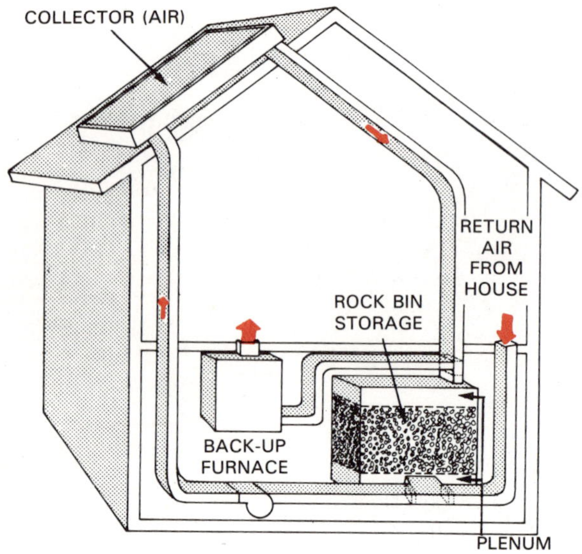

Fig. 8-45. Diagram of an active solar collector with air used as the medium for transferring heat. (HUD)

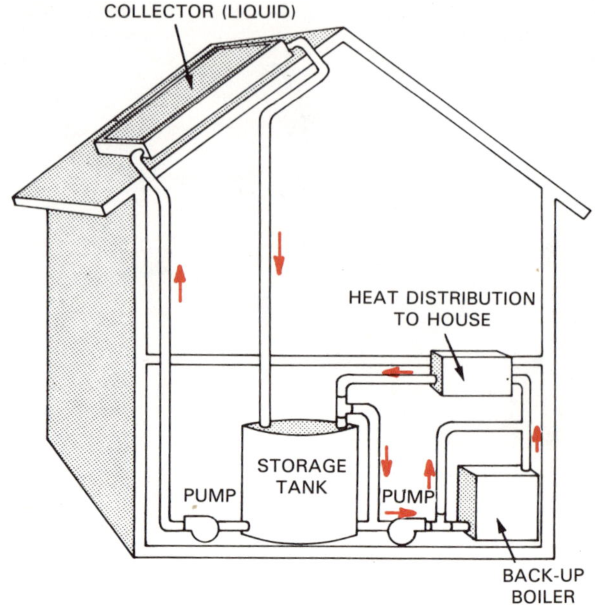

Fig. 8-46. Diagram of an active solar collector using liquid as a medium for transferring heat.

Bins are usually constructed indoors at ground level or in a basement. They may be buried in the ground outside the building. This is not recommended, however. Water may seep into the bin and ruin the insulation. Then too, it would cause moist air to be introduced into the building.

Rocks used for heat storage should be from 3/4 to 1 1/2 in. (19 to 38 mm) in diameter. River gravel is the usual choice. Other storage material such as cast iron, bricks, or ceramic blocks are sometimes used.

Rock bins can be constructed of either masonry or wood. They must be airtight to prevent heat leakage. Wood storage bins must be insulated to a value of between R-19 and R-44.

LIQUID SYSTEMS

A typical liquid system is shown in Fig. 8-46. The heat energy absorbed in the collector is picked up and carried away by water which circulates through tubing attached to the absorber element of the collector. Piping carries the heated water away and brings cool water into the collector to be heated.

The piping is connected with the heating system and with a water storage tank. From 1 to 2 1/2 gallons (4.5 to 11.5 L) of water are needed in the system for every square foot of collector area.

The tank used to store the water may be made of concrete, steel, or fiberglass. Steel tanks should be glass or epoxy lined. Because of its porosity, concrete should be rubber lined.

In cold climates, antifreeze solutions are used instead of plain water. Then a heat exchanger must be used.

HEAT EXCHANGERS

A heat exchanger is a device used in a heating or cooling system. It moves or transfers heat energy from one medium to another.

The radiator of an automobile is one type of heat exchanger that is familiar to almost everyone. Heat caused by the operation of the engine moves into the water. As the water flows through the radiator, its heat transfers to the air.

Heat exchangers in a liquid solar system work the same way. There are several types. The most common are:

1. Air to liquid.
2. Liquid to air.
3. Liquid to liquid.

Fig. 8-47 shows a simple sketch of a liquid to air and air to liquid exchanger. Fig. 8-48 illustrates a liquid to liquid exchanger.

PHASE-CHANGE STORAGE MEDIUMS

Experimental storage units using phase-change materials are being tested. You may recall from earlier discussion that these are materials which change from one state to another as they are heated or cooled. For example, butter is hard when it is cold, but becomes liquid when it is warmed.

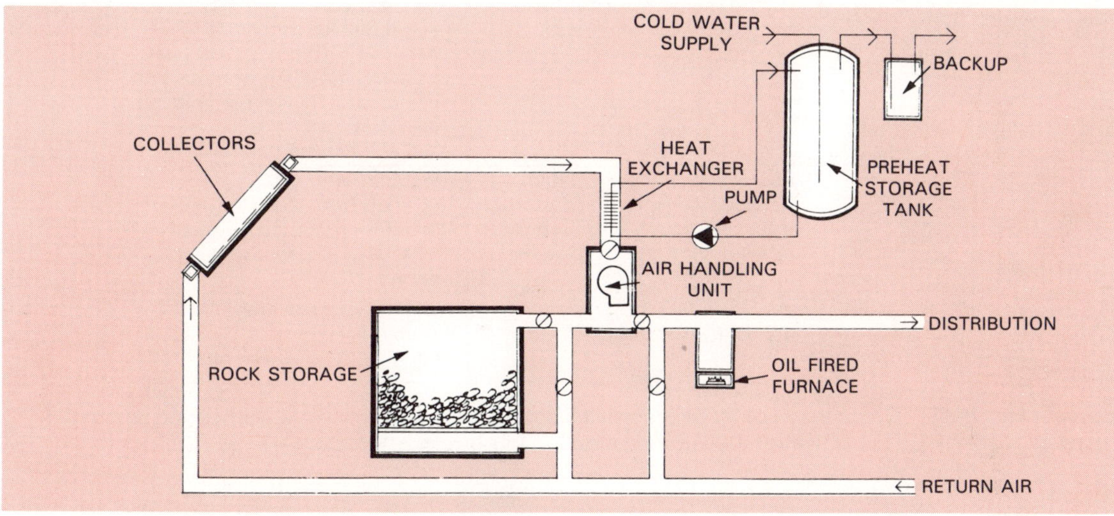

Fig. 8-47. Some systems use heat exchangers so the transfer materials and the medium being heated never mix. This is a diagram of an air to liquid heat exchanger. (HUD)

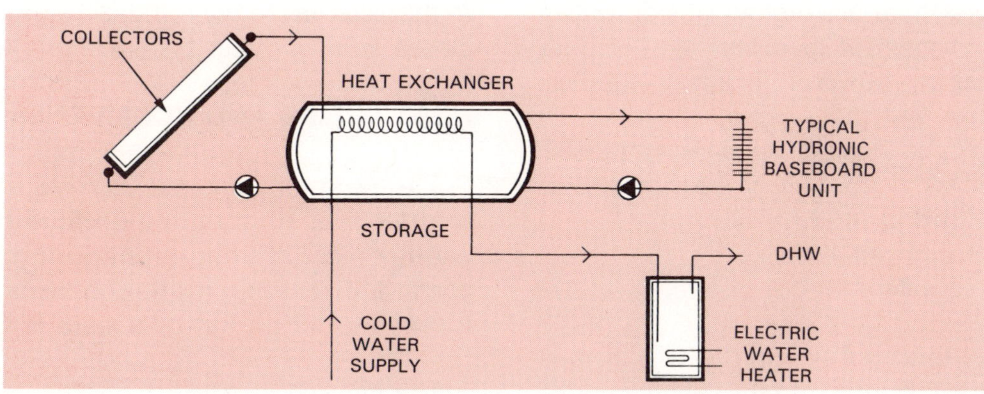

Fig. 8-48. Schematic of a liquid to liquid heat exchanger.

The phase-change materials are generally contained in tubes or trays. Containers must be tightly sealed so the material does not leak out, transfer moisture, or dry out. Large numbers of the storage units are needed to provide a large surface area for heat exchange. This makes phase-change materials very expensive. Fig. 8-49 compares the heat storage capabilities of water, rock, and phase-change materials.

the house supply replenishes the water supply in the solar collector, Fig. 8-50.

An indirect system gets its name from the fact that the solar collector does not directly heat the water you use. The fluid heated by the collector never comes in contact with the hot water. The solution is in a closed loop. The loop runs through the collector and through the supply tank

STORAGE MEDIUM	ADVANTAGES	DISADVANTAGES
Water	—least expensive for retrofitting because of compactness	—possible corrosion —some loss of efficiency because of the need for heat exchangers when nonfreezing liquids or corrosion inhibitors are used —leakage can be destructive
Rock	—no heat exchanger needed between collectors and storage —leakage, though not good for efficiency, is not destructive	—location of any air leakage difficult to detect —retrofitting expensive due to large size storage bin and ducts
Phase-change materials	—very compact storage and good for retrofitting	—expensive container/heat exchanger —Long term reliability not proven

Fig. 8-49. Comparison of three materials used for heat storage in a solar heating system. (National Heating and Cooling Information Center)

SOLAR HOT WATER HEATERS

The most popular and simplest solar energy systems are those used for heating water. The basics of the systems are the same as described earlier in this chapter for space heating. The solar hot water heater, however, is much simpler. There are three systems:

1. The direct open loop system. This is a passive system and is described under the passive solar section of this chapter.
2. The direct pumped system.
3. The indirect (closed loop) system.

In the direct pumped system the solar heated water is emptied directly into the supply tank of a conventional hot water heater. Cold water from

or the hot water heater. The heat is transferred through a heat exchanger. It is usually a coil of tubing inside the water supply tank. As the heated fluid passes through the coil it gives off its heat to the water in the tank. Fig. 8-51 shows a typical closed loop, indirect system.

HEAT PUMPS AND SOLAR ENERGY

The heat pump is a machine which is designed to draw heat from one source and release it at another place at a higher temperature. It can be connected to solar heating systems to extract (draw) more heat out of a solar heat system.

The pump draws heat from the tank and can discharge it in a room at a much higher tempera-

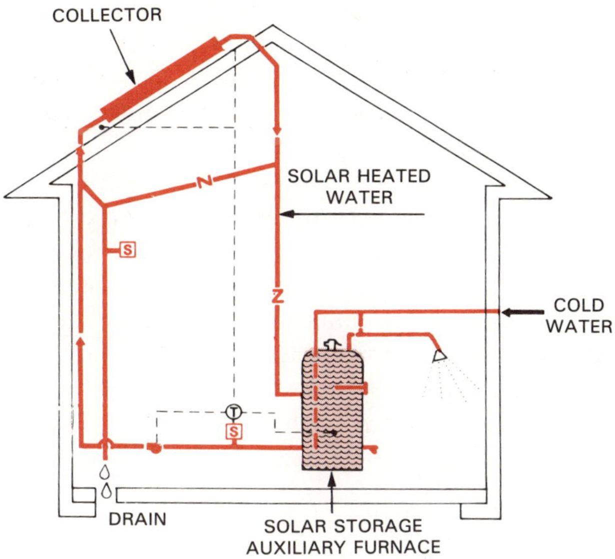

Fig. 8-50. Direct pumped system. A water pump moves water through the collector on the roof.

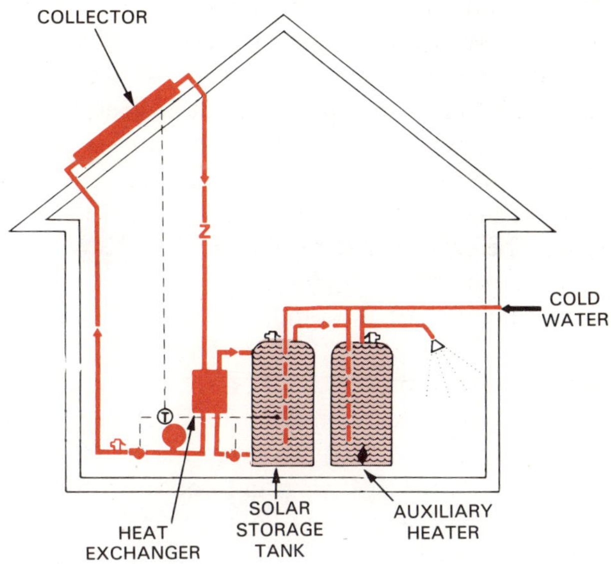

Fig. 8-51. The liquid coming from the roof collector on this system passes through a heat exchanger. It is not used directly for hot water.

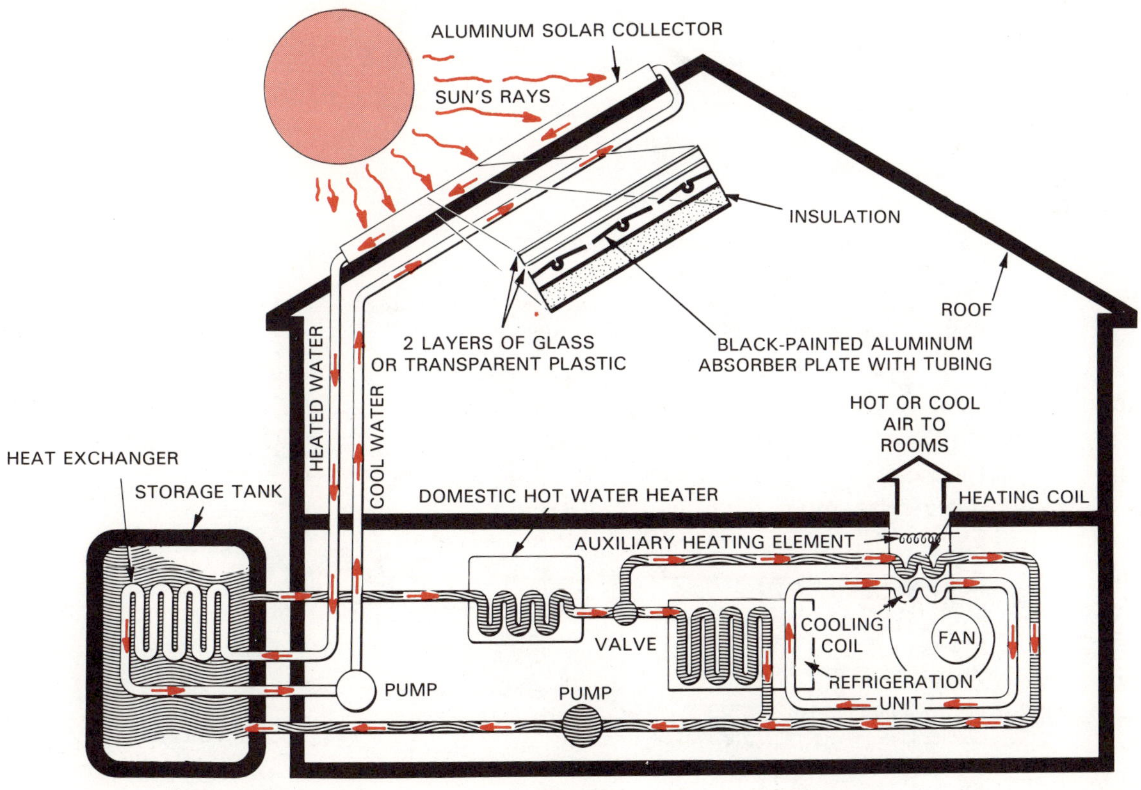

Fig. 8-52. Because heat pumps work better with warmer temperatures, they are often teamed up with solar collectors. In such systems, the heat pump collects its energy from a hot water storage tank.

ture. By turning a valve, the process can be reversed so the heat will be pumped from the room to cool it. The heat is then passed off to the water in the tank where it can be used as hot water supply. Fig. 8-52 is a diagram of a solar-assisted heat pump system.

HIGH TEMPERATURE SOLAR ENERGY

The heating of buildings is done with solar-generated heat at a low temperature. High temperatures are hard, but not impossible, to get through solar radiation. It requires the use of a solar furnace. The furnace gathers the solar energy from a large area and concentrates the rays in a very small area. This is done with an arrangement of many mirrors or heliostats which catch the sunlight and reflect it into a collector.

The principle of high-temperature solar collection has been known for a long time. However, no attempt was made to develop solar furnaces until a few years ago. The first large furnace was designed and constructed at Mountlouis, France in 1952. Later, the French built a second one at Odeillo. A schematic of this installation is shown in Fig. 8-53.

Sixty three mirrors are arranged on a terrace. The mirrors are movable so they can track the sun as it moves through the sky. The sunlight strikes the mirrors and is reflected into the 130 ft. high parabolic collector. It, too, is made up of small mirrors. Because of the shape of the collector, the reflected sunlight is pinpointed on a small part of the target where the boiler is located, Fig. 8-54.

The 9500 mirrors making up the concentrator are flat. Each must be focused by hand so that the area in which the reflected sunlight is gathered measures only about 17 cm (about 7 in.) in diameter. About 270 kW of energy is concentrated in this area.

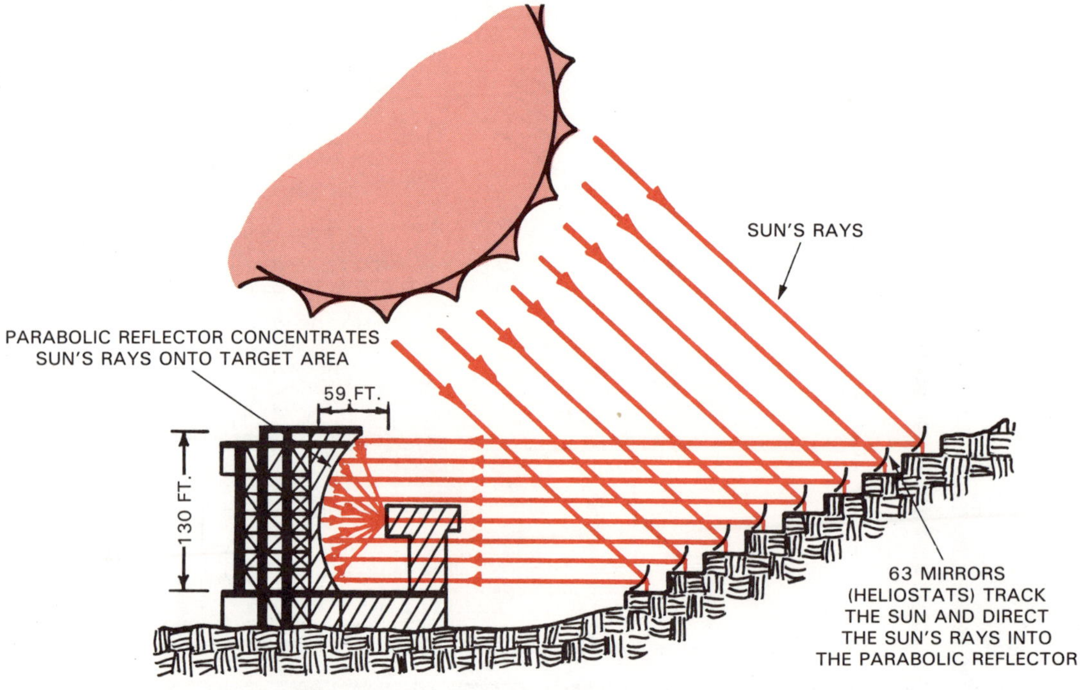

Fig. 8-53. Simple sketch of a solar furnace. The mirrors reflect sunlight back onto a parabolic dish concentrator. It pinpoints all the light to a target area where a boiler is located.

Fig. 8-54. This is a model of a parabolic dish collector. This type can be used for a solar furnace. The surface of the dish is made up of many individual mirrors.

No commercial use has yet been made of solar furnaces. They are still in the development stage. The Mountlouis furnace became the model for others. These were built at:

1. Natick, Massachusetts (later moved to the Sandia Laboratories at White Sands, New Mexico).
2. Sendai, Japan.
3. Odeillo, France.

If the solar furnace proves successful, the heat produced could be used to create steam for generating electric power. The steam might also be used as process heat in manufacturing.

One of the chief disadvantages of the solar furnace is the need to collect the solar energy from a large area. This would require hundreds of heliostats covering many acres of ground.

BLACK BODY RADIATION

Black body radiation is reverse radiation which occurs at night. The earth and other substances which have absorbed heat during the day, give up heat to the atmosphere. That is why in winter, for example, clear nights are cold. There are no clouds to hold back the escape of solar heat.

This principle can be used to provide cooling where electric power is not available. It has been

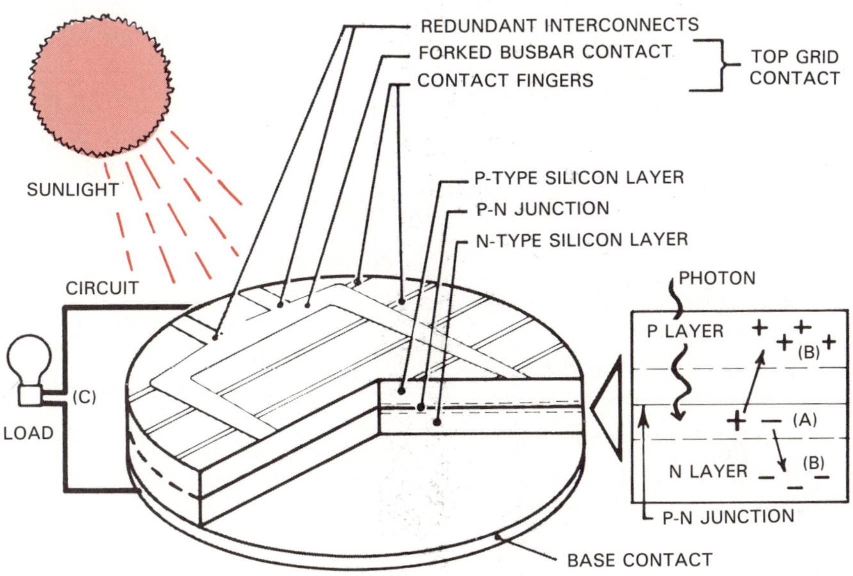

Fig. 8-55. Sketch of a solar cell and how it works. (Solar Power Corp.)

used to keep buildings cool. The palace of the sultan of Zinder in Nigeria, Africa was said to be constructed for this purpose. It was built around a central patio and during the nights the roof radiated heat to the atmosphere while cold air was pulled into the patio and the rooms under the roof to keep the house cooler in the heat of the next day.

French scientists have experimented with this principle. In one experiment they placed a white aluminum plate inside a tub which rested on the roof of their laboratory. A clear plastic sheet covered the tub. When the temperature at the plate was on the freezing mark, 0°C (32°F), on a clear night, the temperature at the film was −40°C (−22°F). Scientists are working to see if black body radiation can be used to cool food and medicine in areas where there is no electric power.

The process has already been successfully used in South America to purify water. Chilean scientists experimenting with black body radiation froze the water at night to separate it from undesirable minerals such as boron and arsenic. They produced water suitable for drinking and for irrigation.

SOLAR CELLS

In Chapter 1 you learned that a solar cell, Fig. 8-55, is a small crystal of silicon that can convert sunlight directly into electricity. Silicon comes from sand, one of the most plentiful materials on earth.

Right now, solar cells, also called PHOTOVOLTAIC cells, are very expensive. This is because of the cost of the materials and the difficulty in making the cells. Some of the first successful solar cells cost about $7000 a square meter and were about 11 percent efficient. The silicon is melted first and then the molten material is slowly drawn from the crucible (pot) so that it forms a single crystal. The crystal, which is shaped like a rod, is very hard. It is sliced into thin discs with a diamond saw. Most of the precious material is wasted since the saw cut is thicker than the disc it is cutting.

MAKING THE CELLS

After the discs are cut they are polished. Then, since silicon is a poor conductor of electricity, impurities must be added while the cells are reheated. The impurities give one side of the wafer a positive charge. The other side gets a negative charge. A thin area in the middle of the wafer is not touched by the impurities. The positive side is called the P layer. The negative side is called the N layer. The neutral area is known as the PN JUNCTION. The junction does not conduct electricity, it insulates the P and N layers from each other.

ETCHING

The cell is not yet ready to produce electricity. An etching process applies a metal grid. This metal coating is needed so the electrical charges can travel out of the cell.

Individual cells are connected to other cells in series (positive terminal to negative terminal) as shown in Fig. 8-56. A group of interconnected cells is called a MODULE. Any number can be connected to get the desired voltage.

Modules are housed in a protective base and covering. A rigid frame provides additional protection as well as a means of mounting the module.

HOW SOLAR CELLS WORK

Sunlight falling on the solar cell delivers bundles of energy called PHOTONS. They dislodge electrons from the solar cell. The electrons flow through the grid and can move out of the module through the conductors. It is thought that one photon of solar energy frees one electron.

Only a very small amount of current can be gotten from a single solar cell, but, when many

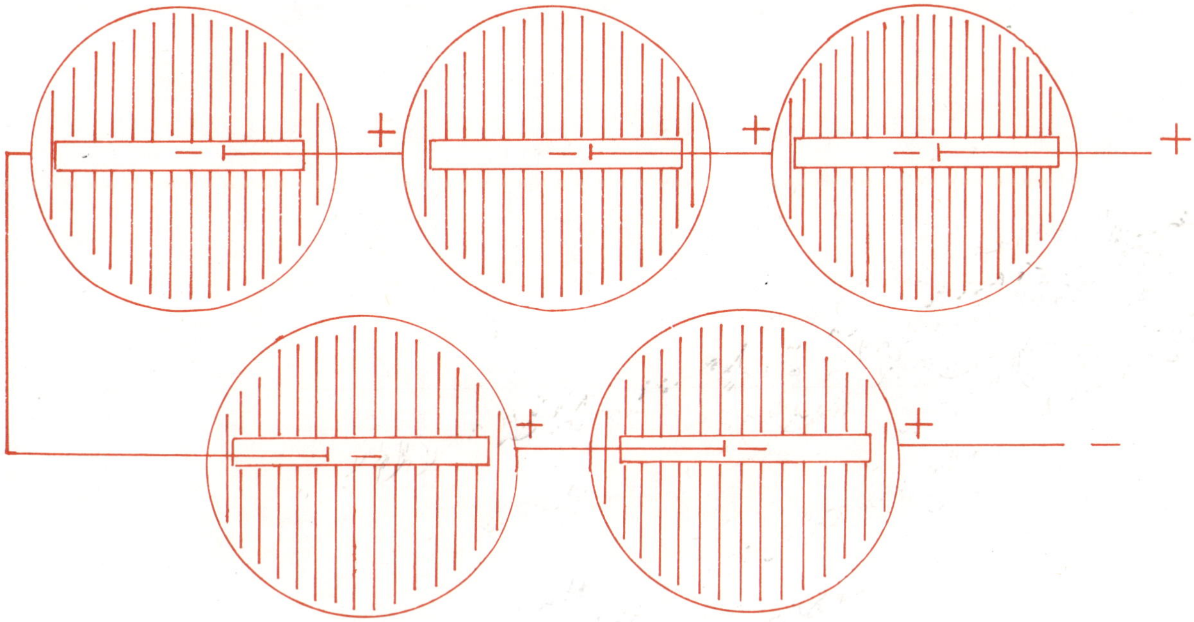

Fig. 8-56. Diagram of a solar cell array connected in series. Total voltage will be the sum of the individual cell voltages.

are connected, considerable amounts of power can be obtained.

A solar cell is sometimes called a solar battery. It does function much like a battery, but there is one big difference. The battery will eventually wear out while the solar cell will produce current indefinitely as long as the sun shines.

APPLICATIONS OF PHOTOVOLTAIC CELLS

Because of their cost and low efficiency, solar cells are not now practical to use. The one exception is to supply electric power in remote areas where other sources are not available. Besides the space program, Fig. 8-57, where solar cells supply electricity 24 hours a day for satellite operation, this exciting power source is currently used in remote areas for:

1. Portable systems for water purification.
2. Operating small battery chargers.
3. Telephone communications, Fig. 8-58.

One exception is the universal application of the solar cell to photography. Small units are found in many light meters and cameras.

An interesting and exciting application of the solar cell was its successful use to power flight. Solar Challenger, a specially built aircraft weighing but 185 lb. (84 kg), in December, 1980, carried a pilot safely over a long distance using only sunshine for power. The plane made about 45 flights. The longest was 18 miles, Fig. 8-59. Fig. 8-60 shows a small automobile with a roof made up entirely of an array of solar cells.

SOLAR CELLS OF TOMORROW

Until solar technology finds a way to improve solar cells and the methods of producing them, we cannot rely on photovoltaics for many uses. For example, they would not be practical now as a substitute for an electric power plant or as a home electric plant.

Like all new products, the solar cell needs to be further developed and refined. Its disadvan-

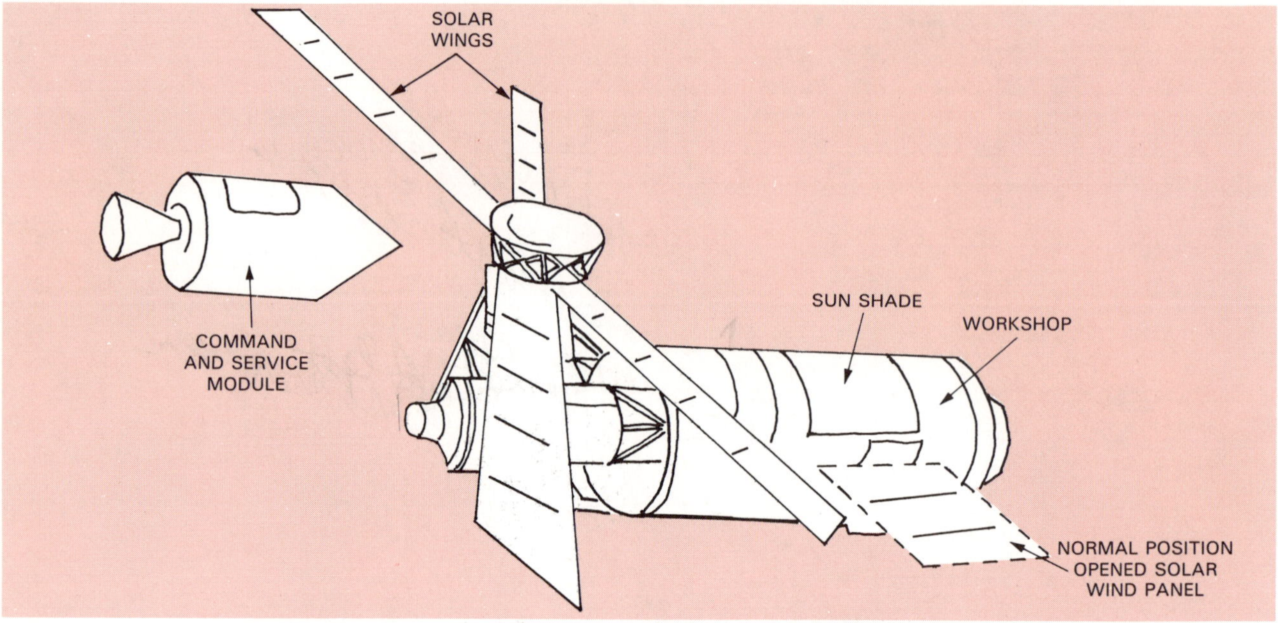

Fig. 8-57. Solar cells provide electrical power to operate space satellites.

tages now can be summed up as too expensive, too large, and not efficient enough.

COST

In 1980, an array of solar cells capable of producing 1 kW of electricity cost $30,000. However, it is predicted that before 1990 the cost per kW will drop to 50 cents. This is still much too expensive when electricity produced by nuclear and fossil fuels is selling for around 7 to 10 cents. Still, experts believe the year 2000 will see the cost of solar cells further reduced to the range of 10 to 30 cents per kW.

SIZE AND EFFICIENCY

The second disadvantage of the solar cell, size, is related to efficiency. A good solar cell will produce about 200 milliamperes of current and around 500 millivolts for every square inch of its surface. At most, the cell is 16 to 25 percent efficient. A 14 volt panel that can produce a watt of power is 6 ft. (1.8 m) square by 1 in. (25.4 mm) thick and weighs around 25 lb. (about 11 kg).

Fig. 8-58. Solar cells supplied telephone power at this booth during the Knoxville, TN, World's Fair in 1982.

Fig. 8-59. First solar flight was made by Solar Challenger in 1980. Weighing only 185 lb., the craft has 15,000 photovoltaic cells in its wings and tail. It was developed by a design team headed by Dr. Paul McCready, a California scientist. (Randa Bishop)

Fig. 8-60. Solar powered automobile exhibited at the 1982 World's Fair.

Just producing one percent of our total electric generating capacity (about 4 million volts) would present a space problem. It would take 80 sq. miles (207 km) of solar panels!

Use of concentrators (magnifiers) to direct more sunlight on the cells (increase insolation) will help. However, the cells themselves would then break down from the intense heat. New materials will need to be developed to stand up to the high temperature.

Modern technology has many challenges facing it to overcome these problems. Experts think they can meet the challenge and solve the problems.

PROMISING NEW METHODS

Even as you read this, research and development is whittling away at the high cost of producing the solar cell. Honeywell Incorporated, for example, is working on a way of coating 2 in. (51 mm) wide ceramic strips with molten silicon. The process is fast, producing an inch of solar cell every 10 seconds. Also, it does away with the old sawing process which wastes so much material.

Another new cost-cutting experiment grows 2 inch wide pure silicon ribbons. Two blocks of graphite, separated by a narrow slot, are dipped into molten silicon. The molten silicon moves up the slot by capillary action. At the top, the silicon cools and crystallizes. This process, called EDGE-DEFINED FILM-FED GROWTH, can make 2 in. of silicon ribbon per minute.

Lasers also show promise as a method of reducing the waste from cutting the crystals into the required wafer thin sections.

PROMISING MATERIALS

Experiments are underway with new materials. While silicon is the time-tested favorite material for solar cells, gallium arsenide is more efficient, transforming 22 percent of solar energy into electrical energy. On the other hand, gallium arsenide cells are 800 percent more expensive than silicon.

A new raw material called amorphous silicon can be produced for a fraction of the cost of older silicon material. Efficiency, however, is very low.

Still another approach being tested is the THIN FILM SOLAR CELL. This is a flat cell thinner than a double edged razor blade. Semiconductor materials such as indium phosphide, tungsten diselenide, and cadmium sulfide are put on bases of glass, plastic, graphite, and metal. The layers are so thin they must be measured in millionths of an inch. The process is simple when compared to the difficulty of growing crystals of silicon. If successful, the cost of thin film cells would be less than a fifth of silicon cells.

None of these processes or materials are perfected to the point where they solve the problems of high cost and low efficiency. Much development and research still remains to be done.

DO YOU KNOW THESE TERMS?

Absorber, convective loop, direct gain solar heating, electromagnetic waves, equilibrium, evacuated tube collector, flat plate collector, fusion, gamma rays, greenhouse effect, heat exchanger, heliostats, infrared rays, insolation, isolated gain solar heating, langley, microwave, parabolic collector, passive solar systems, phase-change materials, photons, photovoltaic cells, plenum, pond, solar constant, thermosiphoning, Trombe wall, ultraviolet rays, volumetric heat capacity, window collector.

SUGGESTED ACTIVITIES

1. Construct a device or toy capable of converting solar radiation to heat.
2. Construct a device which can concentrate the sun's rays on a small area.
3. Build a device which can convert solar radiation to motion.
4. Build, assemble, or demonstrate a device which can convert solar radiation to electricity.
5. Build a device which uses the sun to purify water.
6. Survey your community and report on the number and types of devices which use the sun to save energy. Select one of the types and write a report on how it works. Discuss with the owner the savings experienced in energy.

TEST YOUR KNOWLEDGE

1. Light and heat coming from the sun is caused by ____________ ____________ taking place at its core.
2. If the sun is 93,000,000 miles away and light travels at 186,000 miles per second, how long would it take the light to reach the earth?

3. Describe how electromagnetic waves travel.
4. List the types of electromagnetic waves. Which are visible?
5. Solar constant is all of the radiation falling on an object inside the earth's atmosphere. True or False?
6. The average insolation of (amount of energy reaching) the earth is ____________ kWh per sq. meter per year.
7. The greenhouse effect is caused by: (select correct answer)
   a. Glass protects against the effects of cold winds; therefore it is warmer behind glass.
   b. Solar radiation passes readily through glass but heat energy does not readily pass through.
   c. Glass magnifies the intensity of solar radiation.
8. ____________ ____________ is the amount of heat, in Btus, a substance can hold when its temperature is raised one degree F.
9. What are phase-change materials? Of what use are they to solar energy?
10. On a clear day, the sun delivers about __________ Btu per sq. ft. of radiant energy to the earth's surface. (25, 250, 2500).
11. In an isolated gain type of solar heating system heat is stored and is drawn off as it is needed. True or False?
12. List the three basic elements of an active solar heating system.
13. List three areas where solar cells are used today.
14. At present, the range of efficiency of solar cells is from __________ to __________ percent. (5 to 10, 16 to 25, 18 to 42).

Experimental solar tower being constructed near Barstow, California is designed to heat 1000 homes. Field of sun-tracking mirrors will concentrate sun's rays on boiler mounted on a 250 ft. tower. (Rockwell International-Rocketdyne Div.)

SOLAR TOTAL ENERGY APPLICATION

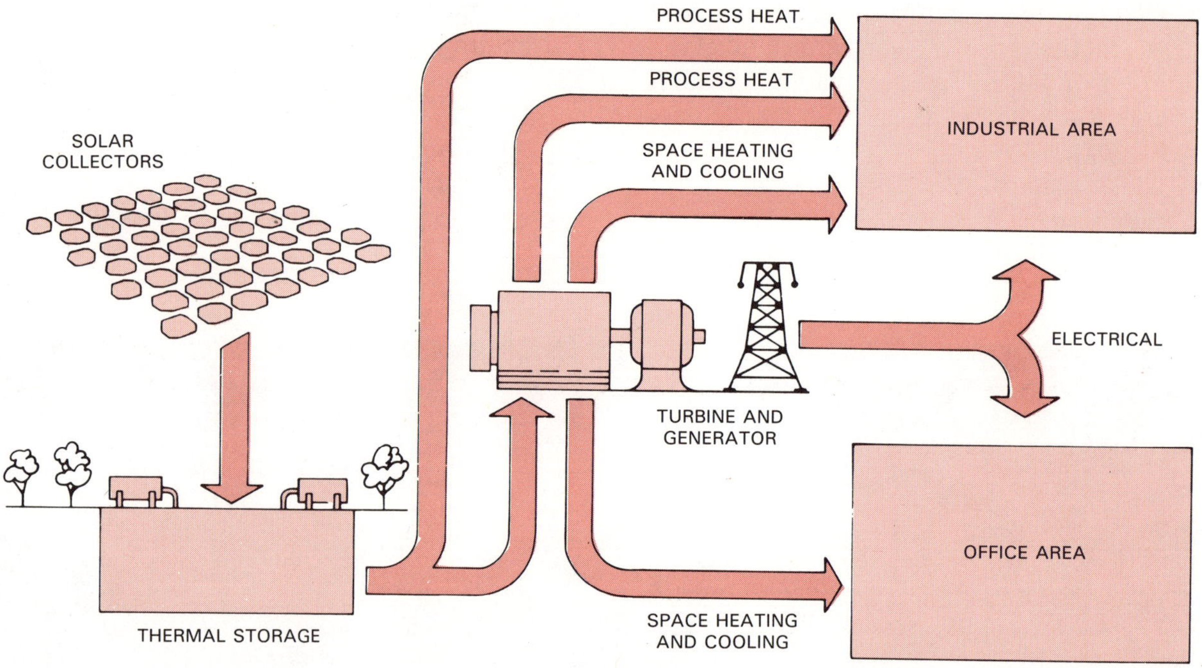

Schematic of a total solar energy system. This system has been designed for a knitwear plant located in Georgia. It has been operating since 1979. Six acres of parabolic collectors concentrate solar rays on a collector. Heat moves from storage area to provide process heat, space heating and cooling, and energy to produce electricity. (U.S. Dept. of Energy)

9 KINETIC ENERGY

The information given in this chapter will enable you to:

- □ List the sources of kinetic energy.
- □ List five sources of kinetic energy in nature which can be harnessed to produce power.
- □ Describe how wind is created and where prevailing winds are located.
- □ Explain the nature of wave energy.
- □ Discuss the origin of tides and list likely sites for tidal power stations.
- □ Discuss how the various forms of natural kinetic energy can be converted to electricity or stored forms of energy.

As you have learned in Chapter 1, kinetic energy is the energy of matter in motion. It is also called MECHANICAL energy. Kinetic energy is made available as a result of:

1. Conversion of other sources of energy.
2. Harnessing of a natural source.

Gasoline is a good example of an energy source converted to produce motion. The gas engine burns the fuel to move a piston. The motion of the piston may move wheels or turn a cutting bar, Fig. 9-1. However, this chapter is really more concerned with the harnessing of natural kinetic forces. These forces are renewable.

KINETIC ENERGY IN NATURE

Most forms of kinetic energy found in nature are caused by the sun. Therefore, we can call them indirect solar energy. Indirect solar sources include the motion of:

1. Wind, Fig. 9-2. Heating action of the sun creates air motion which can sometimes reach great force.
2. Waves, Fig. 9-3. Winds drive water with great energy.
3. Tides. Gravitational attraction of the moon and the sun cause the levels of seas to rise and fall on a regular basis.

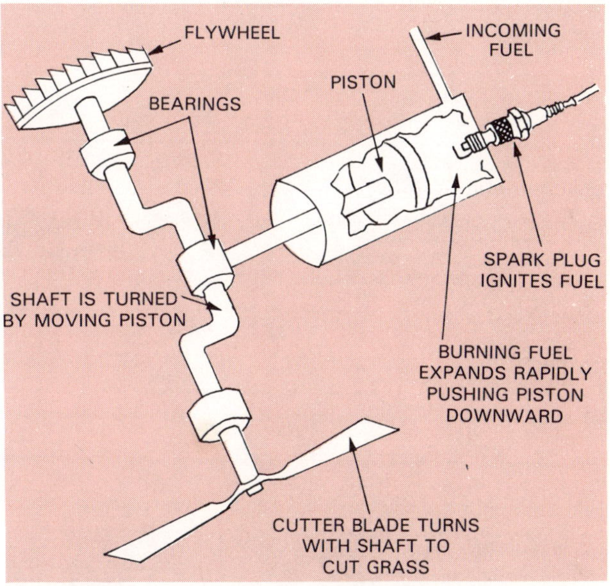

Fig. 9-1. Simplified sketch of gasoline engine. It converts fuel, a chemical energy, to mechanical or kinetic energy.

Fig. 9-2. Wind is kinetic energy resulting from air motion. Its energy can be seen in the swaying of trees.

Fig. 9-3. Ocean waves often have great energy as they batter the seacoast.

Fig. 9-4. Water moves from higher ground to lower ground. An abrupt change in level results in a waterfall which can be used to create usable energy.

4. Falling and flowing water, Fig. 9-4. Rain falls on higher ground and flows to lower levels as the result of the gravitational attraction of the earth.

WIND ENERGY

Wind is air in motion. All winds, from the gentlest breeze to the awesome force of a hurricane, are caused by three natural events:

1. Differences in temperature of the atmosphere.
2. Rotation of the earth.
3. Unequal heating of the oceans and land.

ATMOSPHERIC TEMPERATURE AND AIR PRESSURE

Air is extremely light. However, it has weight and can expand or compress like other gases. One cu. ft. of air (0.28 m^3), on a warm spring day, weighs about 1.22 oz. (35 g).

When air is heated by the sun it starts to move. First it expands and becomes lighter. Then it rises while cooler air sinks and comes in under it, Fig. 9-5.

When warm air rises it causes a lower pressure underneath. The colder air around it is at a higher pressure and so it moves into the space the warm air has left. Thus, when atmosphere at a certain part of the earth is made warmer than atmosphere at other parts, the colder air moves into the warmed region creating wind.

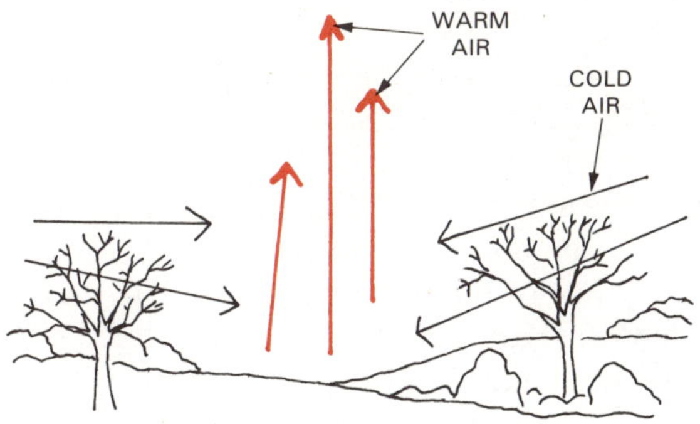

Fig. 9-5. Heated air expands, grows lighter, and rises. Low pressure below the rising air causes descending cold air to rush in to take its place.

World-wide, there are great differences in temperature. For example, the tropics can have temperatures above 100 °F (38 °C) while in the polar regions the temperature is below freezing.

The differences in temperature are mainly the result of how the sun's rays strike the equator and how they strike the polar regions. At the equator the sun's rays fall on the earth from directly above. But at the poles the sunlight strikes the earth at such an angle that the heat is not absorbed nearly as well. As Fig. 9-6 shows, sunshine striking the earth at right angles is better absorbed by the earth's surface. When the sunlight strikes earth at the poles, most of the heat is reflected back into space.

TEMPERATURE'S EFFECT ON WIND

The great differences in temperature between the equatorial zone and the polar regions give rise

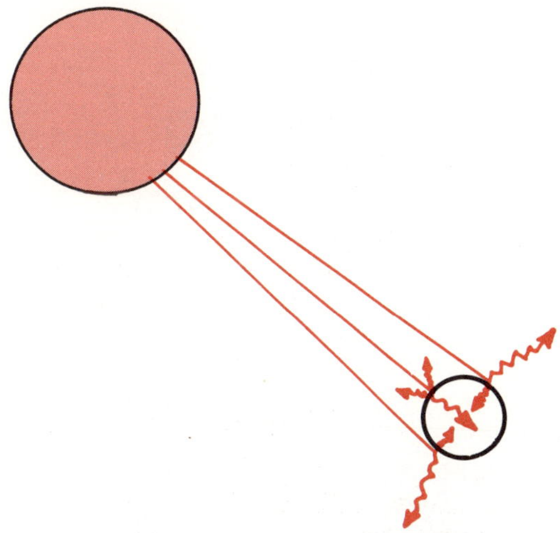

Fig. 9-6. Why some areas of earth are warmer than others. When sun's rays strike earth straight on, much heat is absorbed by the earth's surface. Rays which strike at a glancing angle provide little heating effect because the heat bounces back into space.

to massive movements of air. Hot, light air from the tropics rises and flows toward the poles. At the same time, colder, heavier air from the poles moves under the warm air toward the equator. Fig. 9-7 shows this. This mass movement of air is called a PLANETARY WIND.

WIND FROM EARTH'S ROTATION

Air movement, caused by uneven heating of the earth, is strongly affected by earth's rotation. The speed of rotation is about 1000 mph (1600 kmh) at the equator and 0 mph at the poles.

Strictly speaking, rotation of the earth does not create wind, but it does deflect wind caused by other factors. If the earth did not rotate, the colder winds from the north and south poles would blow toward the equator. Warmer air would rise and flow both north and south toward the poles from the equator.

Because of the rotation, air moving toward the equator is twisted to the west. Air moving toward the poles is turned to the east. See Fig. 9-8.

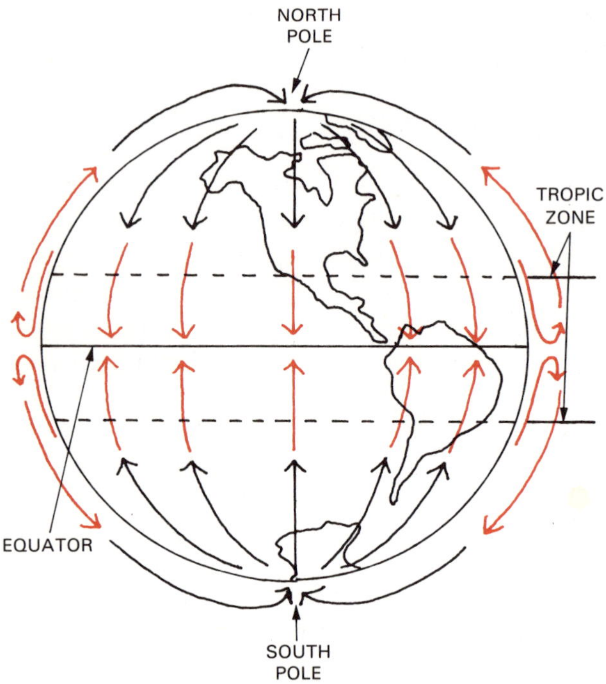

Fig. 9-7. Uneven heating of earth by sun creates a pattern of winds. Low level winds flow toward the equator. Planetary winds bring cooler air from the poles to the tropics. Warmed air, shown in color, rises and flows toward the poles at high altitudes.

WIND PATTERNS

Look at Fig. 9-9. You will notice that there are three distinct areas of wind patterns both north and south of the equator. These areas are also called CELLS. At the equator is the tropical cell. It extends about 30 degrees both north and south. Air at the earth's surface in this cell is rising and is always hot and sultry. There is very little wind but light breezes may come from any direction. The sky is overcast while showers and thunderstorms are frequent. Air pressure is always low.

North and south of the tropical cell are the HORSE LATITUDES. These are high pressure belts where the air is sinking to the surface. The air is dry and warm. Winds are light and changeable or there may be no wind at all for several days. Most of the world's deserts are in the horse latitudes. (The name "horse latitudes" came from crews of sailing ships during the time when colonies were being established in America and other parts of the western hemisphere. Ships were often becalmed (stopped) by the lack of wind and

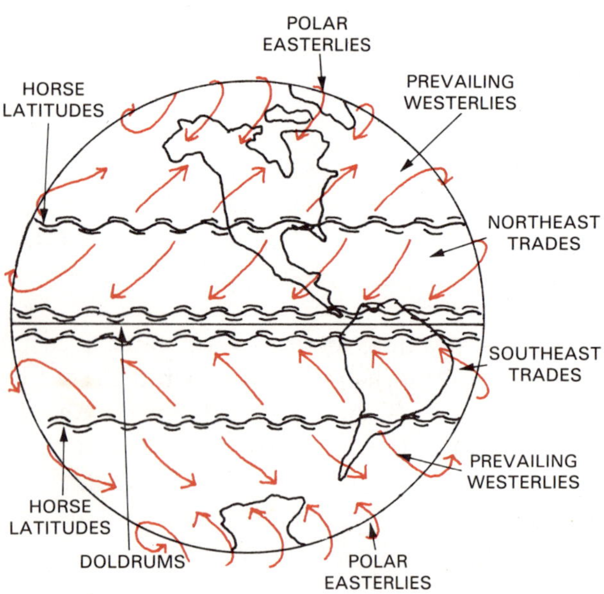

Fig. 9-8. Rotation of the earth deflects planetary winds. The result is bands of wind and bands of calm. Horse latitudes and doldrums are areas of little wind.

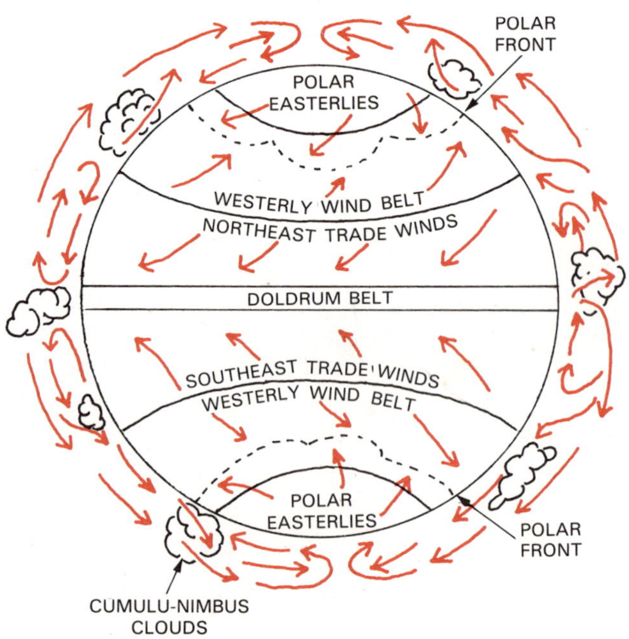

Fig. 9-9. Patterns of wind circulation over whole world.

horses had to be thrown overboard to preserve the supply of drinking water.)

TRADE WINDS

Trade winds are planetary winds that move toward the equator. Their name comes from the old meaning for "trade" which is a "steady track." The trades are the most constant of winds for they often blow for days or weeks with slight change in speed or direction.

EFFECT OF LAND ON WIND

Because land and water heat and cool at different rates, the size and shape of land masses have a great effect on the pattern (direction and force) of winds. Land masses gain and lose heat faster than large bodies of water. Ocean waters are always being stirred up so that heat losses and heat gains at the surface are spread out over a large volume of water, Fig. 9-10. Therefore, the actual change in temperature of the water surface is not as great. The temperature differences between day and night and winter and summer are greatly reduced.

But on land, not much heat penetrates into the soil. The heat that does enter the earth's surface does not become distributed through the soil. As a result, what little heat the earth absorbs during the day is quickly lost during night and during the winter months. The daily and annual difference in temperature is very great.

LOCAL WINDS

Other winds which we encounter from day to day are caused by local temperature differences and topography (lay of the land). These are called local winds. We all know about the sea and land breezes that blow around seacoasts and large lakes. During the day the air above the land warms up much more rapidly than the air above

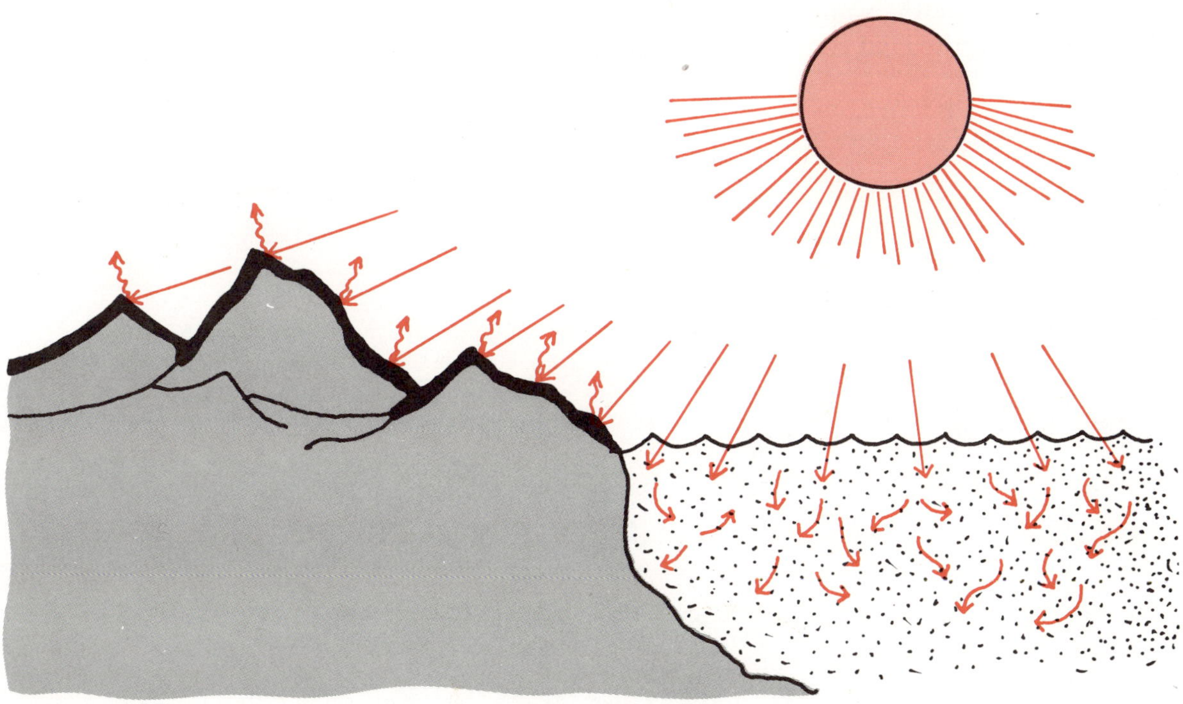

Fig. 9-10. Water and land masses absorb heat differently. Water picks up heat and "stores" it as water circulates. Land, being stationary, absorbs heat to very shallow depth and reflects more of it into atmosphere where it heats up the air.

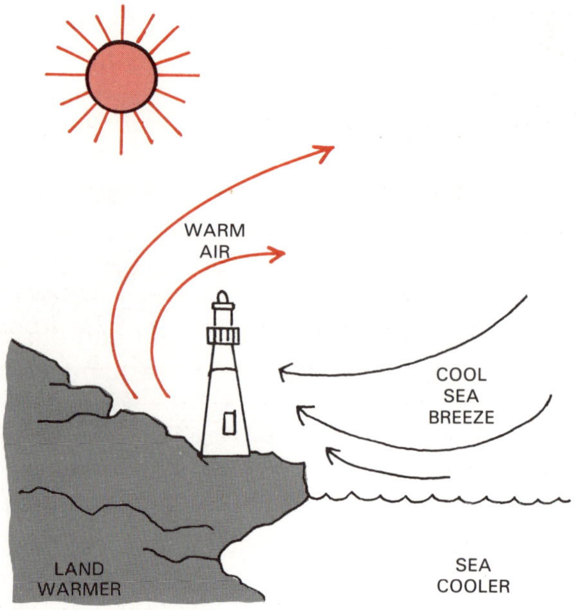

Fig. 9-11. During the day heated air over land rises. Local wind is created as cooler sea air moves over the shore to replace rising warm air.

the water. The heated air over the land rises while cooler air from above the water moves inland, Fig. 9-11. At night, the land cools more rapidly cooling the air above it. Meanwhile, the air above the water is warmed by the stored heat in the water. The air movement reverses. Cooler land air moves in under the rising warm air above the water, Fig. 9-12.

In hilly and mountainous regions, winds are created by unequal heating of the sides and bottoms of the valleys. Down slope winds occur on clear nights because the air on the upper slopes cools and becomes heavier than the air below. It therefore slides into the valley. See Fig. 9-13.

Some local winds occur so often and blow so regularly that they are given names. For example, the name MISTRAL is given the strong cold wind

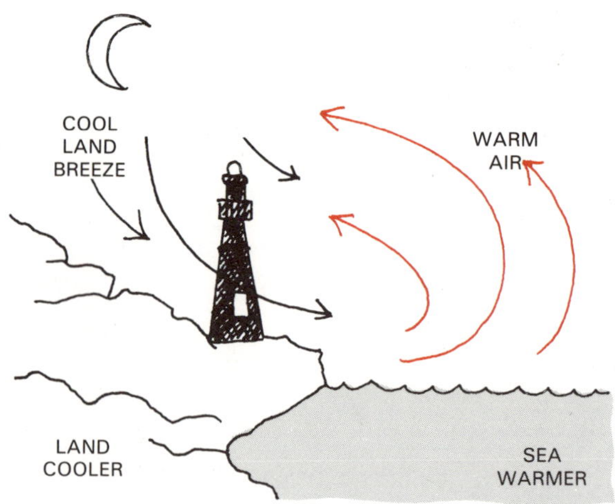

Fig. 9-12. At night the land is cooler. Cooled air over land moves out to sea to replace warmer sea air which is rising.

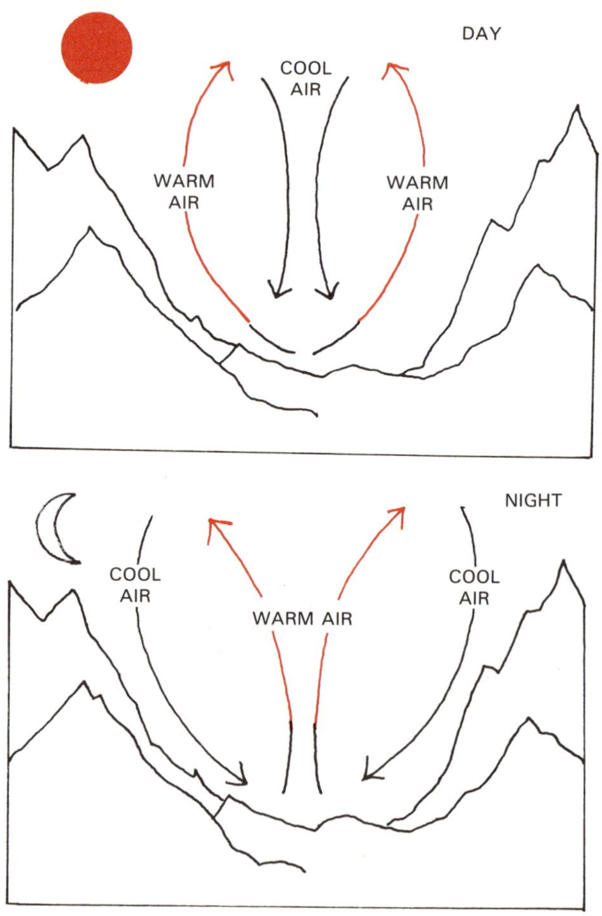

Fig. 9-13. Mountain and valley breezes are mainly updrafts and downdrafts. These are caused by heating and cooling of the land surface.

that blows down the Rhone Valley in France.

When wind blows across a range of mountains the air on the leeward (downwind side) is compressed and heats up. It then blows across the plain as a dry hot wind. An example of this kind is the Santa Ana wind of southern California. It starts in the high desert and spreads across the coastal plain as a dry, dusty wind. Such a wind is also known as a FOEHN.

NORTH AMERICAN WIND PATTERNS

The Rocky Mountains which run north and south along the west coast of the United States, Canada, and Mexico stop most of the strong westerly winds from the Pacific. With no sea winds, inland areas of the United States get cold air moving in from the north pole and warm, moist air from the south. In the Midwest, the prominent winds come from the northwest during winter and from the southwest during summer. See Fig. 9-14.

AMOUNT OF WIND ENERGY IN U.S.

About 2 percent of the solar radiation falling on the earth produces all the wind energy. The total amount of wind energy available in the lower 48 states of the U.S. is about 14 times the energy used by the United States in 1973. About 30 percent of the wind is high up in the atmosphere where it cannot be reached. However, there is still a large amount of wind at and below the 100 ft. level.

TYPES OF WIND

According to weather statistics, there are only a few days a month in most parts of the world when there is no wind. Some, however, are not useful for producing energy. You can classify

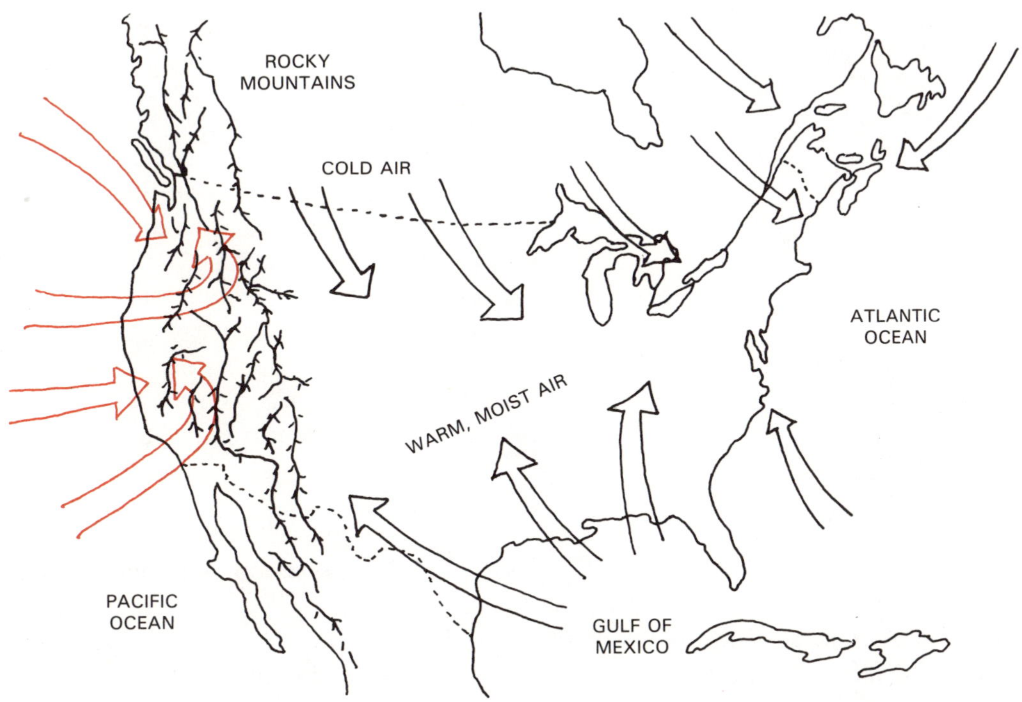

Fig. 9-14. Prevailing wind patterns in U.S. Strong west winds from the Pacific are turned away by the mountains along the west coast. Cold air sweeps down across central plains from Arctic. Moist, warm air moves air from Gulf of Mexico.

winds by their ability to produce energy.
1. Energy winds.
2. Prevalent (prevailing) winds.

Energy winds are those which blow at a speed of 10 to 25 mph (8 to 24 kmh). This type may be available on an average of two days a week. Prevalent winds (from the word prevail) are weak. They travel at about 5 to 15 mph (8 to 24 kmh).

Strong energy winds are found mostly in the temperate and polar regions. Weak or prevalent winds are found in the tropics.

WAVES

When the wind blows over large bodies of water such as lakes or oceans, it drives the water before it in waves. The water moves in huge ripples toward land with great energy. The energy is given up as the water dashes against a shore, Fig. 9-15.

Fig. 9-15. Waves are a result of wind action on large expanses of water.

PROPERTIES OF WAVES

All waves whether traveling in air or water have similar properties. One of these is amplitude. This is the maximum height of the wave. In water the amplitude is the maximum height of the crest of the wave above the still water level.

Another property is called wavelength. This is the distance from one crest to the next one. Still another is velocity or the speed at which the waves are traveling. There are other properties but these are the ones which are important to the harnessing of the energy of ocean waves.

TIDES

Tides are a periodic change in the water levels of seas, oceans, and large lakes. When the water level is high it is known as high tide. When the water is at its lowest level it is called low tide. There is some tidal activity in all large bodies of water but some is so slight that it is hardly noticed.

CAUSES OF TIDES

Rise and fall of tides happens daily on a predictable schedule. Time of a high tide at any place can be established by adding 50 1/2 minutes to the time of high tide for the previous day, Fig. 9-16. This delay coincides (happens at same time) with the daily change in moonrise. It is easy to see, then, that the moon must be the cause. More specifically, it is the gravitational attraction of the moon for the earth.

This gravitational attraction "pulls" at the earth and actually causes a slight bulge in the earth's surface nearest the moon. Water reacts more dramatically to the pull and, thus, collects at where it is nearest the moon. Another bulge forms at the same time on the opposite side and produces a smaller high tide there, Fig. 9-17.

The sun also has some effect on the tides. Its gravitational pull is not as strong because of the greater distance to the sun.

TIDAL RANGE

Tidal range is the difference between low tide and high tide. The greatest ranges occur when the gravitational pull of the sun and the moon are working together. The sun is on the opposite side of the earth from the moon, Fig. 9-18. At other times the two tend to counteract each other. The higher tides are called SPRING tides and the lower tides NEAP tides.

Fig. 9-16. Waters of the seas rise and fall daily because of the gravitational "pull" of the moon and sun. High tidal ranges in Bay of Fundy cause water levels to fall more than 28 ft. at low tide. Effect of low and high side is easy to see at docks. This ship tied up at Port William, Nova Scotia floats at high tide but sits above the water line at low tide. (Nova Scotia Dept., Government Services)

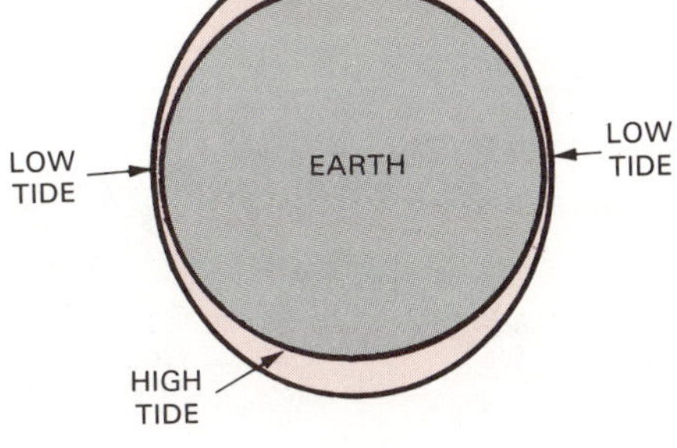

Fig. 9-17. Diagram simplifies explanation of tide. If earth were covered uniformly with water, this is how the gravitational attraction would pull at the water to create high tides.

FALLING AND FLOWING WATER

Falling and flowing water is the action of rivers moving water from higher ground to lower ground. Some of the great rivers of the world may flow for hundreds or thousands of miles before emptying into oceans. It is the action of the sun working as a giant pump that keeps the rivers flowing. It draws up moisture from the oceans and land and releases it over the land. This replenishes the flow of the rivers in a continuous cycle. The flow represents a huge supply of kinetic energy which can be tapped by humankind to do work. See Fig. 9-19.

STORED ENERGY

Some forms of kinetic energy are the result of planned storage. There are serveral types:

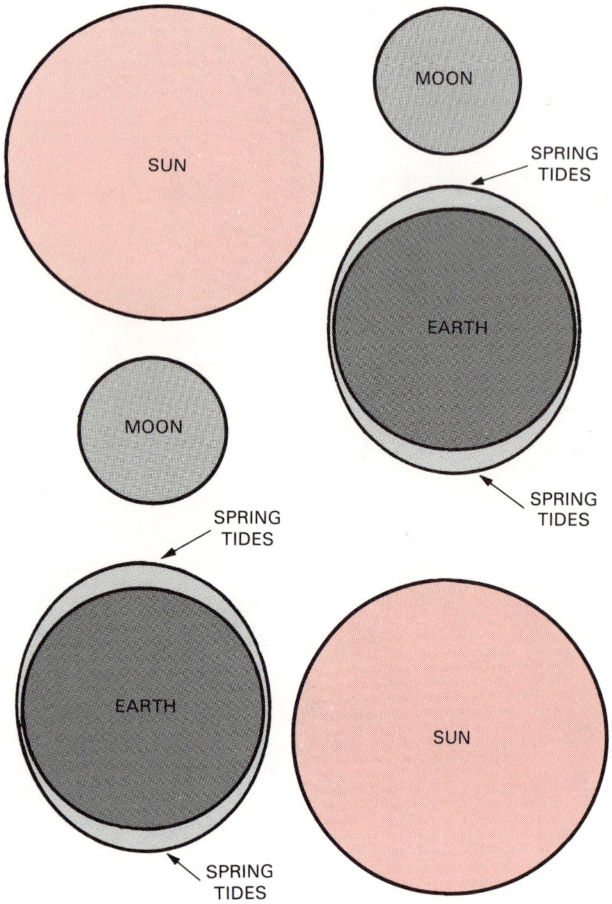

Fig. 9-18. Gravitational pull of sun and earth work together to cause spring (extra high) tides. Left. Sun and moon on same side of earth. Right. Sun and moon on opposite sides.

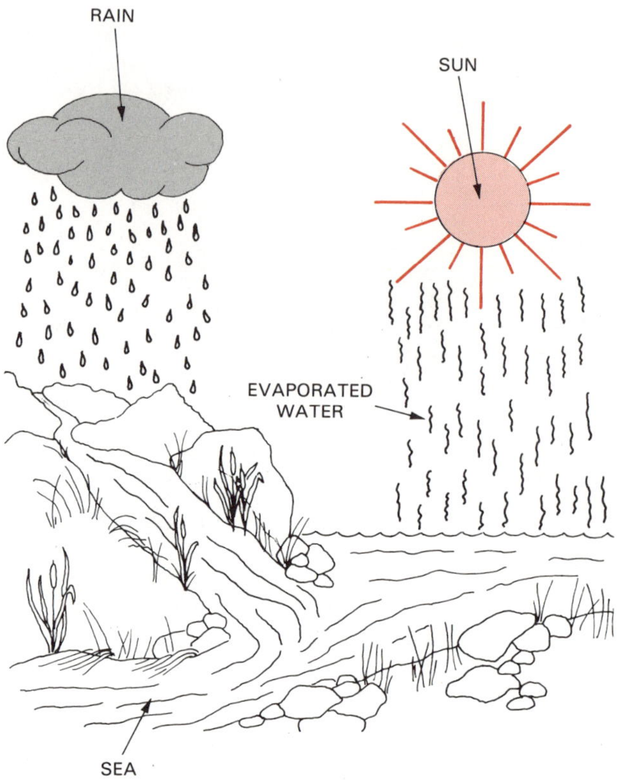

Fig. 9-19. Sun draws water into atmosphere in large quantities. Released in rain, it replenishes flow of rivers.

1. Gravitational storage. The best examples of this are the water towers, Fig. 9-20, which provide the water pressure to feed running water into homes and factories. Another example is the building of dams to store water for generating electricity.
2. Storage by inertia (or inertial storage). This is the storing of energy in flywheels. Inertia is the tendency of a moving or spinning body to continue moving or spinning in the same direction. Such a flywheel can be used to drive electrical generators and propel vehicles or machines such as lawn mowers, Fig. 9-21.
3. Storage of air under pressure. The air, being under pressure, can be used to drive genera-

Fig. 9-20. Planned gravitational storage. Energy of water held at a great height causes the water to flow in home when a faucet is turned on.

Fig. 9-21. Flywheel, because of its weight, will keep on spinning and supplying energy long after force causing it to spin has been removed. This one is part of an old steam engine used to generate electric power.

tors. Compressed air is used to hold up bicycles on pneumatic tires, Fig. 9-22.

4. Storage by deformation. The simplest example of this kind of storage is the winding up of a spring to drive the mechanisms in a clock. The spring tries to return to its original unwound state. As it unwinds it causes the clock wheels to spin and slowly turn the hands. Another example is a toy which moves under power of a rubber band. See Fig. 9-23.

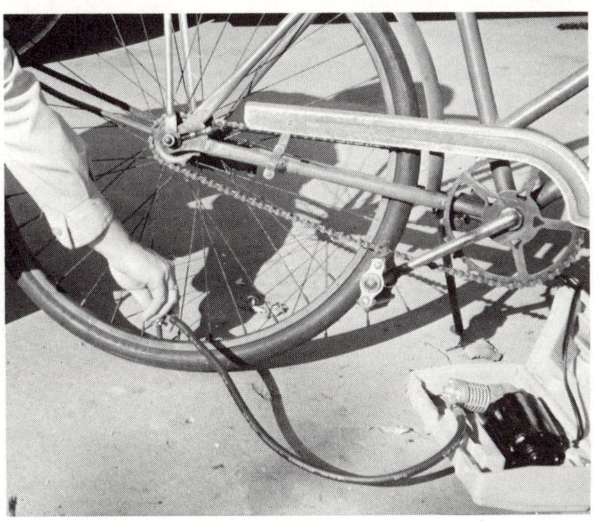

Fig. 9-22. It is storage of compressed air in a tire that makes the tire suitable to hold the weight of bicycle and rider.

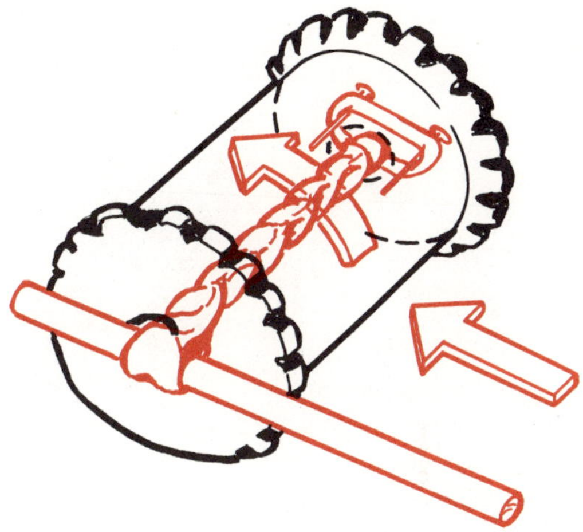

Fig. 9-23. Storage by deformation. Twisted and stretched rubber band in hand made toy tries to unwind to its old shape and length. In doing so it turns the spool.

CONVERTING KINETIC ENERGY

Before any form of energy can do useful work some kind of change must be made. Kinetic energy comes closest to being ready to move loads. Thus, wind can be put to work directly by attaching a sail to a movable object such as a boat. A kite is like a sail. It is kept in the air by the blowing of the wind. So is a hang glider, Fig. 9-24. A river will carry logs to the sawmill or move a boat downstream. Windmills use the motion of the wind to turn a wheel. The wheel drives gears which move a rod up and down to pump water.

In most situations, however, kinetic energy found in nature is transformed (changed) into another form of energy. One of the biggest reasons for changing the kinetic energy is to make it possible to store it or move it to a location where it can be used.

Fig. 9-24. Hang gliders depend on air motion to support the weight of a person.

Fig. 9-25. Ancient windmill from North Falmouth, Massachusetts is patterned after the Dutch mills of the 17th century.

The perfect example of this is the wheel or turbine of a wind machine connected to an electric generator. The electricity can be sent out to distant places over wire conductors. It can also be stored in batteries.

Wind energy can also be stored as a chemical. This is done through electrolysis. Electricity produced by the wind generator separates the hydrogen from water. The hydrogen can be stored or transported in containers.

CONVERTING WIND ENERGY

As you learned in Chapter 2, wind has been used for centuries as a source of power. It pumped water for irrigation, moved ships, and ground flour from grain. In the 17th century, the Netherlands was the world's most industrialized nation simply by using wind power to drive trading ships, run mills, and pump water from lands below sea level. See Fig. 9-25.

Interest in windmills and sailing ships slackened when the steam engine was invented. However, interest in windmills picked up again when Americans began to settle the Great Plains and the West. By 1900 the windmill industry represented a capital investment of more than $4 million. Twenty years later wind power had become a major source of electrical power not only in the rural U.S. but in Europe as well. In the 1930s the Rural Electrification Administration (REA) brought convenient electric power to the U.S. farms. By 1950, small-scale windmills were rare.

WIND GENERATORS

Interest in large wind generators never died. Between 1935 and 1955 a number of large units were built. One of the largest and best known, the 1.25 megawatt Smith-Putnam generator was built in the mountains of central Vermont. Palmer Cosslett Putnam, an engineer who lived on Cape Cod, first conceived the idea of a large generator which would feed electric power into the network of electric power lines normally fed by the fossil fueled electrical power station. He organized a few scientists and engineers to help him. This group designed a wind generator and found a site for testing it. The site chosen was Grandpa's Knob, a 2000 ft. peak in the Green Mountains.

The generator was built by the S. Morgan Smith Company. Its two stainless steel blades had a 175 ft. (53 m) span and sat atop a 110 ft. (33.5 m) tower. It weighed 250 tons and could generate 1250 kilowatts of electricity. This was enough for a small town. The generator was completed in August, 1941. Testing, begun in October of that year, was interrupted frequently because of mechanical failures. In March of 1945, one of the blades fell off. It has not operated since. Though its performance was a disappointment to its designers and builders, it was an important effort by American scientists and engineers to harness wind power.

MORE RECENT DEVELOPMENTS

In 1973, following the oil crisis, the federal government began to sponsor wind generator projects. The first was started at the NASA Lewis Research Center by the National Science Foundation. A wind turbine was developed. The test machine was named "Mod-O." It had a two-blade wind rotor and a gear box which stepped up the speed. The rotor operated at 40 rpm and the gearbox drove the generator at 1800 rpm. In an 18 mph wind it generates about 100 kW of electricity. It began operation at the NASA Plum Brook site in September, 1975. Its purpose was to gather engineering information that could be used in a large-scale wind energy program.

A larger wind turbine, Mod-OA, was developed at Clayton, NM during November, 1977. It generates 200 kW of electricity at 40 rpm., double the power of Mod-O.

In 1976 the federal government added a third wind generator to their test program. It was the Mod-1 which was designed to produce 2000 kW of electric power at a wind speed of 25 mph (40 kmh). This machine, located at Boone, North Carolina, began operating in 1979, but was later shut down. In 1983 it was sold to a private company which planned to convert the generator for use with a water or steam turbine.

MOD-2 WIND TURBINE

Another experimental program in wind power is the Mod-2. Built by Boeing for the Department of Energy, it is a down wind design which can generate 2.5 megawatts of electricity. Three of the units are in operation at Goodnoe Hills near Goldendale, Washington. See Fig. 9-26 and Fig. 9-27.

Fig. 9-26. One of the three Mod-2 wind turbines at Goodnoe Hills in southern Washington state. It weighs 20 percent less than Mod-1 but generates twice as much electricity.
(Boeing Engineering and Construction Co.)

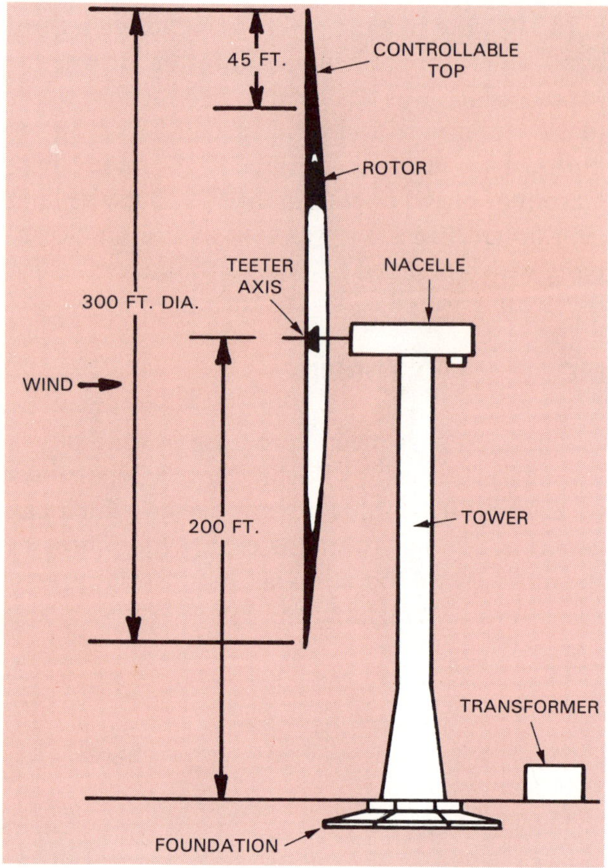

Fig. 9-27. Dimensions of Mod-2. Its tower is made of tubular steel. The blades of the rotor are also of steel.

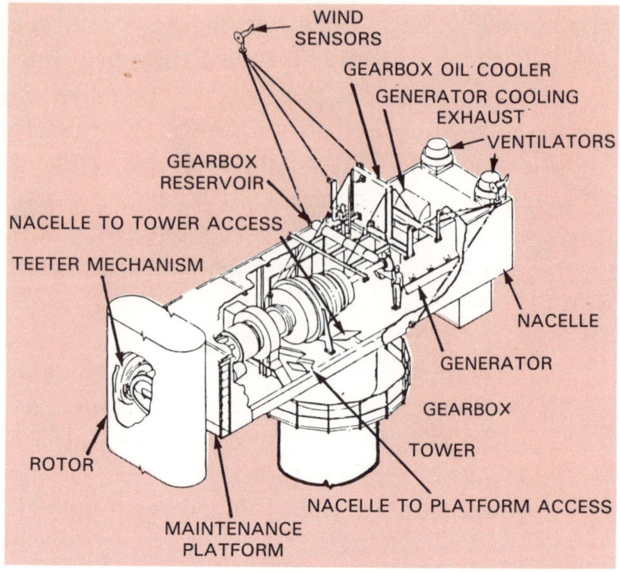

Fig. 9-28. Cutaway of nacelle (housing) for Mod-2 shows gears and other mechanism which control electrical output. Not shown is the microcomputer.

The Mod-2 is controlled completely by a microprocessor (a small computer) in the generator housing, Fig. 9-28. Power from the Mod-2 system is fed into the regional power grid (power lines) of the Bonneville Power Administration.

The units begin to generate electricity in 14 mph winds. At winds of 27.5 mph they reach their rated output of 2.5 megawatts. In stronger winds the blade tips feather (turn slightly to spill the wind). At winds of more than 45 mph the unit shuts down.

At wind speeds of 14 miles an hour a single Mod-2 machine can generate about 10 million kWh of electricity a year. This would supply energy for about 700 homes.

The goal of the Mod programs is to show that large wind machines are practical. Also, researchers want to determine the costs for going into commercial production. Similar testing is underway for small wind machines.

SMALL GENERATOR PROJECTS

Rocky Flats, Colorado has become a test site for small wind generators. The first unit tested there had a 4.1 kW generating capacity. It was used for 2 1/2 months to generate power for a deep well irrigation pump. It was exposed to winds as high as 85 mph (137 kmh).

Other machines have been tested at the site. All have developed defects and failed. Problems have included broken tail vanes, bearing failures, and broken blades.

The tests were originally set up and sponsored by the Energy Research and Development Administration (ERDA). It contracted with a private company to develop the Rocky Flats test site. ERDA had identified a number of needs for wind

power devices in three power ranges:
1. A 1 kW direct current unit for remote locations and where climate is severe.
2. An 8 kW alternating current unit for individual home and farm use.
3. A 40 kW alternating current unit for powering irrigation pumps and to provide electricity for remote communities.

TYPES OF WIND MACHINES

There are two general classifications of wind machines:
1. Horizontal axis.
2. Vertical axis.

HORIZONTAL AXIS MACHINES

In a horizontal axis machine the rotor shaft is horizontal and the rotor or propeller revolves in a vertical plane (path). The farm windmill is a familiar example of this type.

The rotor may have one, two, three, or many blades.

Single bladed rotors, Fig. 9-29, require a counter weight on the opposite side of the rotor shaft to stop vibration. This type is not practical where ice could build up on the rotor and throw it off balance. Two bladed rotors, shown earlier in Fig. 9-26, are most popular because they are strong and less expensive to build. A three blade rotor, Fig. 9-30, spreads stress more evenly, especially when the wind changes direction.

Fig. 9-30. This 25 kW generator has three blade rotor. (Niagara Mohawk Power Corp.)

The bicycle multi-blade, Fig. 9-31, and the sail wing, Fig. 9-32, are new rotor designs. The

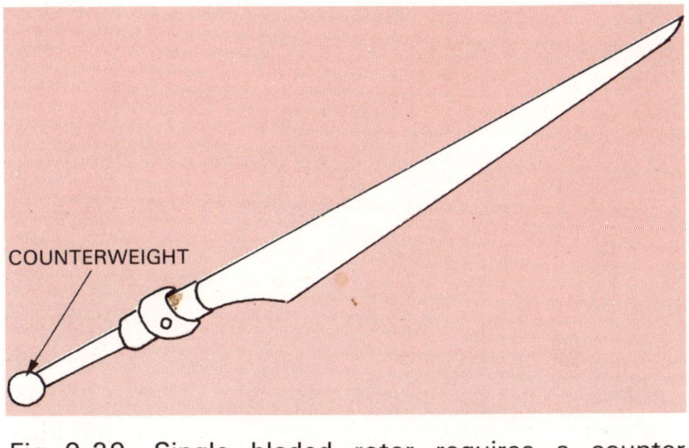

Fig. 9-29. Single bladed rotor requires a counter balance to reduce vibration.

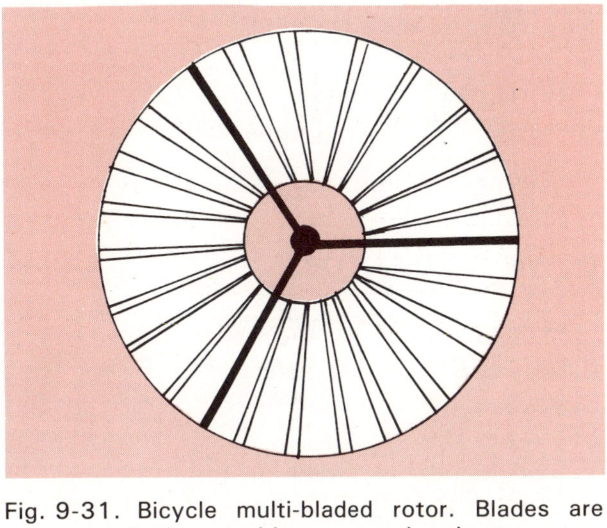

Fig. 9-31. Bicycle multi-bladed rotor. Blades are attached to steel hoop around perimeter.

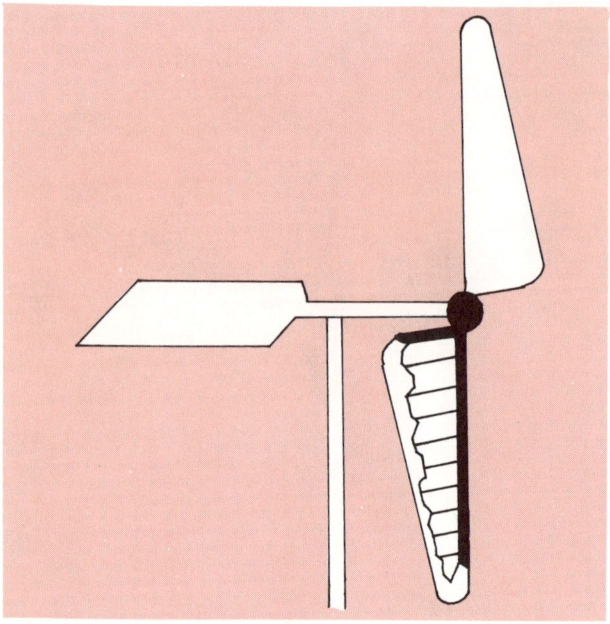

Fig. 9-32. Sail wing has pipe and cable frame covered with cloth sleeve.

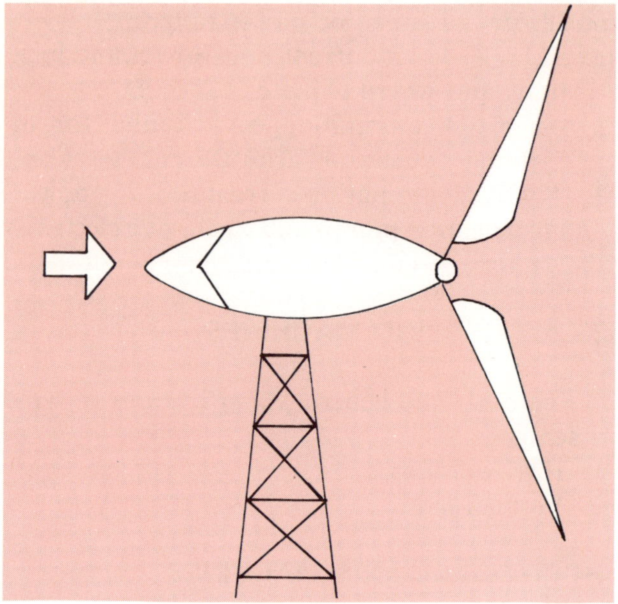

Fig. 9-33. Down wind machines do not need vanes to swing them into the wind.

bicycle rotor has many narrow blades held in tension between the hub and a solid metal rim. It is light and strong.

The sail wing rotor uses a metal tube for the leading edge of the blade. Short bars extend at right angles from the tube to form the rotor blades. Cable is stretched between the tip and the root to form the trailing edge. The blade is then covered with a cloth sleeve.

Horizontal axis wind machines have two other characteristics that separate them into types:

1. Down wind type, Fig. 9-33.
2. Up wind type, Fig. 9-34.

DOWN WIND ROTORS

In down wind types, the rotor is located behind the tower and rotor housing. That is, the wind blows over the tower and housing before striking the rotor. This type has no vane. (This is the fan shaped metal piece keeping the rotor into the wind.) Down wind shapes are preferred for larger wind machines.

UP WIND ROTORS

Up wind rotors are designed to head into the wind. A tail vane keeps it there. Another

Fig. 9-34. Up wind machines face the rotor into the wind.

mechanism pulls the rotor out of the wind when it is blowing too hard or when the machine is shut down.

VERTICAL AXIS WIND MACHINES

In vertical axis wind machines the rotor spins on a vertical spindle. No vanes are needed since the rotor can pick up the wind from any direction. Different kinds are shown in Fig. 9-35.

The Savonius or "S" rotor was patented in 1929 and resembles the Persian windmills used to pump water. It is best used for pumping water and ventilating buildings. It rotates too slowly to generate electrical power.

The Darrieus rotor was patented in 1931. It and the delta Darrieus are vertical axis versions of propeller wind machines. They are usually not self-starting and must be spun by an electric

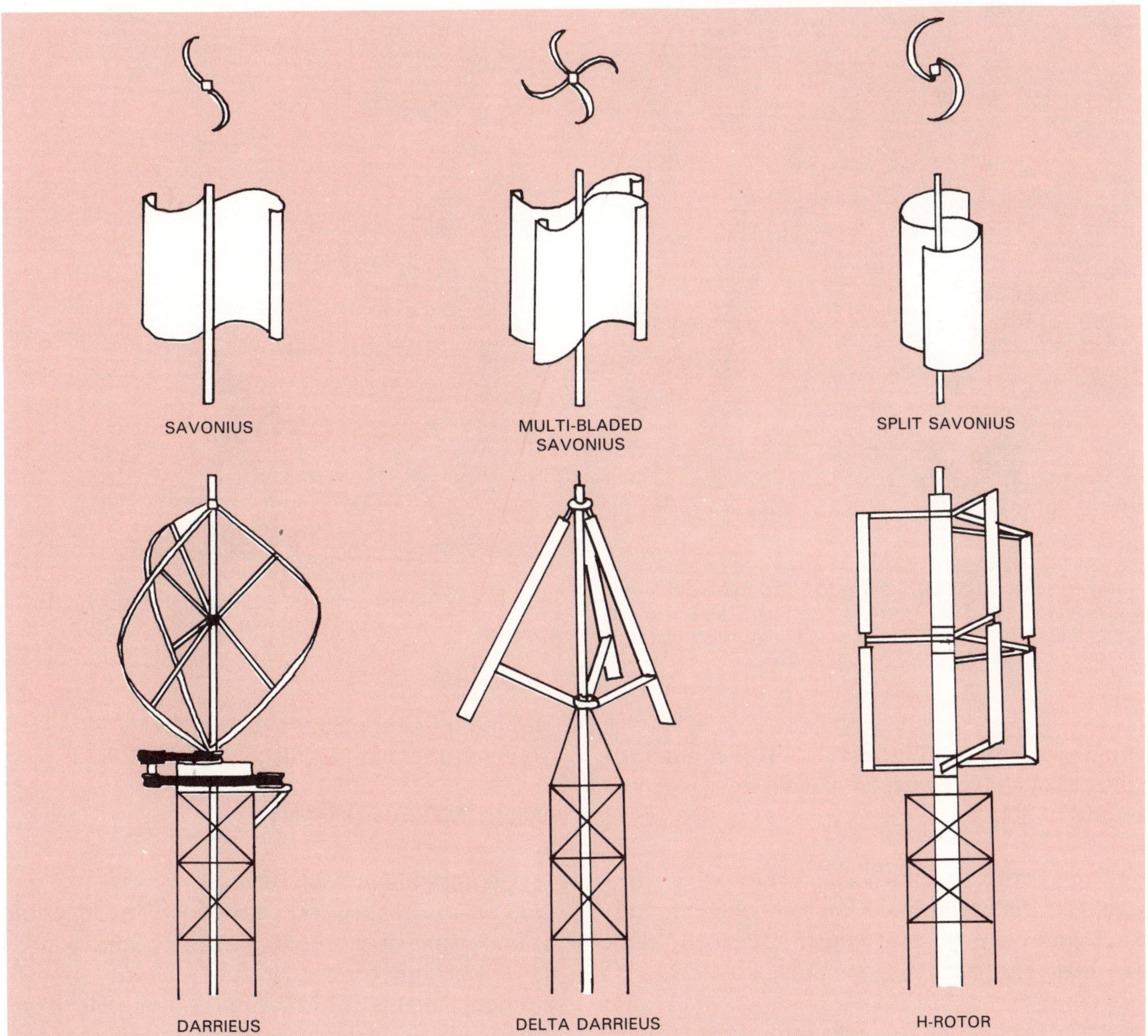

Fig. 9-35. Modern vertical shaft rotor designs.

Fig. 9-36. Darrieus type wind powered generator is installed on a silo in St. Lawrence County, New York. It is used in a wind generator research and development project undertaken by Clarkson College with Niagara Mohawk Power Corp.

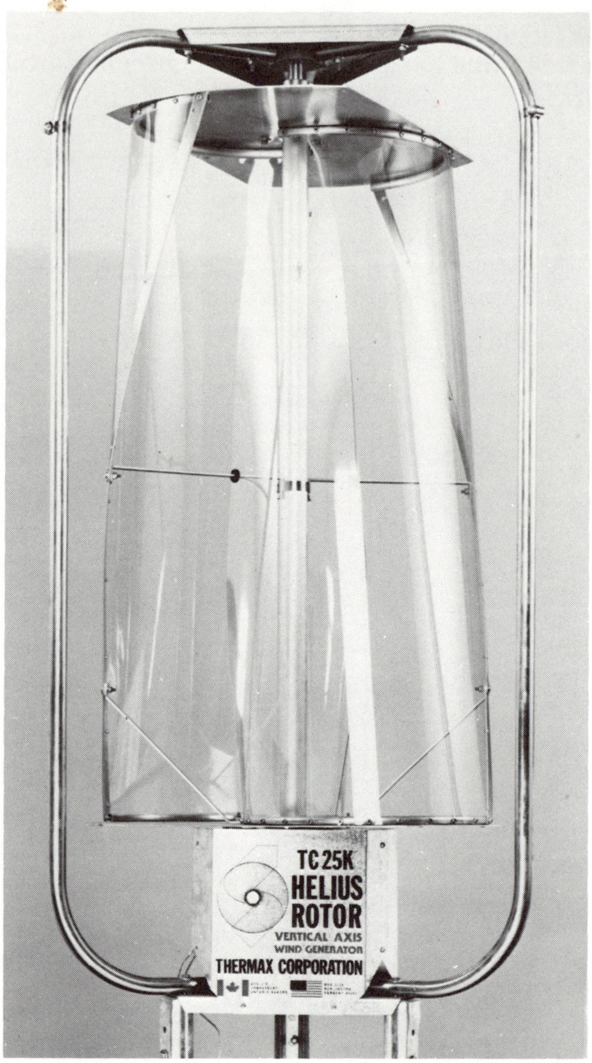

Fig. 9-37. The Helius™ rotor is a variation of the Savonius rotor. (Thermax Corp.)

motor when the wind speeds are high enough for them to operate. An experimental model is shown in Fig. 9-36.

The H-rotor has movable blades which flip twice during each revolution to take greater advantage of the wind. It runs more slowly than the Darrieus, but develops more power.

A variation of the Savonius rotor is the Helius™ rotor, Fig. 9-37. The vanes are shaped and mounted like the Savonius but they are twisted to catch the wind at all angles, Fig. 9-38.

WIND MEASUREMENT

Taking wind measurements is necessary to find a good site for a wind generator. The location is important since a good prevailing wind source is important for generating electric power. Sometimes this information is available from weather stations, airports, or military bases. Many universities, public schools, and private

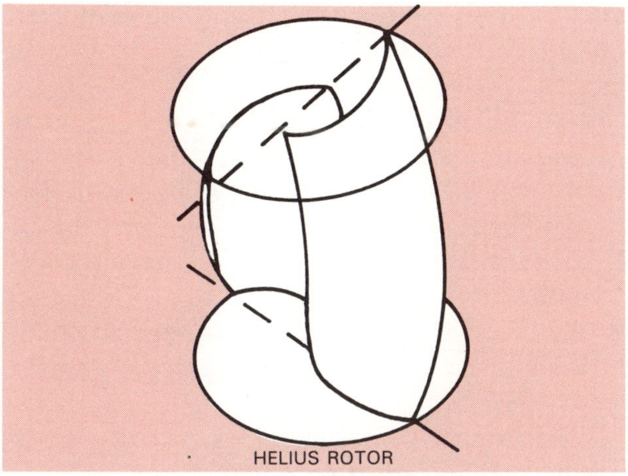

Fig. 9-38. Vanes of the Helius rotor are twisted to catch the wind at all angles.

colleges have weather stations and can supply wind records of your area. If none of these sources of information are available to you, a wind speed indicator can be used to gather information on wind speed in your location.

MEASURING INSTRUMENTS

Instruments which measure wind speed are called WIND SPEED INDICATORS, or ANEMOMETERS. See Fig. 9-39.

A recording anemometer will keep a record of the wind speed over a period of time.

EFFECT OF HEIGHT ON WIND SPEED

The surface of the earth and high structures such as trees and buildings affect the amount of wind available. Wind speeds at low heights are greatest over water and on level plains. Ground level winds are very weak in cities because of the many buildings.

COMPLETE WIND SYSTEMS

The tower, rotor, and turbine or generator make up only a part of the wind power system. Generated electric power is carried away by con-

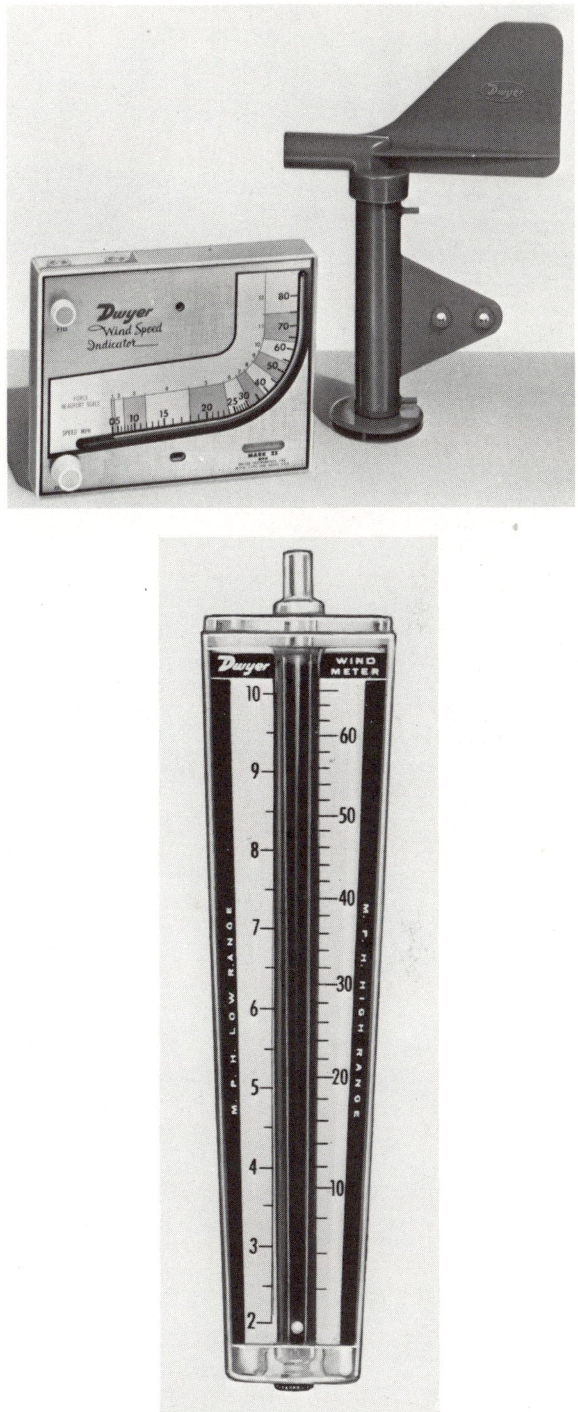

Fig. 9-39. Instruments for measuring wind velocity are called wind speed indicators or anemometers. Top. Instrument intended for a permanent installation. Sensor mounts on a pole or rooftop. Bottom. Portable pocket size meter can be used anywhere. (Dwyer Instruments, Inc.)

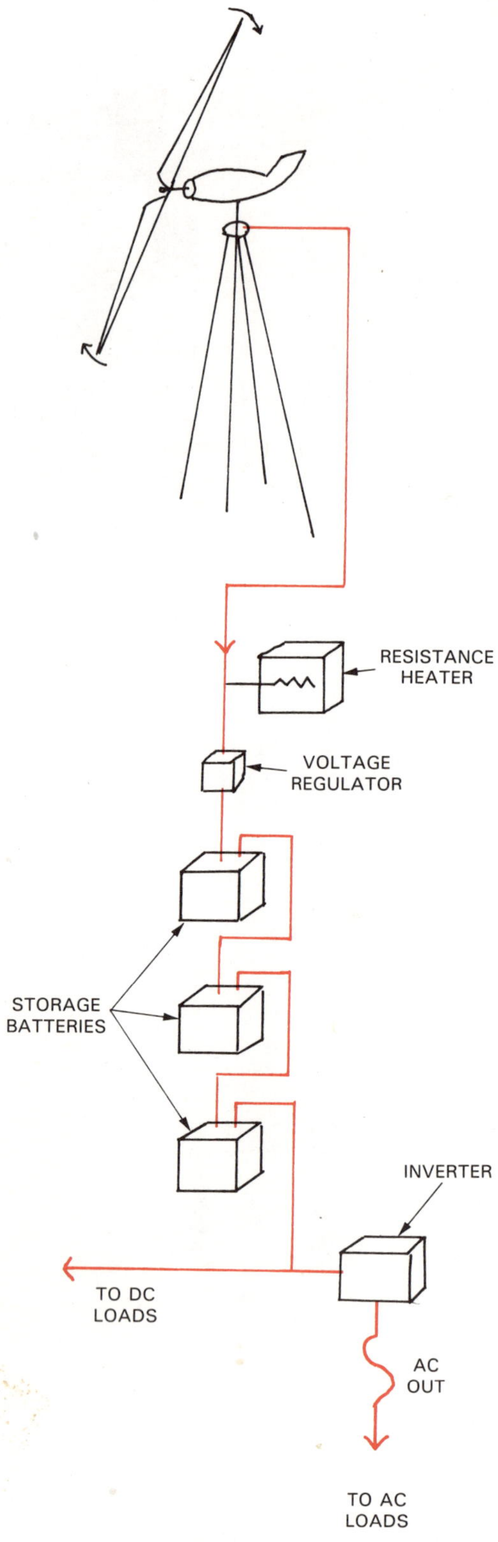

Fig. 9-40. Diagram of a complete wind system. An inverter converts the direct current to alternating current.

ductors. Fig. 9-40 is a diagram of a complete, small scale, wind power system.

A complete system will include the following additional components (parts):

1. Voltage regulator to keep voltage at a certain level.
2. Storage batteries to contain the electrical energy until it is needed.
3. An inverter which steps up voltage and converts direct current to alternating current.

While some wind generators produce alternating current, only direct current can be stored in batteries. Small inexpensive wind generators produce 12 V dc. Larger units produce 24, 36, 120, or even 240 V dc.

VOLTAGE REGULATORS

The job of the voltage regulator is to assure that a constant voltage output is supplied to the load. If the wind generator voltage is not above the voltage for which the regulator is set, a set of contacts in the regulator will be closed. Current will flow through the circuit to the load. As the generator builds up voltage it will reach the rated voltage setting of the regulator. At that point the contacts will separate so that current must go through a resistor. This causes the generator to cut down its output. This action is repeat-

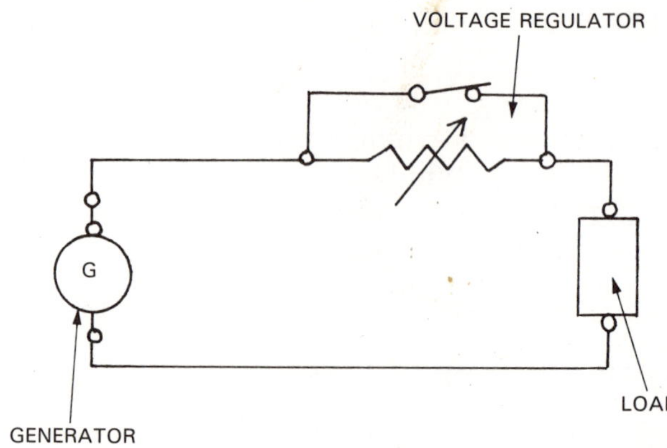

Fig. 9-41. Diagram of a circuit containing a voltage regulator.

ed as often as needed to keep the generator from putting out too high a voltage. See Fig. 9-41.

BATTERIES

Excess electricity generated by the wind machine can be stored in a number of batteries which are connected together. A battery is made up of one or more cells. Each cell includes both positive and negative plates which are separated from each other. Positive plates are connected to the positive post of the battery and negative plates to the negative post. The plates are surrounded by electrolyte. Electrolyte is an acid and water mix which will conduct electricity.

A lead-acid battery, Fig. 9-42, has lead plates and a sulphuric acid-water electrolyte. An alkaline battery has plates of nickel-iron and nickel dioxide. The electrolyte is a mix of potassium hydroxide and water.

Fig. 9-42. Battery suitable for storing excess electricity from a wind power system.

When a lead-acid battery receives electric current the current causes a chemical change to take place in the battery. The water breaks down into hydrogen and oxygen ions. The hydrogen ions attract sulphate ions deposited on the plates when it is discharging. Sulphuric acid is formed.

When current is drawn from the battery the sulphuric acid breaks down into hydrogen and sulphate ions again. Hydrogen and oxygen ions combine and form water. The sulphate ions combine with lead and form a lead sulphate deposit on the plates.

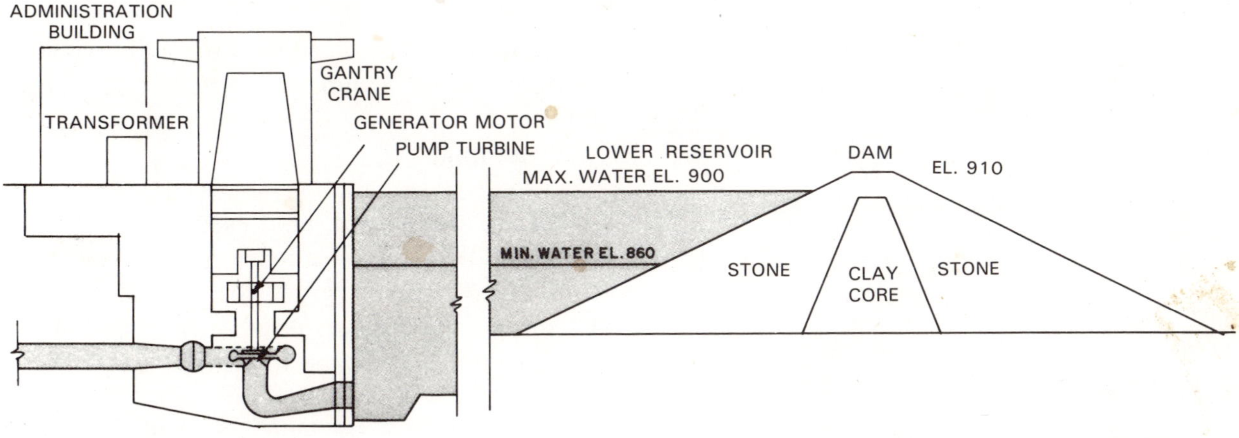

Fig. 9-43. A setup for pumped storage. During off-peak hours, excess power is used to pump water into storage reservoir at right.

PUMPED STORAGE

Because lead-acid and alkaline batteries are an expensive way to store electricity, new storage methods are being explored. One is pumped storage. In this method water is pumped to a high reservoir, Fig. 9-43. Later the water could be released to power hydroelectric generators. This is done at Niagara Falls and other hydroelectric sites.

Another possibility is to store energy as compressed air. Electric motors would run compressors. The compressed air could be stored in underground caverns. See Fig. 9-44. When power is needed, the compressed air would run turbines to generate electricity again.

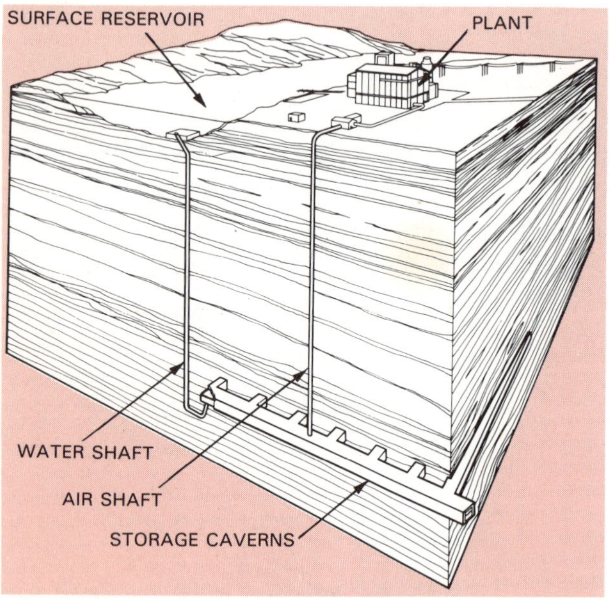

Fig. 9-44. Schematic of an air storage system. Excess electricity is used to compress air into hard-rock caverns. Water is used to keep air at a constant pressure. (EPRI Journal)

INVERTERS

An inverter converts direct current electricity into alternating current. There are several kinds of inverters. A rotary inverter is a dc motor which drives an ac generator. An electronic inverter uses transistors and rectifiers to convert direct current.

When the excess current is to be fed into electric utility lines, a SYNCHRONOUS INVERTER is used. Then when excess electric power is fed back into the utility's power lines, there is no need for batteries.

FUTURE OF WIND ENERGY

While wind is free, the equipment to change its energy into electricity is not. A well-designed wind generator can convert about half of the available mechanical power of the wind into electrical energy. There are other losses so that somewhere between 20 and 50 percent conversion is realized. Experts in this country are experimenting with the idea of large wind energy farms spread across the Great Plains and off our shores.

ENVIRONMENTAL CONSIDERATIONS

There are few undesirable by-products of wind energy. There are no residues to cause smoke or rubbish. There are no health hazards. Still, there are some drawbacks. A single tower does not take up much space, but a giant "wind farm" with many towers and power lines would use up considerable land area. Furthermore, it would not be a pleasant sight.

Large wind generators also cause some annoyance. Residents near a large wind generator in Boone, North Carolina complained of television interference and air disturbance when the unit was running.

On the other hand, land used for wind farming could be used for other purposes at the same time. Cattle could graze around the towers. Farming could be carried on as it is now under and around power line towers.

COSTS OF WIND POWER

Wind power systems have been slow in developing partly because of costs. However, it is thought that once systems are mass produced, wind generators will be able to produce electricity just as cheaply as fossil fueled or nuclear power stations. A major drawback of wind power is that the equipment is often idle for lack of wind. It is estimated that the system would be operating at full capacity only about 6 hours out of every 24. Coal and nuclear power systems can operate at capacity 50 to 60 percent of the time.

On the encouraging side, wind powered systems have low maintenance and no fuel costs. Experts believe that the total cost of the electricity generated by the wind will be about 2 to 3.5 cents a kilowatt hour. This is close to the present average cost of electricity produced by conventional power plants.

It is likely that wind power will be an important supplementary source of electric power.

HYDROELECTRIC ENERGY

As you learned from Chapter 2, one of the oldest methods of using a renewable energy resource is the water wheel. Today when water energy is used, it is used almost entirely to produce electricity.

There are several forms of hydro energy:
1. Hydroelectric power which is derived from the natural force of falling water.
2. Tidal power, which comes from the force of the gravitational pull of the sun and the moon—discussed earlier.
3. Wave power which is caused by the wind driving the water toward the shore.

Nearly all power now taken from water energy is from hydroelectric plants at waterfalls or dams. The usual method is to dam up rivers where the terrain will produce a natural reservoir.

PERCENTAGE OF ENERGY PRODUCED BY HYDROELECTRIC

About 16 percent of all electric power used in the United States comes from hydroelectric installations. This is only 30 percent of what it could be if all resources were developed. It is thought that hydroelectric capacity could be doubled in 20 years.

OBSTACLES

There are reasons why more hydroelectric power plants have not been built:
1. High cost of building dams to form a reservoir of water.
2. Highways, railroads, power lines, and even towns must often be moved.
3. Major equipment such as turbines, generators, and other power station equipment must be designed and built. This takes a great deal of time and money.
4. Farm lands must be abandoned behind the dam.
5. Often, special groups oppose the destruction of other natural resources and wildlife caused by damming up rivers.

HOW HYDROELECTRIC POWER PLANTS ARE DESIGNED

Hydroelectric power plants must have a head of water at least 20 ft. (6 m). Head is the distance from the highest level of the dammed water to the point where it goes through the power producing turbine. Many low-head dams have a head of no more than 100 ft. (30 m). There are high-head hydroelectric plants with a water drop of 100 to 1000 ft. (30 to 300 m). See Fig. 9-45 and Fig. 9-46 for more information on design of dams.

Water storage is another important factor in design. The amount of water storage needed depends on the electrical power demands on the power plant. This demand varies. Electrical loads are greatest during the day and gradually decrease in the afternoon, reaching a low point at night. Usage of electric power is also greater during June, July, and August because of air conditioning.

Water must be stored and released as demand for electricity increases and subsides. When

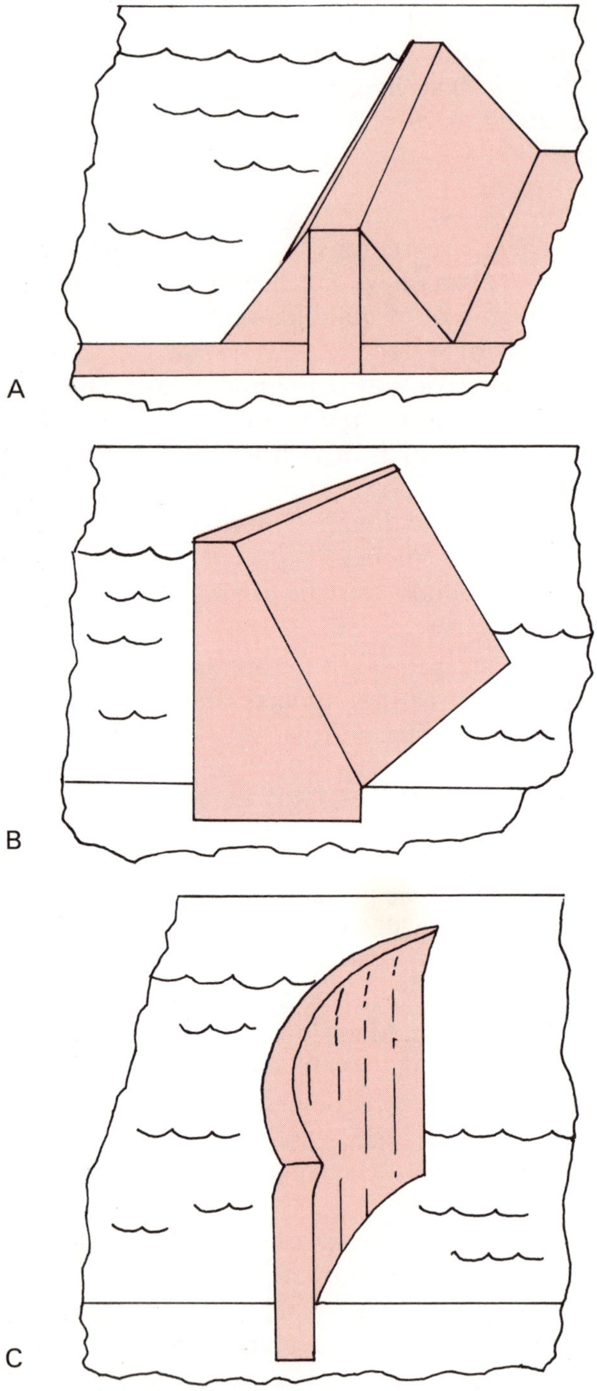

Fig. 9-45. Three types of dams. A—Embankment dam. It uses the weight of earth or rock filling to hold back the water. They are the simplest and cheapest. B—Concrete gravity dams are huge and expensive and usually run across a broad valley. C—Arch dam is thin walled and curved against the stream. Force of water is transferred sideways to steep walls of a valley and downward.

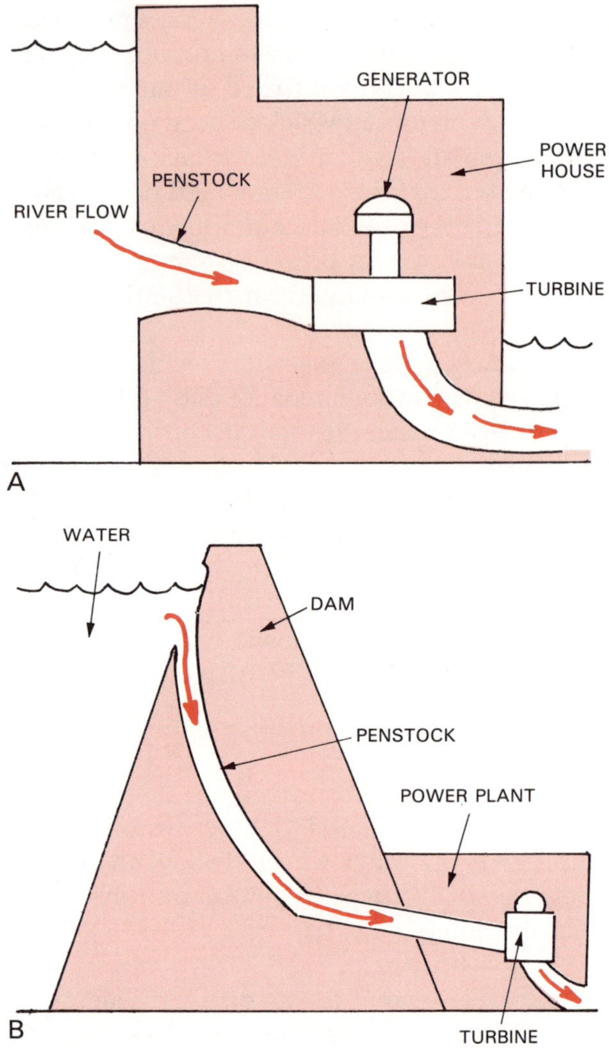

Fig. 9-46. Some hydroelectric power plants operate with a high-head while some have a low-head. A—Typical low-head power plant. B—Typical high-head installation. Low-head type may have less than 100 ft. of head while the high-head type may have a drop of up to 1000 ft.

demand is high, the gates of the dam are opened wider to allow more water to flow through the dam. More power is produced.

HYDROELECTRIC TECHNOLOGY

Fig. 9-47 is a simple schematic of a hydroelectric plant. The water flows from the dam down a large tube called a PENSTOCK. At the bot-

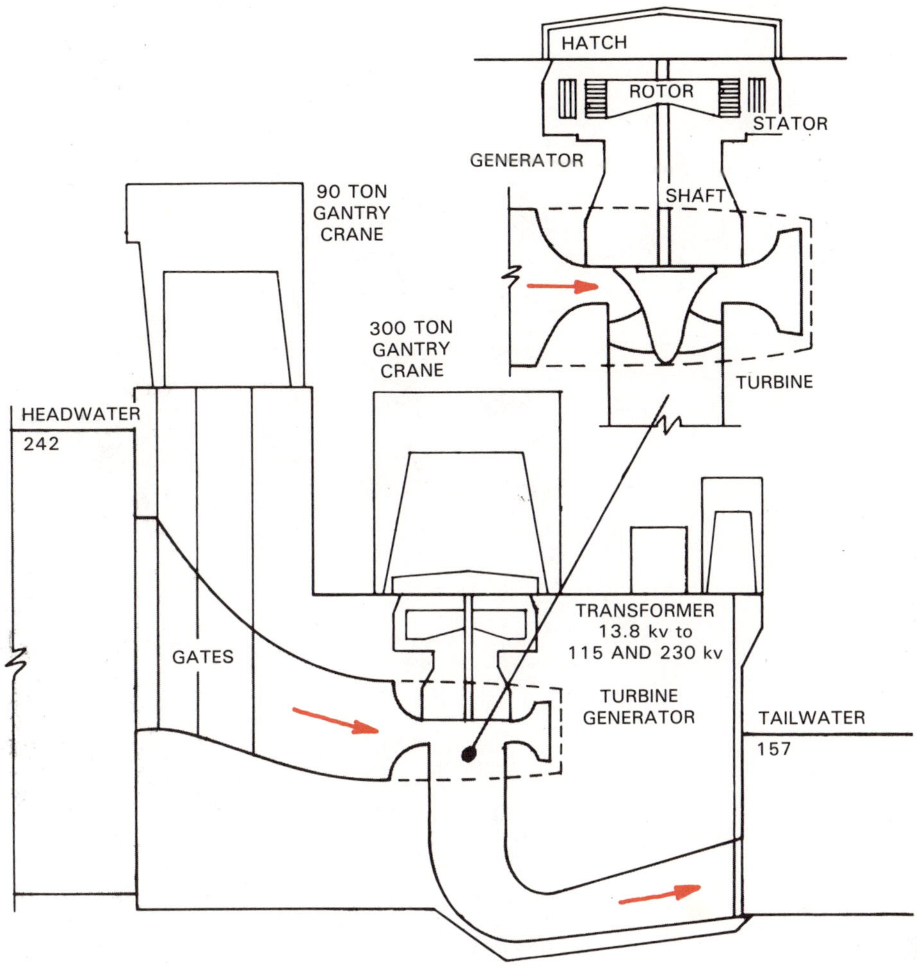

Fig. 9-47. Water flowing through penstock and across the turbine creates power to drive the generator. (New York State Power Authority)

tom of the penstock the fast moving water drives a REACTION or IMPULSE turbine. Reaction turbines work well with low-head hydroelectric plants.

HARNESSING TIDAL WATERS

Rise and fall of tidal waters represents a source of energy which could be converted to useful power. Like wind energy, it is not a constant source. It would have to be captured and stored.

TIDAL MILLS

Use of tidal power is not new. Hundreds of years ago tide waters were trapped at the mouths of rivers emptying into the sea. It was used to run tide mills.

Dams were constructed to form a reservoir upriver. Incoming tide was allowed to flow through gates in the dam to flood the reservoir. At high tide the sluice gates were closed. Several hours later when the tide had fallen, the trapped water was released to turn water wheels.

Between 1581 and 1822 one of the arches in the London Bridge had such a mill. It was used to pump water from the Thames River to the center of London.

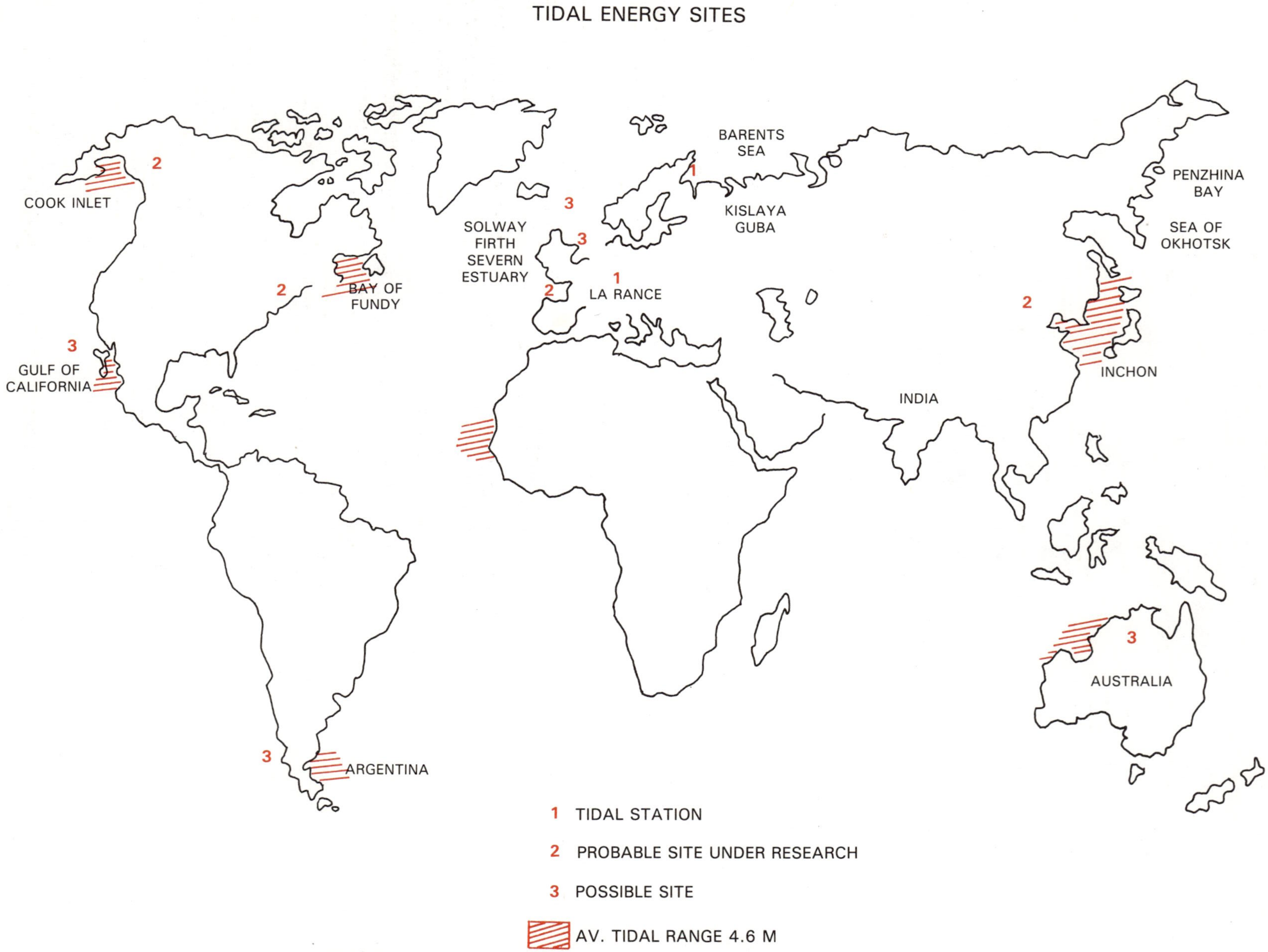

Fig. 9-48. Many potential tidal power sites in the world remain to be researched and developed. Rising costs of fossil fuels make them more attractive as sources of electric power.

A tidal power station operates much like the ancient tidal mills just described. Trapped tidal water flows through an opening in a dam to spin a turbine. Certain coastal areas of the earth would be well suited for trapping the tidal waters, Fig. 9-48.

The coastal feature best suited for a tidal power station is a narrow estuary (place where a river empties into the sea). The difference between high tide and low tide should be 30 ft. (about 10 m) or more. Such features are found at:

1. The Bay of Fundy in eastern Canada.
2. The Pacific coast of Alaska.
3. The Severn and Solway estuaries in England.
4. The Barents and Okhotsk Seas in Russia.
5. Northwestern Australia.
6. Northeastern Brazil.
7. Southern Argentina.
8. Western Africa.
9. The eastern fringe of the West Indies.
10. Northwestern France.

It is thought that tidal plants built on all these sites could produce nearly 2000 billion kilowatts of electricity every year.

Tidal power stations need a plentiful supply of water to keep the generators producing power. Since the tide waters are available only for a short time every day, the water must be captured by huge dams across the mouth of the estuary. It is very expensive to build such dams and to construct the turbines needed to turn the generators. The estuary must be able to hold a great deal of water behind the dam. Likewise the incoming tide must bring a great deal of water into the estuary.

EXISTING AND PLANNED TIDAL POWER STATIONS

The first tidal power station was opened at the Rance River in northwestern France in 1962. A second station, Kislaya Guba, has since been con-

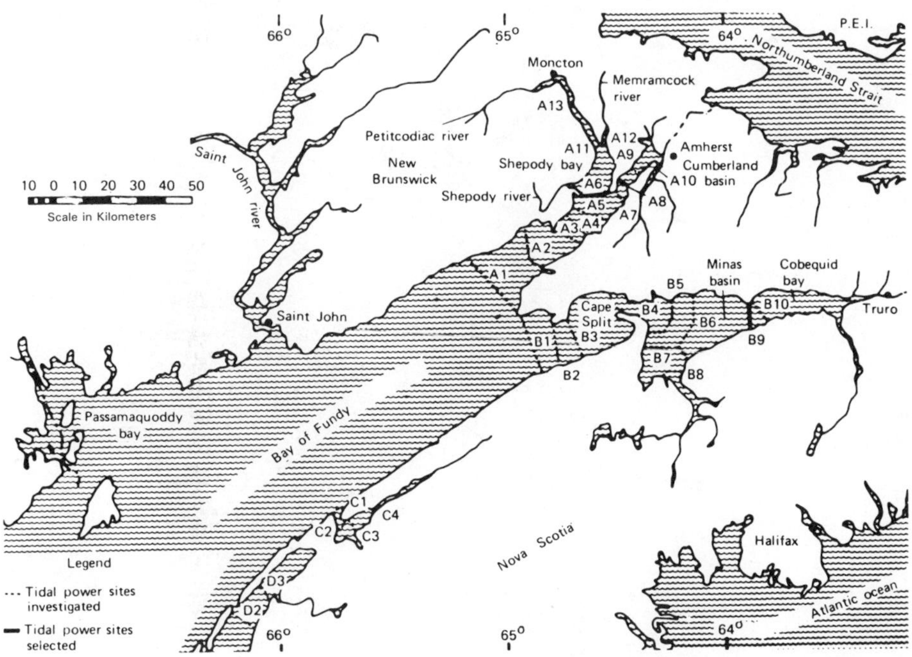

Fig. 9-49. Location of first North American tidal power project. It became operational in 1983. (Tidal Power Corp.)

structed in Russia on the Barents Sea.

Turbines at the Rance River tidal power station can generate electric power as the tidal waters move in either direction. The estuary forms a large natural reservoir. The tidal range (difference between low and high tide) can be as much as 28 ft. (8.5 m).

BAY OF FUNDY PROJECT

The Bay of Fundy tidal power project is designed to produce electrical power for Nova Scotia, Fig. 9-49. The first phase of construction is a pilot project, a generating station which has been built near Annapolis Royal on Hog's Island in northwest Nova Scotia. The tidal waters are

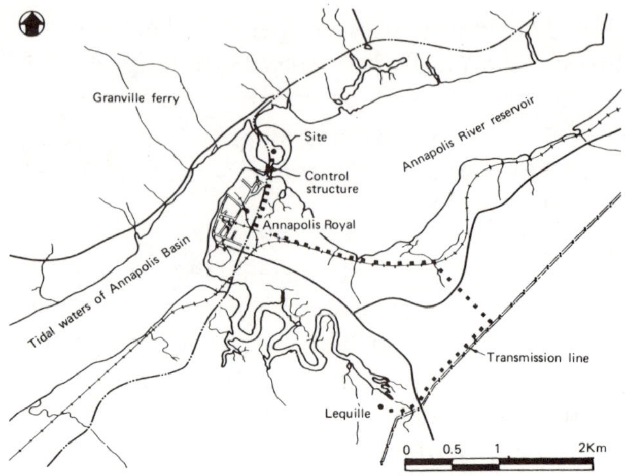

Fig. 9-50. Small area map of power station. The installation takes advantage of a barrage (like a levee) which protects agricultural marshlands.

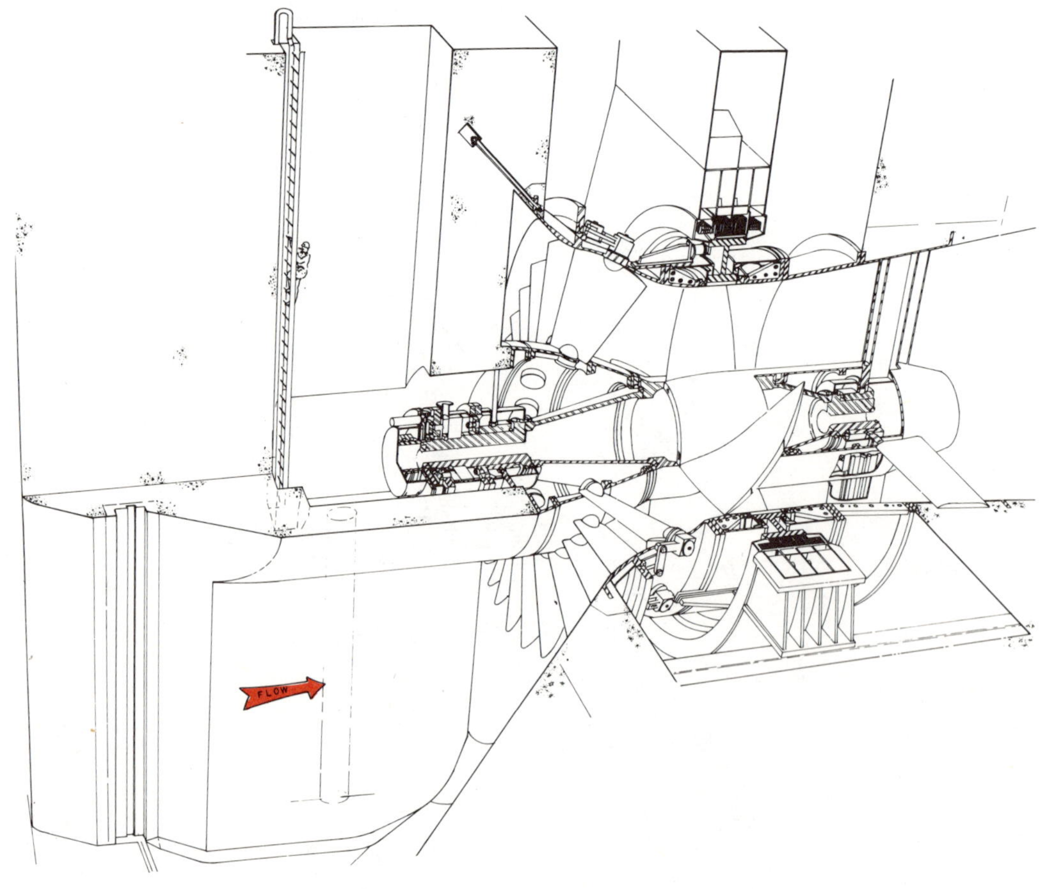

Fig. 9-51. Cutaway of Straflo turbine. Generator poles are arranged around the outside of the propeller fins. (Tidal Power Corp.)

held in a natural reservoir or basin formed by the mouth of the Annapolis River, Fig. 9-50. The project will test the suitability of a new turbine design. The unit is a modern adaptation of an axial flow turbine with a rim type generator. It is mounted horizontally to capture the energy of a horizontal flow of water. Fig. 9-51 is a cutaway drawing of the turbine and generator. The propeller consists of a hub and four blades. It rotates inside the huge generator stator. Its operating head range is 4.6 ft. (1.4 m) to 22 ft. (6.8 m). It is capable of producing 50 GW of electric power annually.

The tidal range at Annapolis Royal and other points along the coast of Nova Scotia is as high as 28.5 ft. (8.7 m) during spring tides and as low as 14 ft. (4.4 m) during neap tides. The average range is 21 ft. (6.4 m).

ENERGY FROM WAVES

Ocean waves hold 10 times more energy than tides. Wave action in the Atlantic, Pacific, and Indian Oceans is capable of producing 40 to 70 kilowatts of electrical power for every meter of wave front. Great Britain is experimenting with devices which will trap the wave energy. See Fig. 9-52. While the cost of construction of the devices is not a problem, a system to carry the electrical power ashore would be expensive. The devices being built are for generating electricity. However, they could be used to power water electrolysis. The hydrogen gas produced could then be shipped to fuel cells located on shore.

DO YOU KNOW THESE TERMS?

Amplitude, anemometer, cells, centrifugal force, Darrieus rotor, down wind rotor, electrolysis, estuary, gravitational storage, Helius rotor, horizontal axis wind machine, horse latitudes, inertial storage, inverter, local winds, microprocessor, mistral, neap tide, planetary winds, rotor, Santa Ana wind, Savonius rotor, spring tide, storage battery, tidal power, tidal range, topography, trade winds, up wind rotor, vertical axis wind machine, voltage regulator.

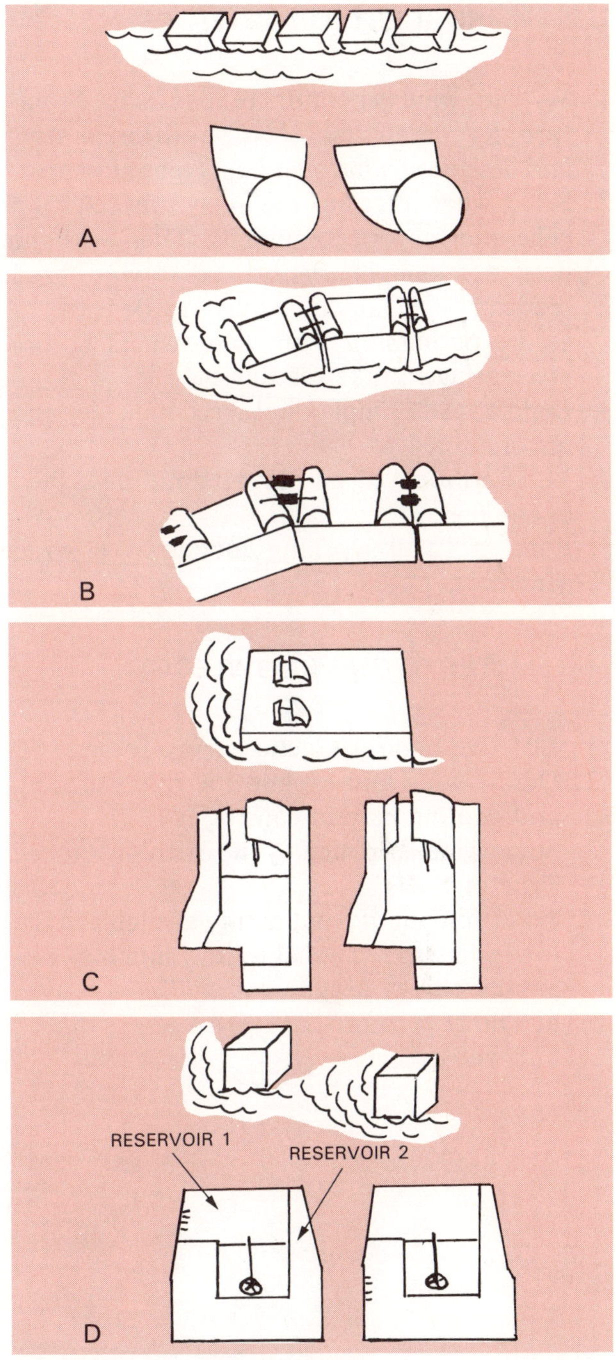

Fig. 9-52. Many devices are being tested to capture the energy of waves. A—Air pressure ring buoy. Water action builds up air pressure to operate air turbines in the buoys. B—Cockerell contouring raft. Wave motion operates hydraulic motors or pumps. This energy is converted to high pressure in a fluid. In turn, the fluid would operate a turbine. C—Oscillating water column (foreground) creates air pressure to drive an air turbine. D—Russell wave rectifier captures water in its reservoirs. The water drives turbines.

SUGGESTED ACTIVITIES

1. Write to wind generator companies for literature. Report to your class on different sizes and designs of wind generators manufactured.
2. Select what you think are likely sites for wind generators in your community. Then, with the help of your instructor, take wind level readings with an anemometer. Be sure to secure permission from landowners before attempting to take readings.
3. Build a small, simple wind speed indicator and test it.
4. Build a model of an old windmill described in the text.
5. Build a small working model of a wind generator.

TEST YOUR KNOWLEDGE

1. List five natural kinetic energy forces.
2. Define wind and explain the role warming and cooling play in causing it.
3. Some wind is created by the earth's rotation. True or False?
4. Total amount of wind energy available in the lower 48 states is equal to how much of our annual energy usage:
   a. Thirty percent.
   b. Two percent.
   c. Two times.
   d. Fourteen times.
   e. Forty times.
5. Name three properties of waves and define them.
6. What keeps rivers flowing?
7. Kinetic energy found in nature is usually converted to another form before use. The reason for this is (pick best answer or answers):
   a. It is not needed in its natural state.
   b. There is little for which kinetic energy can be used.
   c. Conversion makes it easier to store.
   d. Conversion makes it easier to transport to where it can be used.
8. A down wind generator needs a vane to keep it facing into the wind. True or False?
9. A ____________ is a device for measuring wind speed.
10. What items make up a complete wind-powered electrical system?
11. Hydroelectric power comes from the natural force of____________________________.
12. Explain the difference between a low-head and high-head dam.
13. What two natural features of a coast would make it a desirable place to build a tidal power station?
14. Ocean waves hold ________ times more energy than ocean tides.

10 NUCLEAR ENERGY

The information given in this chapter will enable you to:

- Describe the basic principles of nuclear fission and nuclear fusion.
- Explain the process of extracting, refining, and fabricating a fuel assembly for a fission reactor.
- List four basic types of nuclear reactors.
- Discuss the advantages and disadvantages of nuclear power.
- Describe current experimentation with fusion power systems.

Present-day nuclear power plants, Fig. 10-1, use the heat from the fission of uranium fuel to heat water. (As you learned in Chapter 1, fission is the splitting of atoms of uranium.) The steam from the heated water drives a turbine which is connected to an electrical generator.

To understand the splitting of an atom, it is necessary to consider the atom and its parts. Remember, it is the smallest particle of any element that has all the chemical properties of that element.

All matter is made up of atoms. Water, wood, rocks, and even ourselves are made up of these tiny particles. So tiny are they that billions of them can rest on the point of a pencil with room to spare.

Each atom is made up of even smaller particles called protons, neutrons, and electrons. Protons are positively charged; neutrons have no charge at all. Electrons are negatively charged.

In the atom the protons and neutrons are held in the center or nucleus. The electrons, however, are in motion. They circle or orbit around the nucleus. See Fig. 10-2.

Every atom of an element has the same number of protons in the nucleus. However, there can be different numbers of neutrons in different atoms of the same element. These different forms of the same element are called isotopes, Fig. 10-3. Later, when we talk about different uranium isotopes, we will give them numbers to identify them.

An atom is mostly empty space. Most of the mass (weight) of an atom is in the nucleus. Electrons orbiting (spinning in a circular path) the nucleus are much lighter than protons and neutrons.

The orbits of electrons are arranged in "shells." Each shell holds only a certain number of electrons. The shell closest to the nucleus may contain two electrons, the next can hold up to eight and the third one, up to 18. The fourth can contain up to 32. Each electron, regardless of how many are in the shell, has its own separate orbit.

At the beginning of nuclear fission, a neutron

Fig. 10-1. Indian Point 3 nuclear power station in New York state has been operating since 1976. A pressurized water reactor, it develops 965,000 kW of electrical power. (Power Authority of the State of New York)

Fig. 10-2. An atom is made up of a nucleus of neutral and positively charged particles surrounded by orbiting negatively charged particles. (Westinghouse Nuclear Energy Systems)

Fig. 10-3. Isotopes are atoms of the same element having different numbers of neutrons in their nuclei. (Westinghouse Nuclear Energy Systems)

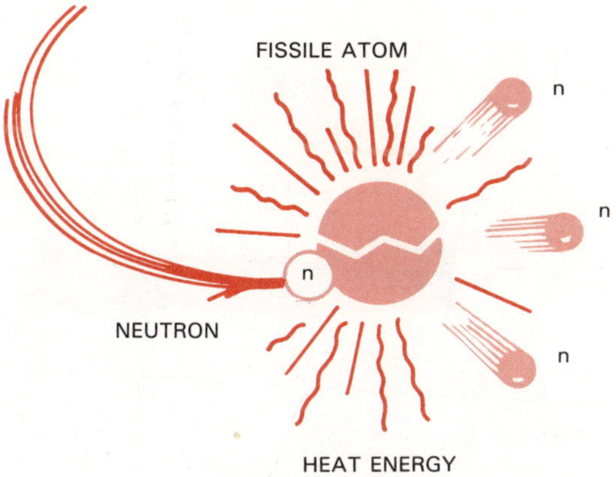

Fig. 10-4. Fission is the splitting of a uranium atom. It occurs when a free neutron from one atom strikes and splits another atom of uranium. Two new atoms are formed along with great amounts of heat energy. (Westinghouse Electric Corp., Advanced Power Systems Div.)

from one atom strikes another atom with great speed and causes the atom to split into two lighter atoms. See Fig. 10-4. If conditions are right, this sets up a chain reaction.

CHAIN REACTION

Nuclear power depends upon one reaction causing other reactions. Each neutron freed by fission can go on to split other uranium atoms so that the process continues, Fig. 10-5. This series of events is called a CHAIN REACTION. The process of splitting atoms causes a tremendous amount of heat.

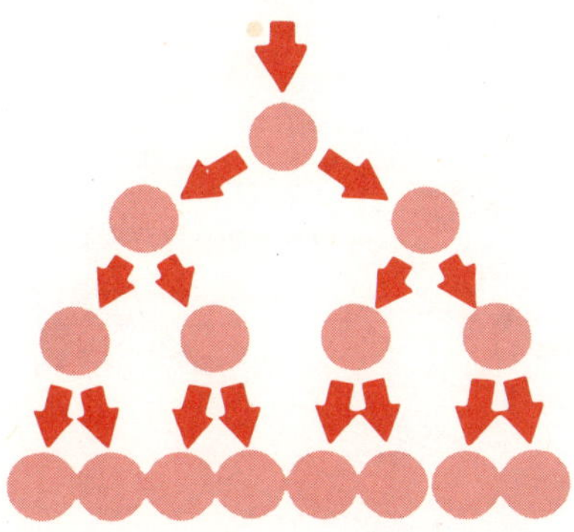

Fig. 10-5. A chain reaction occurs when a neutron splits an atom and releases enough neutrons so that many other atoms are split in the same way. (Westinghouse Nuclear Energy Systems)

HEAT OF FISSION

A very small amount of uranium fuel will produce a great deal of energy. It comes out of the chain reaction as heat. There is about as much heat in a thimble sized pellet of nuclear fuel as there is in nearly a ton of coal, Fig. 10-6.

The splitting of one atom releases about 190 MeV (million electron volts) of energy. (An electron volt is but a tiny fraction of a calorie. There are 2.6×10^{16} MeV in each calorie.)

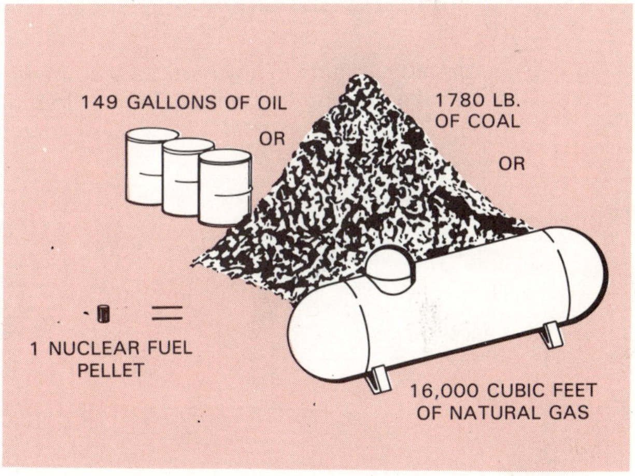

Fig. 10-6. A thimble-sized pellet of uranium fuel contains about as much heat as a ton of coal. (Westinghouse Electric Corp., Advanced power Systems Divisions)

NUCLEAR FUELS

Uranium is thought to be one of the substances in volcanic ash which was spewed onto the earth's surface millions of years ago. Rains dissolved the uranium out of the ash and carried it back underground where it hardened into an ore.

Since it is found near the earth's surface, like

Fig. 10-7. Uranium mining in northern Saskatchewan, Canada. Left. An aerial view of an open pit mine at Rabbit Lake, Canada taken during the winter. Right. Uranium-bearing rock is loaded on a truck for transport to the mill. (Gulf Oil Corp.)

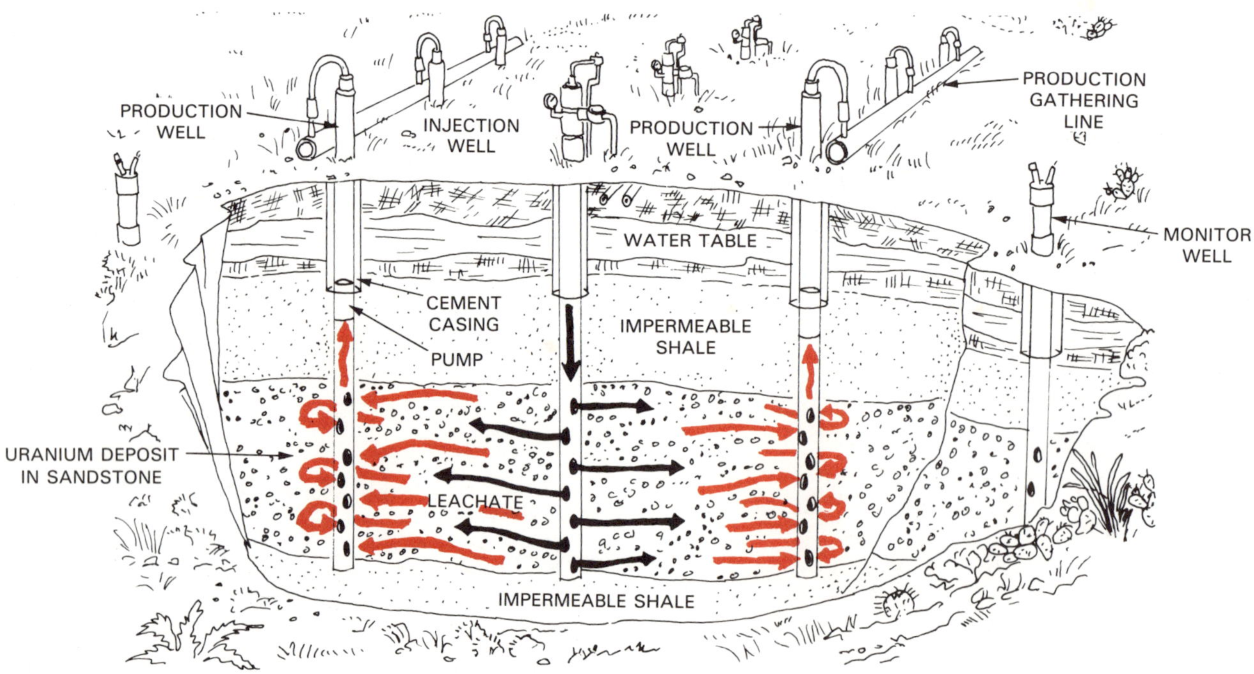

Fig. 10-8. Diagram of an in situ leaching operation. A mixture of water and chemicals is pumped down injection wells to dissolve uranium ore. A producing well recovers the dissolved ore. (Mobil Energy Minerals Div.)

coal and oil, it can be mined in much the same way:

1. By open pit mining, Fig. 10-7.
2. By underground shaft mining.

Recently a new method of mining has been used. Known as "in situ" (in place) leaching, it reverses the process which originally separated the uranium from the volcanic ash. Water is pumped into the ground dissolving the uranium. It can then be pumped to the surface. Fig. 10-8 is a diagram of an in situ leaching operation.

URANIUM SUPPLIES

Like our fossil fuel supplies, our supply of uranium will run out one day. Only one out of every 140 uranium atoms can cause fission. Thus, large amounts of uranium ore must be processed. A ton of uranium yields only about 14 lb. (6.4 kg) of fission fuel.

U.S. supplies of uranium could be as much as 4 million tons. This includes known and possible reserves. This reserve would be about 286,000 lb. (8 107 963.5 kg) of nuclear fuel. In 1982 there were 77 nuclear reactors in full operation and 60 more under construction. Uranium supplies for that many reactors would last about 200 years.

However, there is a way to get more fuel from other atoms of uranium that will not go into fission in their natural state. Remember the isotope discussed earlier? You will recall we said that different atoms of the same element can have different numbers of neutrons in their nucleus. We call such atoms isotopes. One of these isotopes will produce fission if it is given another neutron for its nucleus. Scientists have discovered that this isotope, already present in nuclear fuel, will automatically absorb a stray neutron during nuclear fission. So the process of making new fuel (called BREEDING) goes on all the time in a nuclear reactor.

To classify them, scientists have given certain isotopes of uranium names or numbers. The isotope which is naturally able to be split is called U-235. (The "U" stands for uranium.) The isotope which can be made to produce fission is labeled "U-238." When it gains a neutron it becomes U-239 or plutonium. If all of these isotopes could be converted, the supply of nuclear fuel would be extended to supply 1000 large reactors for 30,000 years.

PROCESSING URANIUM

Uranium which is mined in the traditional way must be ground to the size of fine sand. Leached uranium is separated from the water-chemical solution. Chemical treatment produces a mixture known as YELLOW CAKE. It may be in the form of a bright yellow slurry (soupy mix) or a powder.

Before it is used as a nuclear fuel the yellow cake is converted into a gas called URANIUM HEXAFLUORIDE. It is enriched to increase the percentage of U-235 to about 3 or 4 percent.

FUEL FABRICATION

The enriched uranium hexafluoride gas is then formed into pellets. This is done by milling, pressing and sintering (baking in a furnace like pottery), and grinding. Each pellet is about the size of a fingertip.

Pellets are packed into specially made tubes. Loaded tubes are welded shut and bundled together in metal frames. The frames maintain carefully engineered spacing between the rods. The bundles are called FUEL ASSEMBLIES. They will make up the core of a nuclear reactor. See Fig. 10-9.

FISSION

The chain reaction discussed earlier is started when a neutron "bullet" is fired at a U-235 atom. As the nucleus splits it shoots out two or three neutrons. They, in turn, strike more U-235 atoms and split them. Once started, the process will continue without any further help.

Fig. 10-9. Quality control engineers inspect a newly built nuclear fuel assembly at a fabrication plant. (Exxon Corp.)

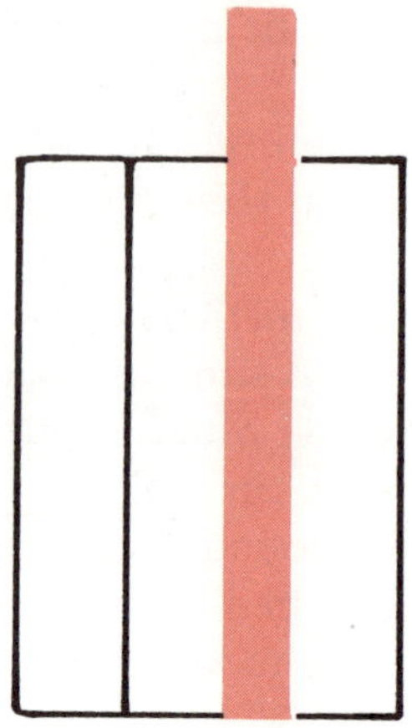

Fig. 10-10. Rods are used to slow down or stop the chain reaction of nuclear fission. The rods are lowered into the core where they can absorb some of the bombarding neutrons. (Westinghouse Nuclear Energy Systems)

The flying neutrons may be speeded up by MODERATORS to increase the chance of their bumping into other U-235 atoms. CONTROL RODS are used to slow down the rate of a nuclear reaction. The rods are movable so that they can be inserted when the reactions need to be slowed down and removed when an increase in reaction is needed. See Fig. 10-10.

The reaction takes place in a strong, closed container called a REACTOR.

NUCLEAR REACTORS

The nuclear reactor is to a nuclear power plant what a furnace is to your home. It contains the nuclear fission so that it does not harm its surroundings. Then it distributes the heat of fission so that the heat can be used to generate electric power. It is similar to the operation of the furnace which contains the burning fuel and circulates heated air or water to heat rooms in the house.

A steel pressure tank, called a VESSEL, holds several thousand fuel rods which make up the fuel core (supply). The vessel, shown in Fig. 10-11, has walls 9 in. thick. Around the vessel is another structure called a "steel containment." It is made of steel plate which is leakproof. See Fig. 10-12. Around the vessel and the steel containment is a shield building. This building has walls of reinforced concrete three or more feet thick.

REACTOR TYPES

Different types of reactors are being used or considered. They include:

1. Light water reactor (LWR).
   a. Boiling water reactor (BWR).
   b. Pressurized water reactor (PWR).
2. Heavy water reactor (CANDU).
3. High temperature gas reactor (HTGR).
4. Fast breeder reactor (FBR).

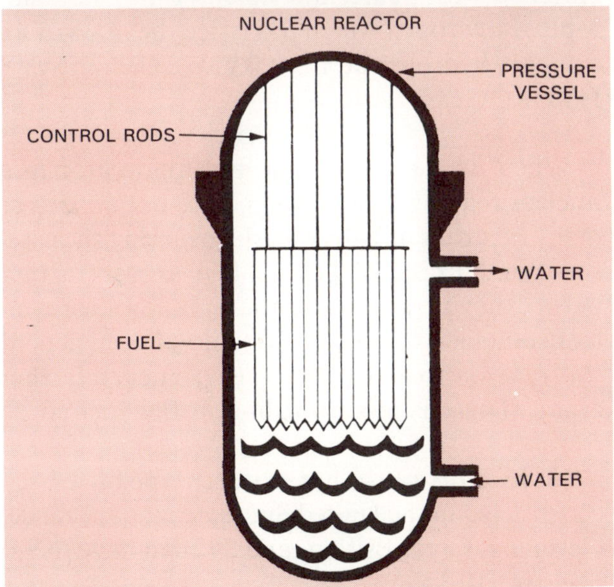

Fig. 10-11. A pressure vessel is the part of the nuclear reactor which holds the control rods and fuel. It is where fission takes place. Top. Vessel, weighing 900 tons, is carefully lowered into a shielding wall at a nuclear generating station. (Niagara Mohawk Power Corp.) Bottom. Diagram of pressure vessel. (Edison Electric Institute)

Fig. 10-12. Steel containment vessel like this holds the pressure vessel. (Niagara Mohawk Power Corp.)

a. Liquid metal fast breeder reactor (LMFBR).
b. Gas cooled fast breeder reactor (GCFBR).

LIGHT WATER REACTORS

Most operating nuclear power plants of today are light water reactors. They use ordinary water as a coolant and to transfer heat energy to the turbine. The water flows around the uranium fuel assemblies carrying away the heat. It turns to steam so that it can drive a turbine. The water also acts as a moderator.

BOILING WATER REACTOR

Of the two types of light water reactors, the boiling water reactor (BWR) is the simplest in design. The water, in cooling the reactor, comes to a boil. Heat of fission brings the water to 545°F (285°C) under a pressure of 1000 psi (6894 kPa). The steam is piped directly to the turbine. A simple schematic of the BWR is shown in Fig.

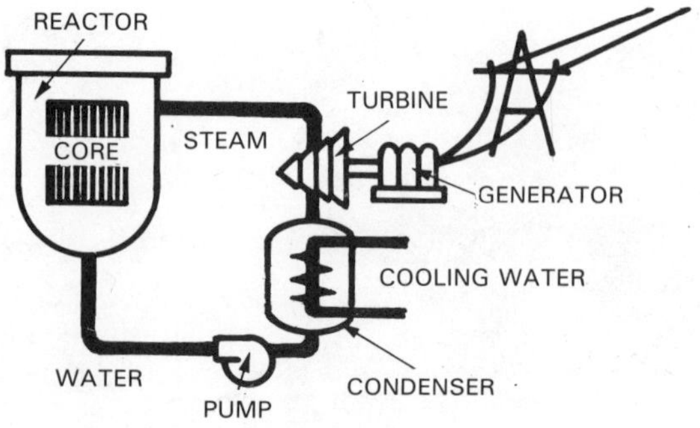

Fig. 10-13. Boiling water reactor creates steam inside the reactor vessel. The steam is piped directly to the turbine. (Westinghouse Electric Corp., Advanced Power Systems Div.)

10-13. Spent steam is cooled back to liquid and returns to the reactor and repeats the cycle.

PRESSURIZED WATER REACTOR

The pressurized water reactor uses water under high pressure to cool the reactor and to move the heat produced by fission. However, unlike the boiling water reactor, the heated water or steam does not go directly to the turbine. It moves through its own piping system, called the PRIMARY LOOP, to a steam generator. The coolant water gives up its heat to water in a SECONDARY LOOP. The radioactive water in the primary loop never mixes with water in the secondary loop. It, thereby, prevents water in the secondary loop from becoming radioactive. Heat, picked up in the steam generator, moves as steam to the turbine and causes it to turn.

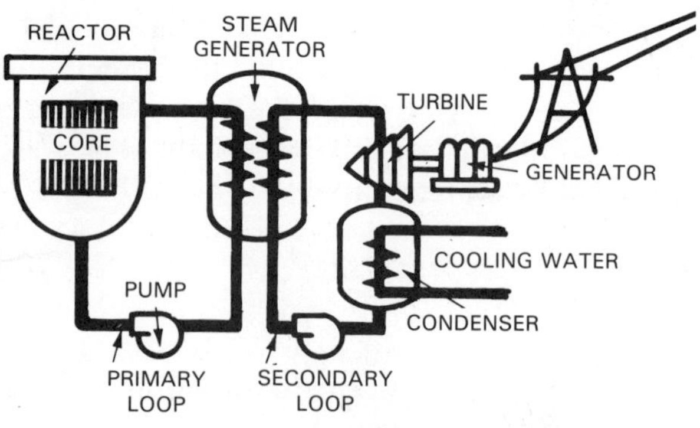

Fig. 10-14. Pressurized water reactor uses two loops carrying water to transfer heat of fission from the core to the turbine. Cooled water is returned to repeat the cycle over and over again. (Westinghouse Electric Corp., Advanced Power Systems Div.)

Water from the primary loop returns to the reactor after it has lost much of its heat to the steam generator. Likewise, steam in the secondary loop, after it passes through the turbine, is returned to the steam generator. However, it first passes through a condenser to cool it to a liquid. Fig. 10-14 is a simple schematic of a PWR.

The PWR operates with water at a pressure of 2250 psi (about 15,500 kPa) and a temperature of 600°F (315°C). Because of the high pressure in the primary loop, the coolant never boils. The boiling occurs in the secondary loop as it picks up the heat of the coolant in the steam generator.

HEAVY WATER REACTOR

Heavy water is ordinary water which has been enriched with deuterium. (Deuterium is an isotope of hydrogen which has twice the mass of an ordinary hydrogen atom. It is also called HEAVY HYDROGEN.

When used as a coolant and moderator in a nuclear reactor, heavy water does not absorb as many neutrons as light water. This being the case, the bombarding neutrons can travel farther before they slow down to the right level for controlled nuclear reaction. This allows designers of the reactor to place the fuel elements farther apart. There is less crowding in the fuel core.

No reactors of the heavy water type have been constructed in the United States. Canada, France, Germany, Great Britain, and Sweden have them. See Fig. 10-15.

HIGH TEMPERATURE GAS REACTOR (HTGR)

If you look at the diagram of the high temperature gas reactor in Fig. 10-16, you will not see much difference between it and the light

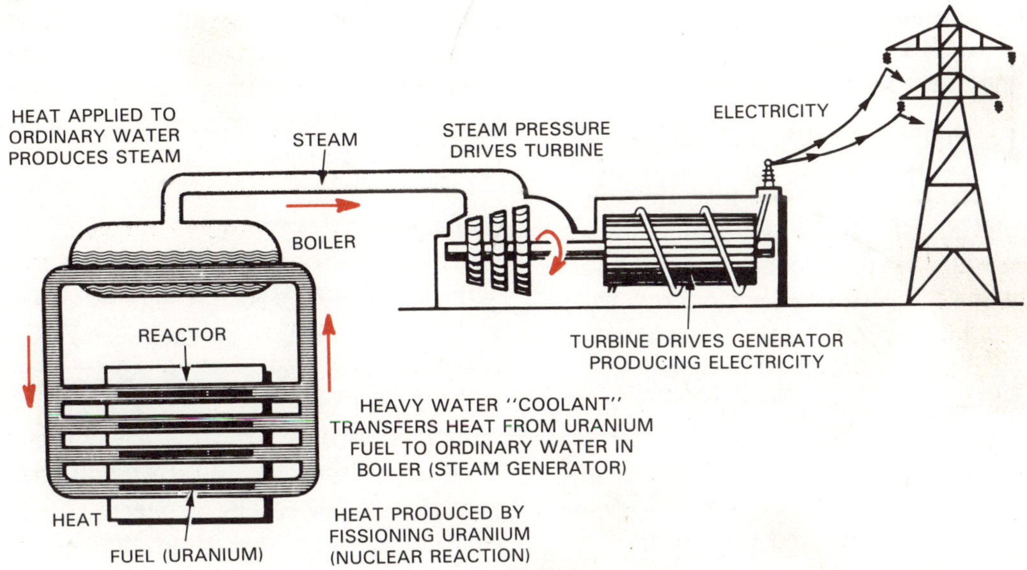

Fig. 10-15. Simplified sketch of Canada's heavy water reactor. It uses ordinary water enriched with deuterium as a coolant. (Nuclear Power in Canada/Canadian Nuclear Assoc.)

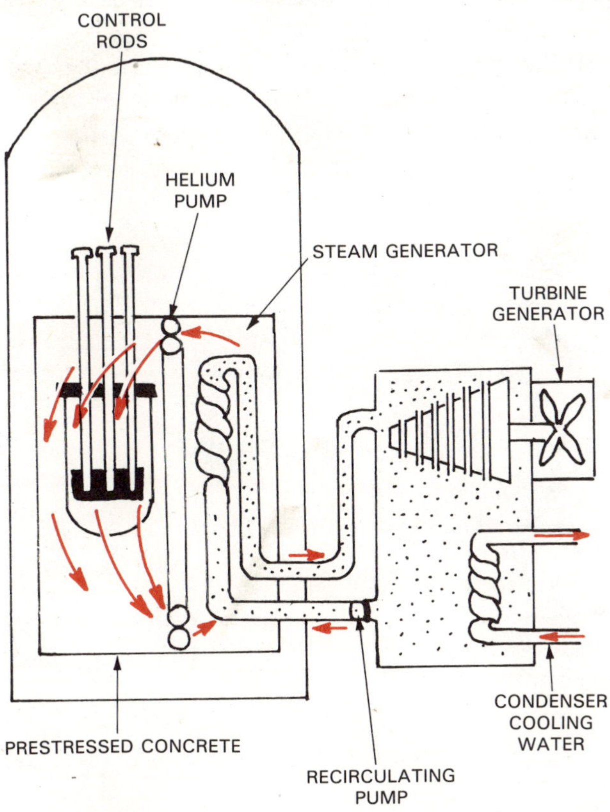

Fig. 10-16. The high temperature gas reactor (HTGR) uses helium gas to carry thermal energy from the pressure vessel to the steam generator.

water reactor. They are constructed much the same except that the HTGR uses helium gas to carry the heat away from the reactor core to the generator. There are several advantages in the use of helium:

1. Heat is transferred much better.
2. The system can operate at higher temperatures and pressures.
3. There is less danger from radiation since helium is an inert gas. (This means it does not mix well with other chemical elements.)

Controlling fission

As with the light water reactor, the rate of fission must be controlled. Graphite (a material like the "lead" in a lead pencil) is used as the moderator. It is applied as a coating to tiny particles of the fuel.

The HTGR uses three isotopes of uranium as fuel:

1. U-235 which is used to start the nuclear reaction.
2. Thorium 232 (Th-232) which is used during normal reactor operation. It is plentiful and

less expensive than U-235. Not in itself a fuel, it converts to U-233 during reaction and this isotope becomes the fuel.

Preparing the fuel

The three isotopes are formed into tiny particles called MICROSPHERES. One microsphere contains U-235 and the other contains U-232 and U-233. Then the microspheres are coated with graphite and blended into a pellet. Each pellet is about 1/2 in. (12.5 mm) in diameter and 1 1/2 in. (38 mm) long.

The pellets (or fuel rods) are loaded into holes in a machined graphite block. A single fuel element is made up of about 2000 fuel rods. The reactor will need up to 800 fuel elements.

Like the pressurized water reactor, the HTGR uses two separate heat transfer systems. Super heated helium in the first loop circulates around a steam generator inside the reactor vessel. Water flowing through the second loop turns to steam in the steam generator and is piped, under high pressure, to the turbine.

Pressure and temperature are very high. The helium reaches temperatures above 1425°F (775°C) and a pressure of 700 psi (4825 kPa). In the secondary loop, temperature reaches 1000°F (535°C) and pressure, 2400 psi (16,545 kPa).

BREEDER REACTOR

The breeder reactor is usually called the fast breeder reactor because there is no moderator to slow down the neutrons which are creating fission. It is an advanced nuclear power plant design which can produce more nuclear fuel while it is producing electricity. In fact, it will produce more fuel than it uses. The process, as you learned earlier, is called breeding because it adds a neutron to U-238 and changes it to a type that will become part of a nuclear chain reaction, Fig. 10-17.

Use of the breeder reactor would produce fuel from useless isotopes making up over 99 percent of the mined uranium. This would cut down greatly the amount of fuel that must be mined, supporters of the breeder reactor program say.

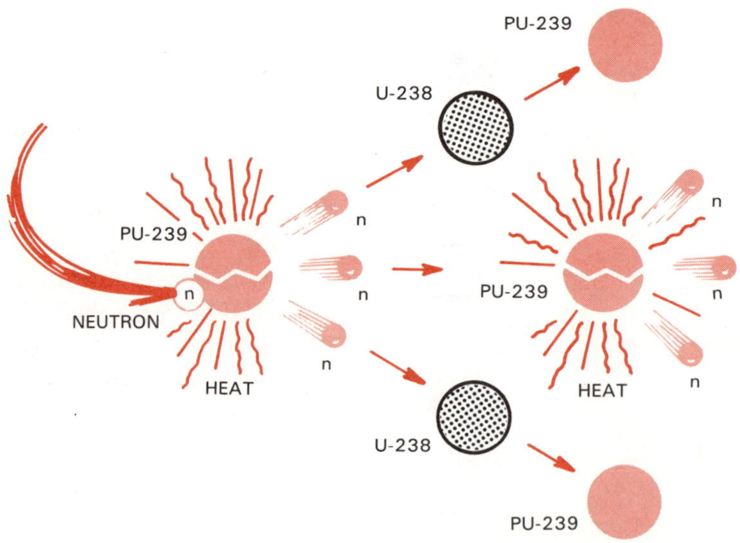

Fig. 10-17. Creating plutonium from U-238. An atom of U-238 absorbs a neutron, becomes fissile, and splits when struck by a second neutron. This starts a chain reaction.
(Westinghouse Electric Corp., Advanced Power Systems Div.)

Types of breeder reactors

There are two types of breeder reactors. One is the liquid metal fast breeder reactor (called LMFBR). The other is called the gas cooled fast breeder reactor (GCFBR).

Liquid metal breeder reactor

The only commercial liquid metal breeder reactor design in the United States is shown in Fig. 10-18. It is being built on the Clinch River in eastern Tennessee but construction has been halted. The reactor was designed to develop 400 megawatts (MW) of electrical power.

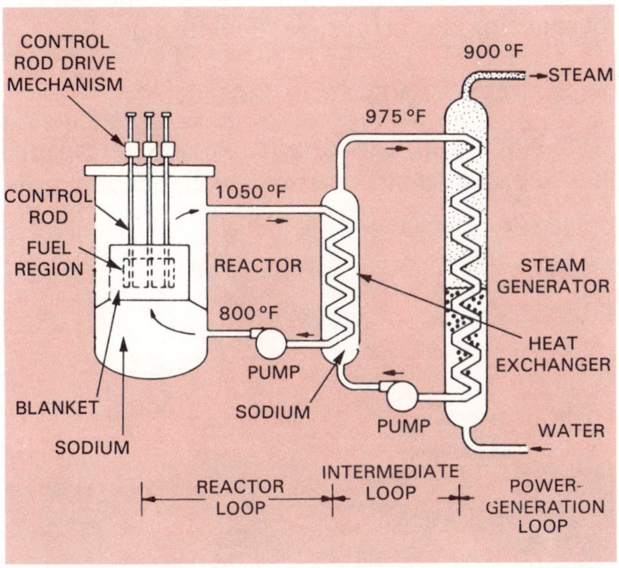

Fig. 10-18. Diagram of Clinch River Liquid metal breeder reactor. It is under construction in Tennessee. (Breeder Reactor Corp.)

One way in which the LMFBR is different from other reactors (besides its ability to make new fuel) is its use of liquid sodium as a coolant. Sodium is a soft, waxy, silver white, metallic alkali. It is valuable as a coolant because it can carry more heat than water and helium gas. It melts at 210°F (98°C) and boils near 1600°F (880°C).

The primary loop carries the heated sodium to a heat exchanger inside the containment structure. This heat exchanger must be housed inside the building because the sodium becomes radioactive.

A second sodium loop takes the heat from the first exchanger and delivers it to a second exchanger. A third loop filled with water is heated by the second exchanger and delivers steam to the turbine.

Gas cooled fast breeder reactor

The gas cooled fast breeder reactor is still in the experimental stage. Its first fuel charge would be developed from the spent uranium discarded by the light water reactors and from the stockpile of U-238 left after the refining of uranium. From each pound of fuel used it will produce 1.45 lb. of new fuel. Fig. 10-19 is a simple drawing of the inside of a GCFR. Fig. 10-20 is a cutaway view of the fuel element.

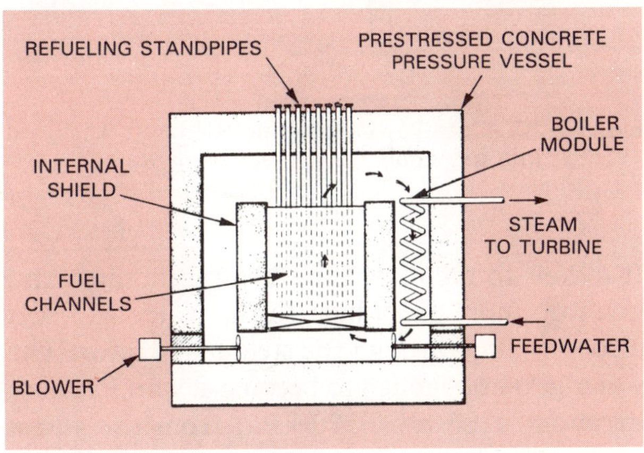

Fig. 10-19. Simple schematic of a gas cooled fast breeder reactor. Helium is used as the coolant.

The GCFR is like the high temperature gas reactor. It will use helium as the coolant. Unlike molten sodium coolant, helium will not cause corrosion in metals it contacts. Helium sealed in the GCFR heats up to about 1022°F (about 550°C).

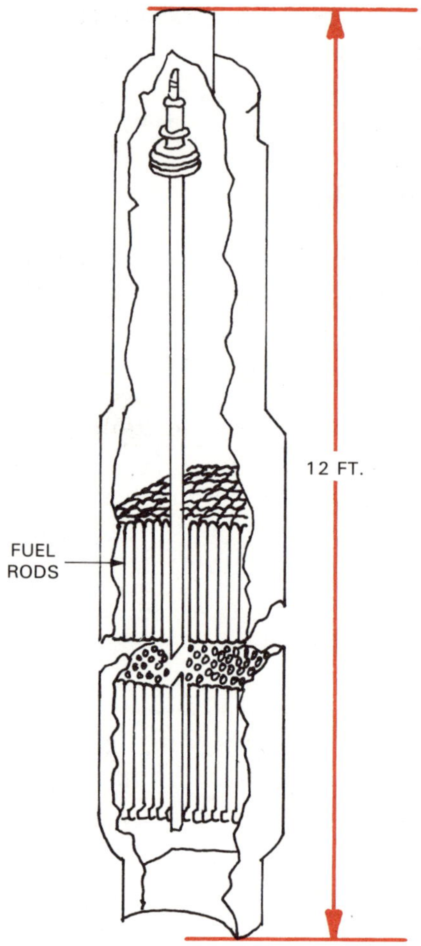

Fig. 10-20. Cutaway view of fuel element. It is used in a gas cooled fast breeder reactor.

It travels to the top of the steam generator and gives up most of its heat to the water which is also traveling through the steam generator. The water gets hot enough to become steam. Pressure of about 1200 psi (8280 kPa) from the steam powers the turbine.

Helium leaving the steam generator still has heat energy. It is circulated through a second heat exchanger generating enough additional steam to run a pump which circulates the helium.

Most nuclear reactors use a heat exchanger. This is a device that transfers heat from one fluid (liquid or air) to another without mixing them. A car radiator is a simple example. Hot coolant flows through tubes as air flows around the outsides of the tubes picking up heat. The heat from the coolant transfers to the air.

NUCLEAR PROPULSION

Nuclear propulsion means using a nuclear reactor to power vehicles. Nuclear fuel is required in very small amounts and does not need to be replaced often. The small nuclear reactor, thereby, has some advantages for moving people and materials.

It is suitable for ships, aircraft, and rockets where cost is not as important as the need to go long distances and for long periods without refueling. It is not practical for small land vehicles because of the protective shielding needed.

NUCLEAR POWERED VESSELS

A Navy ship, the aircraft carrier Enterprise, has eight nuclear reactors. A guided missile

Fig. 10-21. The U.S.S. Enterprise was the first U.S. nuclear powered ship. (U.S. Navy)

cruiser and a destroyer are also nuclear powered. See Fig. 10-21.

Powering ships requires a reactor similar to, but smaller than, the reactors used in electric power production. Heated coolant (steam) drives a turbine connected directly to the propellers. Some of the steam is used to drive a generator which supplies electrical power.

Nuclear power was first used in submarines, Fig. 10-22. The reactors are the lightwater pressurized type. Molten sodium was tested in one submarine but was later replaced with a pressurized water reactor.

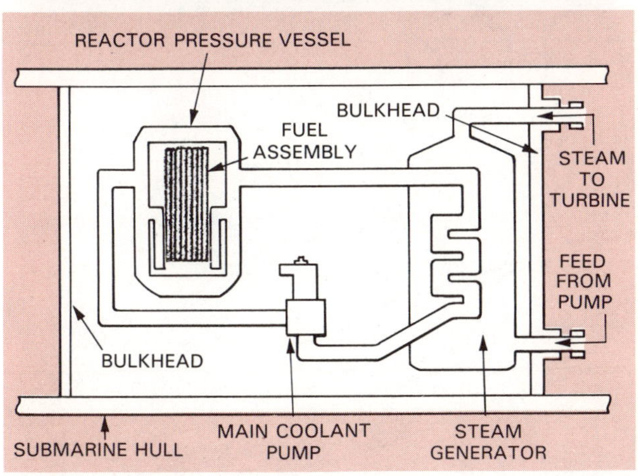

Fig. 10-22. Nuclear power for propulsion was first installed in submarines. Top. Bow view of nuclear propelled submarine, U.S.S. Ohio. Bottom. Schematic of typical sub nuclear power plant. (U.S. Navy)

NUCLEAR AIRCRAFT ENGINES

Experimental models of two nuclear aircraft engines have been built:

1. A direct-cycle engine.
2. Indirect-cycle engine.

In the direct-cycle engine, compressed air is piped into the reactor core. It is heated and carried back to the engine. Here it is expelled to create thrust like that of a jet engine.

The operation of the indirect-cycle is similar. A liquid metal coolant enters the reactor core, picks up heat, and transfers it to compressed air outside the reactor. Then the heated compressed air is exhausted in the engine to drive the aircraft.

NUCLEAR POWERED ROCKETS

During the 1960s the United States developed a nuclear reactor suitable for propelling rockets. Known as the Kiwi, it was a low-powered reactor using hydrogen gas as a propellant.

BY-PRODUCTS OF FISSION

By-products are materials which are produced or left over after the making of another product. Fission, which produces heat for energy, has several by-products. Some are useful, some are not, Fig. 10-23. As U-235 atoms are splitting, fragments of materials are left over. One of the fragments is atoms of elements that are lighter in weight than uranium. A few are gases; others are solids. Some are not radioactive while in others the radioactivity lasts fractions of a second. Still others are radioactive for thousands of years.

Some of the atoms of uranium capture a neutron without splitting. They become elements heavier than uranium. The most important of these heavy elements is plutonium.

Every nuclear reactor builds up these by-products during fission. After a while the by-products become so plentiful that the efficiency

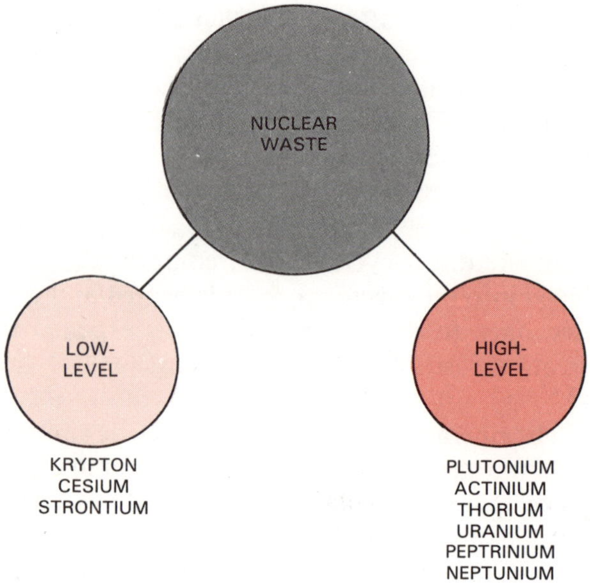

Fig. 10-23. Fission produces low-level and high-level waste. Half life of low-level waste is 25 to 30 years; high-level wastes have half-lives of 25,000 years or more.

of the reactor is reduced. When this happens, the fuel rods containing the pellets of fuel must be replaced. This usually happens after about three years of operation. Every year about one-third of the fuel rods are replaced.

REPROCESSING AND STORAGE

Old rods removed from the reactor, in addition to usable by-products, still contain about 30 percent of the original fuel. Large numbers of by-products would have stopped fission had the rods been allowed to remain longer in the reactor. Both the leftover U-235 and the plutonium by-product can be reclaimed and reprocessed into new fuel.

TEMPORARY STORAGE

The spent fuel rods are first stored in pools of water at the reactor because of the radioactivity.

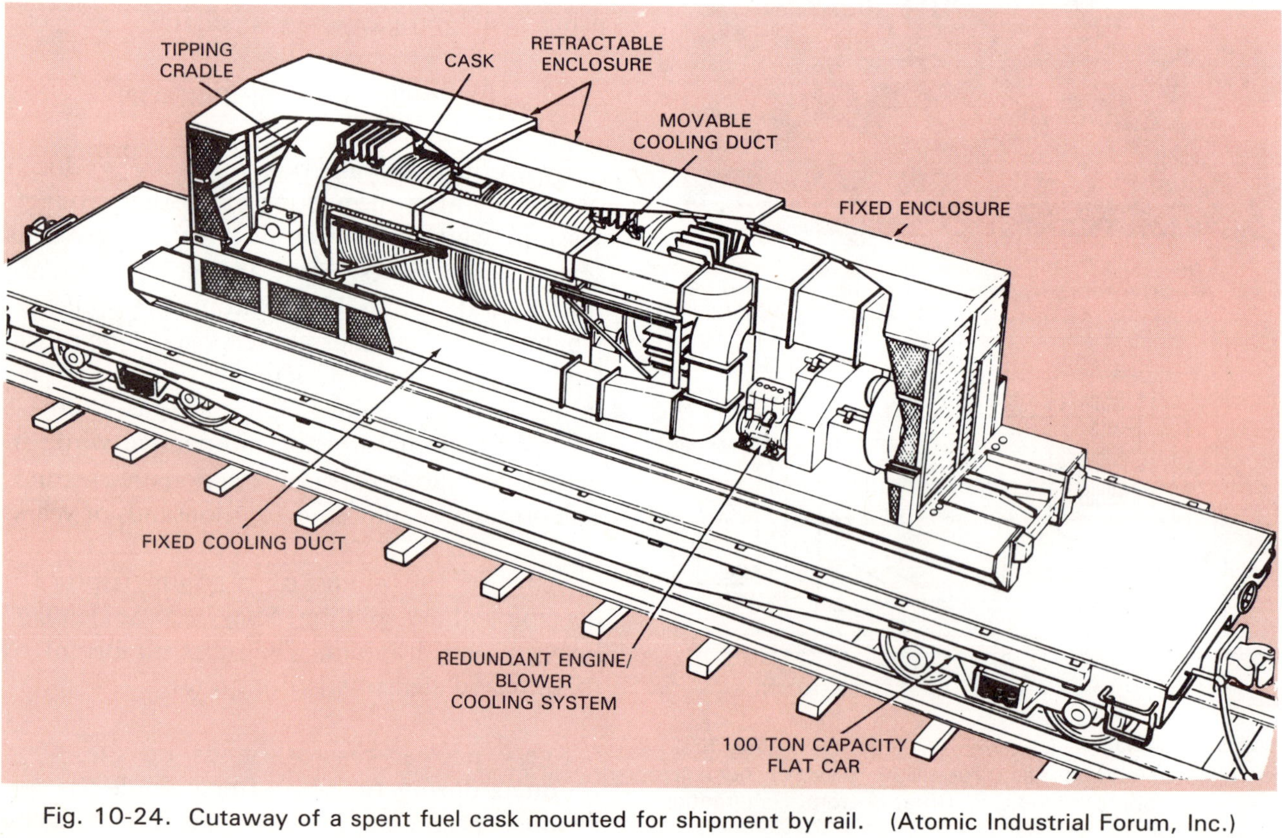

Fig. 10-24. Cutaway of a spent fuel cask mounted for shipment by rail. (Atomic Industrial Forum, Inc.)

After four months, when the radioactivity has been reduced by more than 90 percent, the rods can be shipped in special containers to be reprocessed. See Fig. 10-24.

Again, the fuel is stored. The shipping casks are opened under water. The fuel rods are removed and stored in pools once more.

REPROCESSING

About 99 percent of the uranium and plutonium is recovered from the spent fuel, Fig. 10-25. The rods are moved to concrete-walled work areas for reclamation. Here they are cut into small pieces. The work rooms, called "hot cells," have walls several feet thick to protect workers from radiation. The fuel must be handled entirely by remote controlled machinery.

Next, the pieces are treated to a nitric acid bath to leach out (dissolve) the spent fuel. Another chemical separates the plutonium and the uranium from the acid solution.

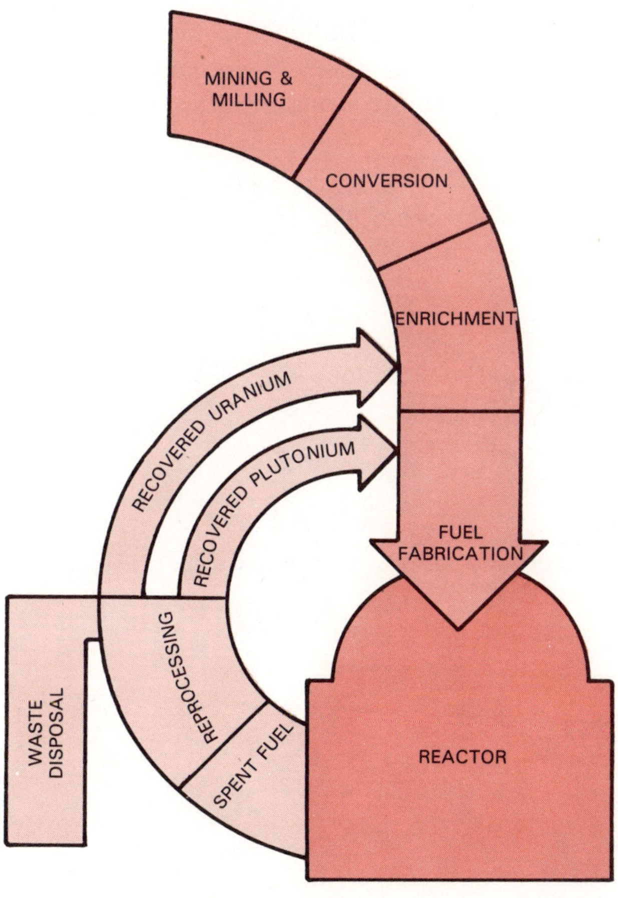

Fig. 10-25. Reprocessing of spent fuel recovers 30 percent of the original load of U-235.

WASTE TREATMENT AND STORAGE

Leftover waste must be safely stored so that it does not present a danger to health. Since it is in liquid form, its volume will be reduced through evaporation. The waste is stored for up to five years in double-walled stainless steel tanks. Finally, the waste, still liquid, will be solidified and stored indefinitely.

It is estimated that the yearly volume of waste from one nuclear reactor will occupy 70 cu. ft. (1982 L), or a space roughly 4 ft. square and 4 1/2 ft. high. This solid waste would be contained in 12 stainless steel canisters 1 ft. in diameter and 10 ft. long.

For a fee, the federal government will take custody of the wastes. Final storage would be in deep geological formations. Most promising, according to federal experts, are salt beds. These formations have been undisturbed for several hundred million years. They are expected to remain stable. Since salt is soluble in water, its existence indicates that there is an absence of ground water. (Water is the only means by which the solidified buried wastes could be leaked into the environment.)

It is estimated that there are over 256 million acres (103 hectares) of underground salt formations in the United States. About 2000 acres would be needed as storage. Fig. 10-26 shows how an underground storage vault might look.

TRANSMUTATION

Scientists are exploring still another method of disposing of the radioactive waste through trans-

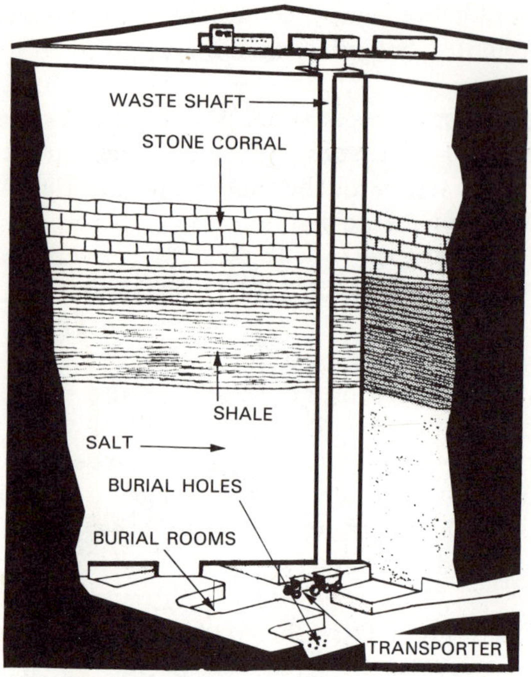

Fig. 10-26. One method of storing radioactive materials is to bury them in underground salt formations. (Atomic Industrial Forum, Inc.)

mutation. This is a process which separates radioactive chemicals into materials with a shorter radioactive life.

THE FUTURE OF FISSION

Nuclear power stations now account for about 5 percent of our electric power generation and could eventually produce 45 percent or more. The prospect of having a fuel source, plutonium, which lasts 30,000 years is attractive and exciting. This would allow technology the time to develop other energy sources, especially solar and nuclear fusion.

However, the future of nuclear fission as a power source is uncertain. The problem is in the nature of uranium and plutonium. They are a danger to us in two ways:

1. They are highly radioactive because of the very thing that makes them good fuels—an unstable atomic structure. Exposure to this

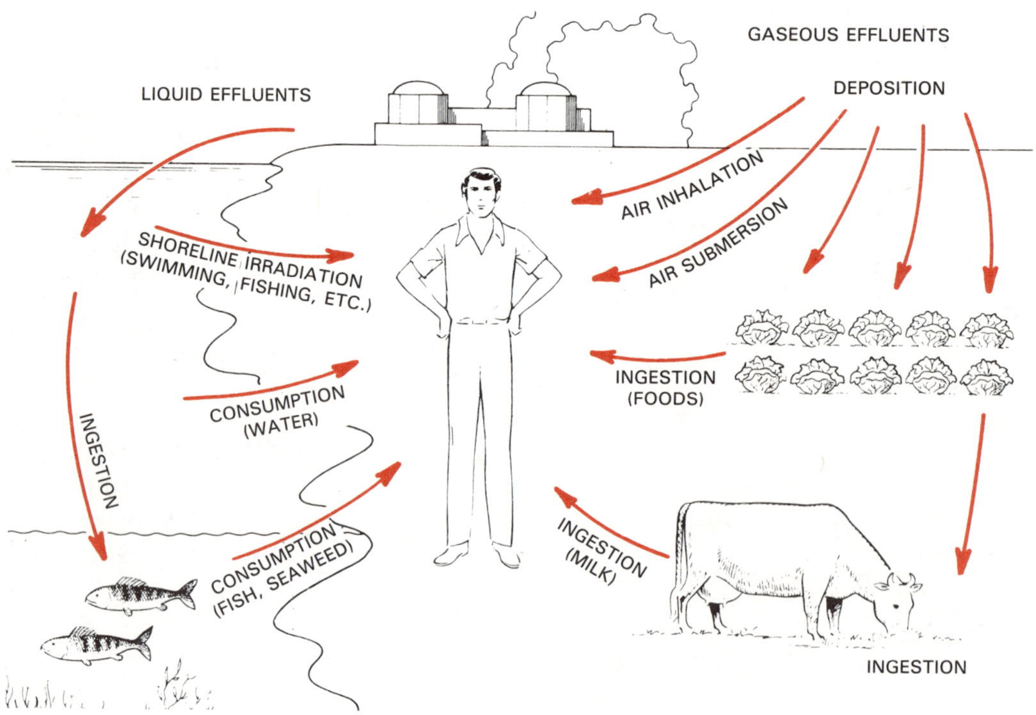

Fig. 10-27. Radiation exposure can reach humans through many pathways. (Nuclear Power in Canada/Canadian Nuclear Assoc.)

radiation at high enough levels can cause cancer as well as great genetic damage to all life forms.

Exposure can reach humans through a number of ways, Fig. 10-27. Plutonium, especially, is one of the most toxic substances known.

2. Plutonium is also the "stuff" from which atom bombs are made. There is concern that, as more and more plutonium is transported, it could fall into the hands of terrorists or blackmailers and be used to produce bombs.

Safeguards used at nuclear power stations reduce the danger of exposure to radiation. In fact, the operating stations have a good safety record. Still the irreversible (cannot be undone) effects of just one nuclear accident (a meltdown at a power station, or release of radioactive gas) disturb many people including scientists and environmentalists.

The only incident in the years nuclear power stations have been operating apparently caused little environmental damage. It did demonstrate, however, that it can happen and may happen again, environmentalists say.

Another concern is that stored wastes will accidentally leak radiation into the water supply. Once there, it would be introduced into the entire food chain.

Science and technology are working to build in ever better safeguards against accidental or malicious discharge of radiation materials into the air or water. Eventually society must weigh these safeguards and the need for energy against the potential side effects and decide for or against.

NUCLEAR FUSION

Another source of nuclear energy is called fusion. Unlike fission which is the splitting of atoms, fusion is the combining of atoms. This is how the sun produces radiation while using hydrogen as a fuel.

Getting two atoms of any kind of matter to fuse is very difficult since the nucleus of the one atom repels the nucleus of another atom. You can think of the nuclei as similar poles of a magnet. If you try to push the positive poles together the magnets will resist. Nuclei are positively charged with protons. Therefore one nucleus will have to hurled at another with great force to cause them to fuse. The more protons the nucleus has the more it resists fusion. It is, therefore, necessary to use nuclei with the smallest number of protons. See Fig. 10-28.

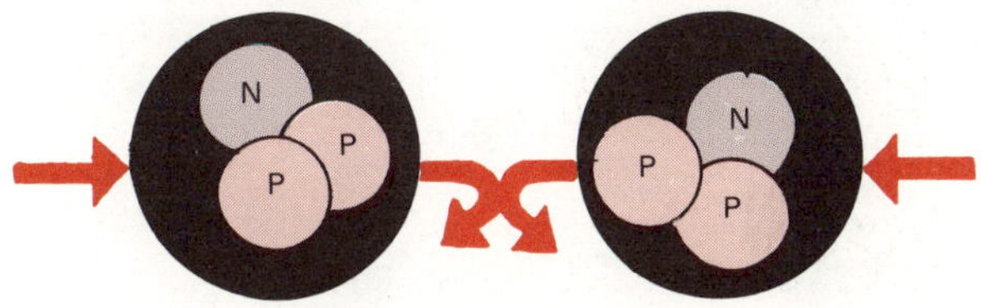

Fig. 10-28. Nuclei of atoms tend to repel each other. This makes fusion difficult to achieve. The fewer protons in the nuclei the easier fusion becomes.

HYDROGEN ISOTOPE

Two isotopes of hydrogen are the most suitable fuel for fusion. One is the isotope DEUTERIUM (D) also known as "heavy hydrogen." The other is the isotope TRITIUM (T). It can be extracted (separated) from water and from lithium, a soft, silver white metal. Though more rare than deuterium, it is abundant enough that supplies from lithium alone would provide sufficient energy for 50,000 years.

Heavy hydrogen is in plentiful supply on the earth since it is present in water. For every 6500 atoms of ordinary hydrogen in water there is one atom of deuterium. It is calculated that the earth's oceans, rivers, and lakes hold about 10 trillion tons of deuterium. Between six and eight trillion tons of this are recoverable.

DEUTERIUM AS AN ENERGY SUPPLY

The amount of energy available from deuterium is staggering when compared to any other energy source. When two deuterium particles fuse, the new mass (weight) of the combined particles is less than the sum of their masses before the fusion. The lost mass is converted into energy. The burst of energy from even a tiny amount of deuterium is gigantic. A look at the four reactions in fusion indicates how much energy is released, Fig. 10-29.

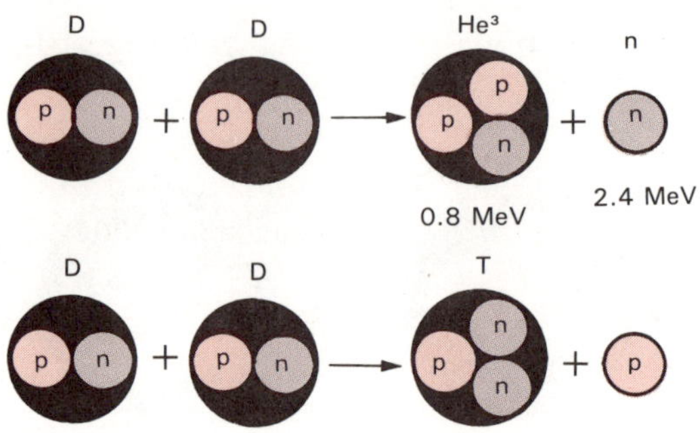

Fig. 10-29. The fusion of two nuclei release from 2.4 to 3 million electron volts of energy. (U.S. Atomic Energy Commission)

1. In the first two types of reactions, deuterium particles react with other deuterium particles:
   a. A deuterium particle fuses with another deuterium particle. This forms a tritium particle and releases one proton and 3.3 Mev (75 million Btu per gram) of energy.
   b. A deuterium particle fuses with another deuterium particle forming a helium particle and releasing a neutron and 4 Mev (92 million Btu per gram) of energy.
2. In two additional reactions, deuterium forms energy while fusing with products of the first two reactions, Fig. 10-30.
   a. Deuterium fuses with tritium (T) forming a new isotope of helium He^4, while releasing a neutron and 17.6 Mev (313 million Btu per gram) of energy.
   b. Deuterium fuses with helium (He^3) forming a new helium isotope and releases a neutron. Energy released equals 18.3 Mev (334 million Btu per gram) of energy.

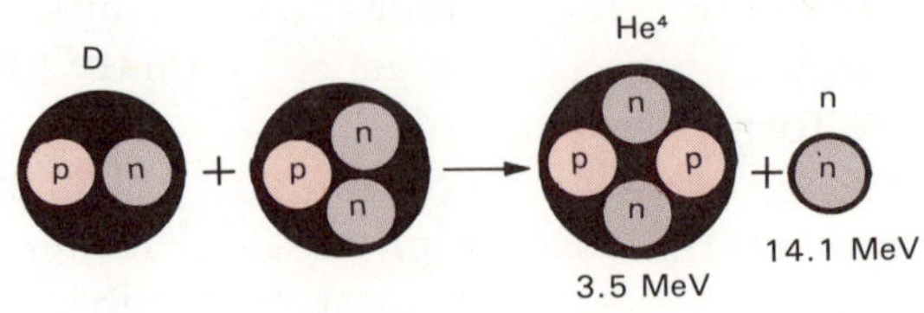

Fig. 10-30. When fusion takes place between deuterium and tritium, the release of energy equals more than 14 million electron volts.

By comparison, a ton of coal produces only 25 million Btu when burned. Each ton of deuterium produces 60 million times more energy. At present rates of energy consumption, the world could be using fusion energy for the next 50 billion years!

CREATING FUSION

As you already learned, getting two particles of matter to fuse is not easy. Something must be done to get them to move with such force that they overcome their electrical repulsion. They must be hurled at each other with great force.

In early experiments, scientists used a PARTICLE ACCELERATOR to hurl deuterium particles against a stationary target made up of deuterium or tritium. This caused fusion but no chain reaction. Much more energy was consumed in causing the fusion than was released.

Later experiments proved that it is much more practical to use heat to speed up the motion of the particles. If a mixture of deuterium and tritium is heated to a temperature of 90,000,000 to 180,000,000°F (50,000,000 to 100,000,000°C), the particles will collide and fuse.

PLASMA

Matter subjected to such high temperatures as are needed for fusion will not remain in their normal state (solid, liquid, or gaseous). They turn into a fourth state called PLASMA or IONIZED GAS. In this state the electrons leave the nuclei and form a mix of charged electrons and the stripped nuclei. The nuclei, called ions, are positively charged while the electrons are negatively charged.

Although plasma contains free positive ions and free negative electrons, the numbers of positive and negative electrical charges are in balance. The plasma as a whole, then, is electrically neutral.

CONFINEMENT

Maintaining high temperatures to create plasma is hard to do. It calls for special containers. Ordinary bottles, cans, or tanks will not do. The particles have to be contained in a vacuum where they will not touch any matter which will take away their heat. There are two problems then:

1. Finding a way to contain the mixture. It must be confined for about one second.
2. Suspending the mixture in a vacuum where it can be kept free of any foreign substance and out of contact with the sides of the container.

Two ways of containing and protecting the mixture are:

1. Magnetic confinement.
2. Inertial confinement.

MAGNETIC CONFINEMENT

Since the plasma must literally float in a vacuum, a method is needed to suspend it in its containing vessel. If it were allowed to touch the sides, the plasma would cool off so rapidly that fusion would stop altogether. This is done by surrounding it with a magnetic field which presses

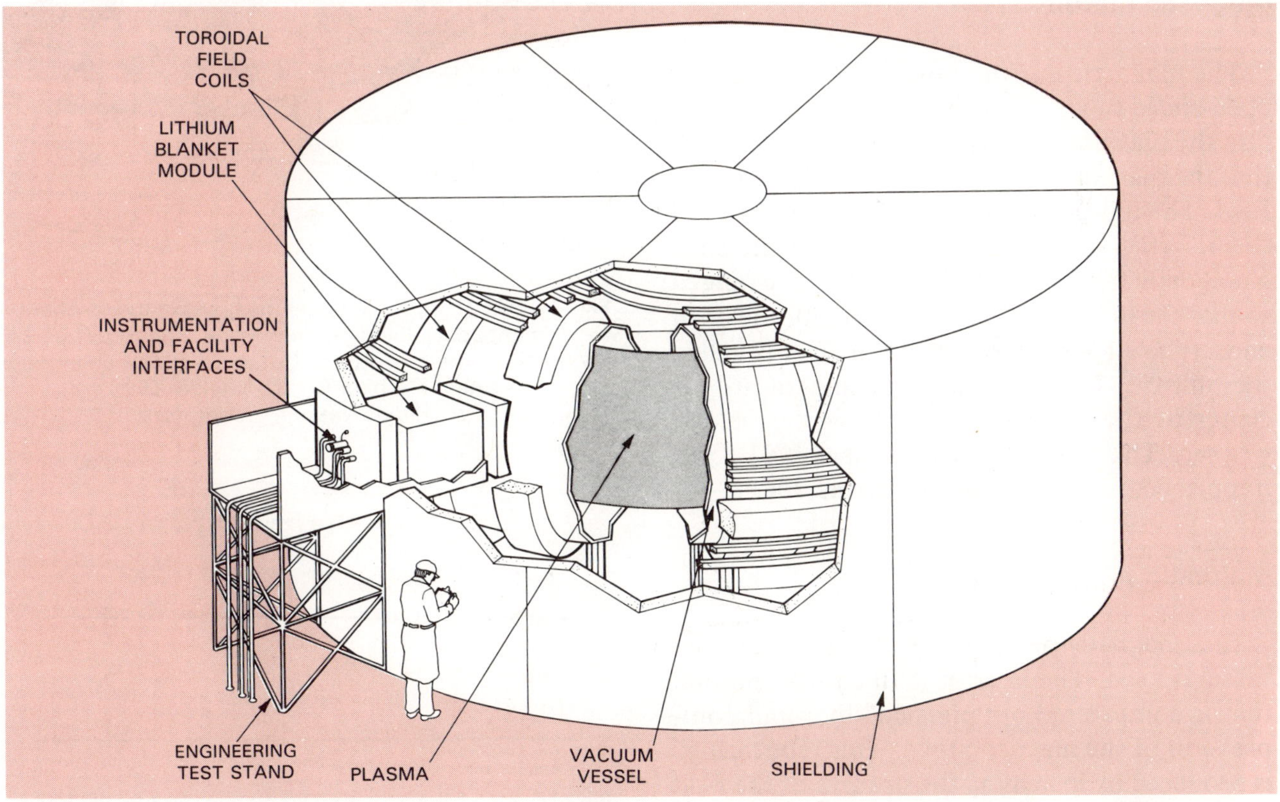

Fig. 10-31. Cutaway of the Princeton University fusion test reactor. Hot plasma is suspended in the chamber by magnetism. (EPRI Journal)

inward on the plasma keeping it away from the vessel sides. This magnetic action is often referred to as a "magnetic bottle."

Experiments are underway using three different systems of magnetic confinement.

The "doughnut" chamber

The toroidal-shaped (doughnut-like) chamber is hollow. Fig. 10-31 is a cutaway of the chamber. The plasma travels around in this vacuum chamber. The Russian version, known as "Tokamak," is the most successful. Most U.S. experiments are being done with this type. Chambers in this country include the PLT (Princeton Large Torum), Alcator at Massachusetts Institute of Technology, Doublet IIA at La Jolla, California, and the Elmo Bumpy Torus at Oak Ridge, Tennessee. See Fig. 10-32.

Magnetic mirrors

The magnetic mirror system, Fig. 10-33, uses open-ended tubes rather than closed tubes to contain the plasma. The system is based on the fact that the plasma cannot pass through a magnetic field. The tubes have field coils wrapped around them. When a current passes through the coils a magnetic field is set up. The coils are arranged so that the magnetic field is strongest at the open ends. The stronger field of magnetism is called the "mirror." It turns back the plasma toward the center of the tube where the magnetic field is weaker. The plasma will loop back and forth trapped between the magnetic mirrors, Fig. 10-34.

Magnetic pinch device

In a magnetic pinch device, the torus shaped vessel is used. The interior is filled with plasma hat is compressed or "pinched" by rapid com-ression of the magnetic field. This "squeezing" produced by increasing the strength of the field forcing the plasma toward the center of the . See Fig. 10-35.

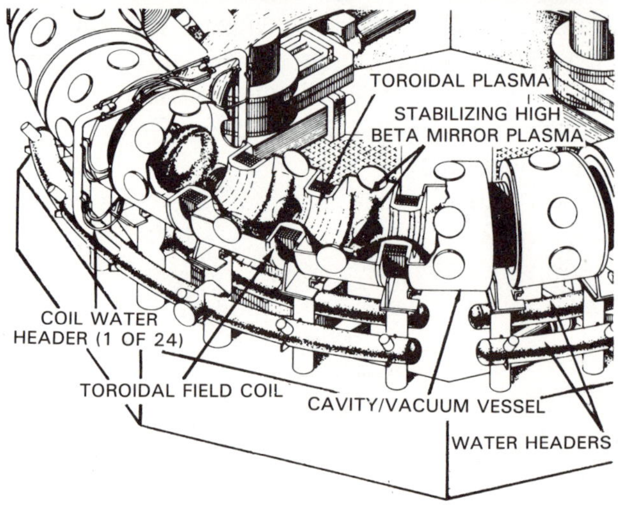

Fig. 10-32. Top. Fusion chamber used in fusion energy experiments at Oak Ridge, Tennessee. The torus has 24 identical cavities. Bottom. Cutaway of the chamber. (Union Carbide Corp., Nuclear Div., Oak Ridge National Laboratory)

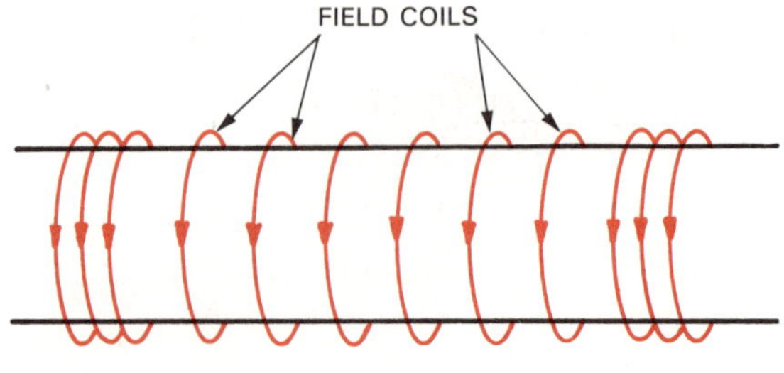

Fig. 10-33. Magnetic mirror system. Field coil circling the plasma chamber sets up a magnetic "girdle" to contain the plasma.

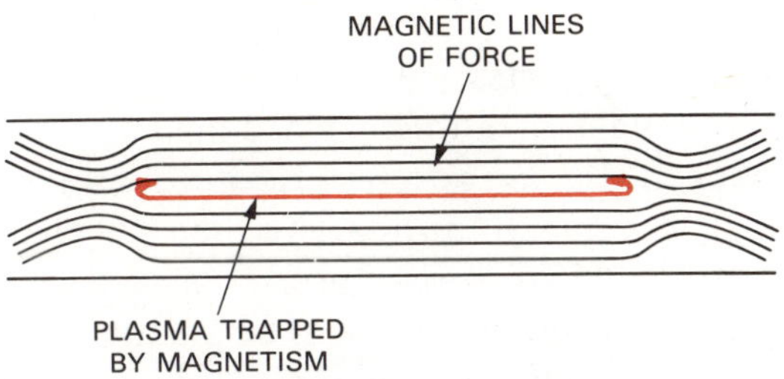

Fig. 10-34. Strong magnetic lines of force at the ends of the chamber trap the plasma turning it back to the center of the chamber. (Union Carbide Corp., Nuclear Div., Oak Ridge National Laboratory)

Heating the plasma

Heating of the plasma can be done in one of three ways:

1. Electric currents, induced into the plasma, can provide the necessary heat. The plasma heats up like the heating element of an electric toaster. Toroidal vessels rely, to some extent, on this kind of heating.
2. A nuclear beam could be injected into the plasma. The beam is made up of very energetic deuterons (deuterium isotopes). They are fired into the plasma inside confinement vessels.
3. Compression of the plasma will cause it to heat up enough to start the fusion reaction. This technique for heating can be used in the magnetic pinch confinement vessels. Pinching heats up the plasma while confining it.

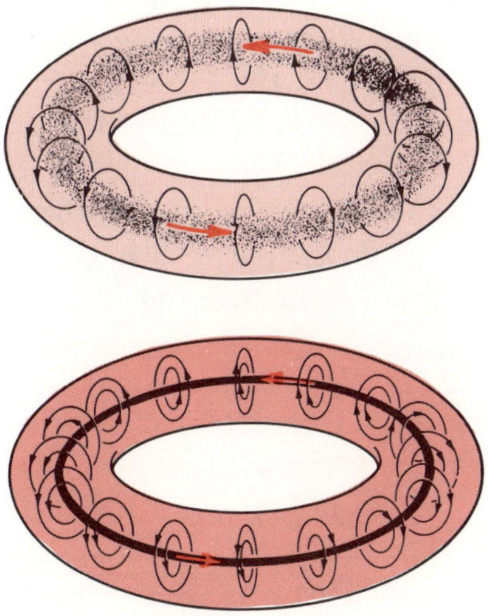

Fig. 10-35. How plasma is "pinched" by magnetism. Top. Plasma takes up much space in the toroidal (ring shaped) tube at first. Bottom. But when heavy current flows through the coil surrounding the tube, the magnetic lines of force squeeze the plasma into a smaller space and make it more dense. (Union Carbide Corp., Nuclear Div., Oak Ridge National Laboratory)

INERTIAL CONFINEMENT

The inertial system does not use magnetism to confine the plasma. Rather, the fusion fuel, a mixture of deuterium and tritium, is frozen into small, round droplets less than 1 mm (about 1/25 in.) in diameter. One of the droplets is located in the middle of a vacuum chamber. Then it is hit from all sides by an energy source. This can be a laser beam or a beam of charged nuclear particles. The particles may be ions or electrons.

LASER BEAM BOMBARDMENT

A powerful laser beam may be focused with great accuracy on the pellet. A system of mirrors breaks the beam into several smaller beams that hit the pellet from all sides.

The bombardment must take place in a split second. This heats the fusion fuel rapidly without expanding it. The outside layer of the pellet vaporizes and rushes outward. As it does so it creates a force against the remainder of the pellet causing it to implode (fall inward with crushing force). The great density and heat ignites the fusion reaction. The pellet explodes, Fig. 10-36. Most of the energy is carried by the released neutrons.

THE FUTURE OF FUSION ENERGY

Many engineering problems need to be solved before nuclear fusion can be a success. So far, a self-sustaining fusion reaction has not been possible. For fusion, three conditions must occur at the same time. There must be exact:

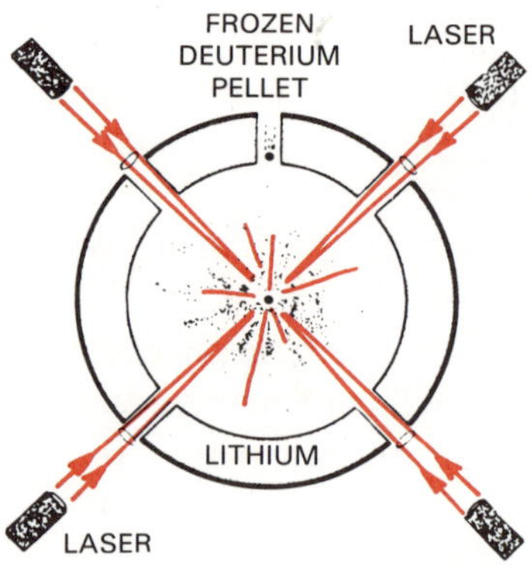

Fig. 10-36. One way of causing fusion is by directing a laser at a frozen pellet of deuterium inside an evacuated (airless) chamber. The pellet is violently compressed causing it to produce heat and neutrons. A layer of lithium recovers the heat and breeds tritium. (Nuclear Power in Canada/Canadian Nuclear Assoc.)

1. Temperature.
2. Confinement time.
3. Plasma density.

Only then will fusion reaction continue on its own.

Despite almost 30 years of work on fusion, scientists and engineers have not been able to produce all three of these conditions in one machine. Should these researchers succeed tomorrow in demonstrating a method of sustaining a fusion reaction, other problems would have to be solved. One of them would be the construction of a chamber that could withstand the fusion reaction for a long period of time without disintegrating (crumbling). Another would be the tremendous cost of designing and building such a reactor. Should they succeed in overcoming the obstacles, humankind will have an energy source for millions or even billions of years.

FUSION ENERGY AND THE ENVIRONMENT

Successful fusion reaction does not threaten the environment to the same extent as fission reaction. Some writers contend that fusion could be as nonpolluting as any energy we have except solar. The advantages of fusion energy as an environmentally acceptable energy are based upon the following facts:

1. The danger of a nuclear accident is sharply reduced. Fusion reactors do not contain the huge amounts of radioactive material characteristic of fission reactors.
2. Danger of meltdown would be greatly reduced if not altogether eliminated. If any part of the system fails, the plasma is destroyed and the reaction stops.
3. Storage of waste material from spent fuel is less of a problem for two reasons. The waste is not so long-lived as fission by-products. The amount of waste is small compared to fission.
4. Since plutonium is not a by-product of fusion, the danger of nuclear materials falling into unfriendly or irresponsible hands is not a threat.
5. Gathering the fusion fuel does not destroy the environment. Removing one hydrogen atom in 6500 from sea water will have no measurable effect on the water.

FUSION "TORCH"

It is possible that once developed, fusion power could be put to other uses than for propulsion and generation of electric power. One proposal is the development of the "fusion torch." The plasma coming from the exhaust of the fusion reactor might be used to break down solids or liquids. The extremely high temperature would vaporize the material and then ionize it. (This means to break up the particles into atoms.) Fig. 10-37 shows a simple schematic for a fusion torch.

The torch could be used for disposing of all kinds of waste materials, processing of ores, and production of liquid fuels. It is believed that the fusion torch could also be used to generate ultraviolet or X rays. This would be done by putting small amounts of heavy atoms into the plasma exhaust.

Energy created in this way could then be used for various purposes:

1. Desalinization (taking salt out) of seawater.

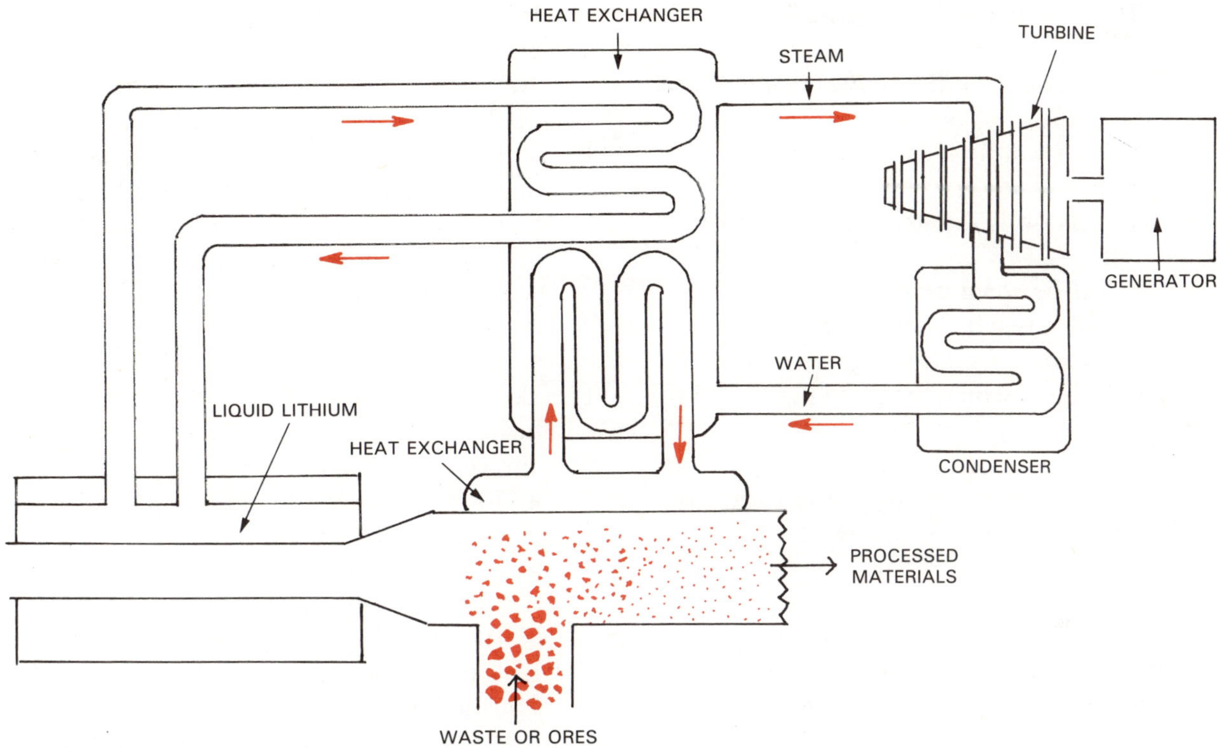

Fig. 10-37. How a "fusion torch" might work. Some of the energy from the superheated plasma would be used to vaporize waste material or ores and reduce them to atoms. Then the atoms could be reconstituted (put back together) into useful raw materials.

2. Production of hydrogen.
3. Processing of chemicals.

Because plasmas, not chemicals, would be used to separate elements there would be less pollution of the environment.

The idea of a fusion torch, being untested, may not be practical in every case. It does appear that exciting industrial uses may be found for fusion energy if the scientific and engineering problems can be solved.

DO YOU KNOW THESE TERMS?

Breeder reactor, by-products, centrifuge, chain reaction, control rod, deuterium, electron, electron volt, fission, heavy hydrogen, in situ, ionize, isotope, laser, leaching, microspheres, neutron, nucleus, plasma, plutonium, proton, radiation, reactor, reactor vessel, sintering, sodium, thorium, Tokamak, toroidal, transmutation, U-235, uranium, uranium hexafluoride, yellow cake.

SUGGESTED ACTIVITIES

1. As a class, debate whether development of nuclear energy by fission should continue.
2. Develop a report on whether the United States should continue research and development of a commercial fusion reactor.
3. Construct a model of one of the following:
   a. Containment vessel for a nuclear reactor.
   b. Nuclear fission chain reaction.
   c. Underground storage vault for nuclear waste.
   d. Fuel assembly for a nuclear reactor.

TEST YOUR KNOWLEDGE

1. Different forms of the same element are called ____________.
2. Describe, in simple terms, how nuclear fission takes place.
3. Heat of fission is due to (select the correct answer):
   a. Friction between neutrons.
   b. Some atoms which burn up.
   c. Conversion of matter into energy.
   d. The speed at which the freed neutrons are traveling.
4. ____________ ____________ a mixture of oxides of uranium may be in the form of a bright yellow slurry or a powder.
5. Name the four basic or main types of nuclear reactors.
6. The __________ reactor will actually produce more fuel than it uses.
7. A fuel element is a bundle of hollow rods containing pellets of uranium fuel. True or False?
8. Plutonium is an atom of uranium which has gained a ____________.
9. Nuclear ______________ means using a nuclear reactor to power vehicles.
10. Nuclear fusion (select correct answers):
    a. Is the combining of two atoms into one.
    b. Is easiest using two isotopes of hydrogen.
    c. Is an attractive energy source because the supply of fuel is virtually unlimited.
    d. Is already commercially feasible.
    e. All of the above.
11. To create fusion, the deuterium must be heated to temperatures in excess of (select correct answer):
    a. 5000°C.
    b. 10,000°C.
    c. 25,000°C.
    d. 50,000°C.
    e. 100,000°C.
12. List the three ways in which fusion fuel (plasma) may be heated.
13. List four advantages of fusion energy.

Turbine generator for nuclear power station. This one has a capacity of 965,000 kW and is driven by a pressurized water reactor. (Power Authority of the State of New York)

11 BIOMASS ENERGY

The information given in this chapter will enable you to:

- Define biomass and list its various forms.
- List and describe processes by which biomass is converted to energy.
- List and describe fuels produced through biomass conversion.
- Describe the densification process for producing solid biomass fuel.
- Discuss the potential for a biomass fuel industry.
- Discuss problems connected with growing and collecting biomass materials.

Biomass is live organic material such as trees, farm crops, seaweed, or algae, Fig. 11-1. It is also wastes such as manure, trash, sewage, and garbage. All biomass is a form of indirect solar energy. It represents material that once grew and thus captured sunlight. The process by which plants store energy is called PHOTOSYNTHESIS.

The term, biomass, has only recently become a common part of our energy language. It comes from the words BIOLOGY and MASS. Biology is a study of living organisms, which includes plant life. Mass refers to the body of organic mat-

Fig. 11-1. Our forests are a huge biomass resource which produces energy as well as lumber and other wood products.

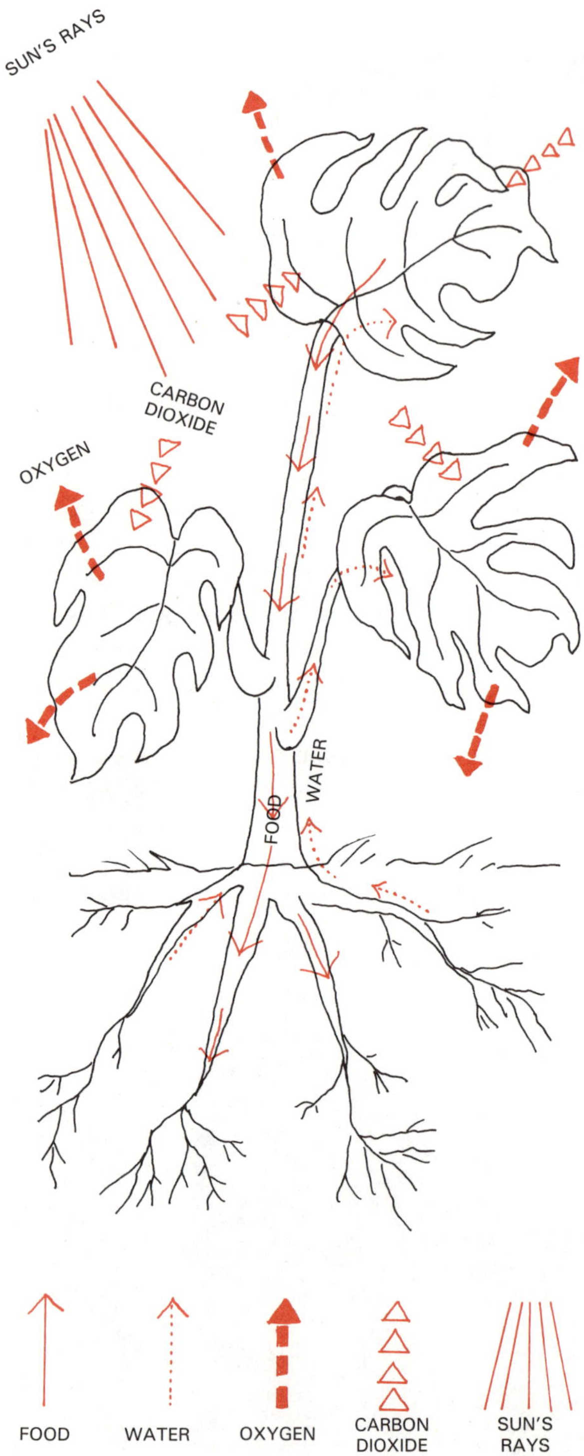

Fig. 11-2. Plants take in sunlight, water, carbon dioxide, and nutrients. They combine the carbon with hydrogen from the water and store it. The oxygen released from the water is passed on to the atmosphere.

ter available in a region, country, or in the whole world.

HOW BIOMASS IS PRODUCED

As you learned in Chapter 1, plant life is a form of potential energy. The energy is stored in the cells of plants.

The plants use radiant energy to convert carbon dioxide, water, and nutrients from the soil into sugar and oxygen. Certain enzymes within the leaves of the plant help make the process work. (An enzyme is a type of protein which causes chemical action to take place within plant and animal cells.) Fig. 11-2 illustrates the process of photosynthesis.

Carbon is an important element found in the air and in all kinds of organisms living and dead, Fig. 11-3. The amount of carbon stored every year through photosynthesis is about 10 times the world's total annual energy use. Only a half percent of this carbon is consumed as food or used to produce fibers for manufactured products. The rest is available as a source of largely untapped energy.

CONVERTING BIOMASS ENERGY

Before it can be used as any kind of energy or food source, biomass must be converted to a more usable form. Plant life is converted in several ways:

1. Metabolic conversion. This means being consumed (eaten) by animals, Fig. 11-4, or humans. The metabolism (digestion) of the animal and human body is such that is reverses the process of photosynthesis. Heat, at a very low level, oxygen, and enzyme action break down the plant cells to create food for energy and growth. The by-product is carbon dioxide and fecal wastes.
2. Thermochemical conversion. This is an industrial process which depends upon higher levels of heat to burn the biomass or to convert it to a combustible gas or liquid.
3. Biological conversion. This, too, is an industrial

LOCATION	QUANTITIES (in billions of tons)
Atmosphere	700
Dead organic matter land	700
Land plants	450
Marine zooplankton	5
Marine phytoplankton	more than 5
Dead organic matter marine	3000
Fossil fuels	10,000
Sedimentary rock	20,000,000
Sea surface layers	500
Deep sea	34,500

Fig. 11-3. Distribution of carbon in the world. Large quantities are found in the atmosphere and in dead organic matter.

process. It uses enzymes, fungi (molds), or microorganisms (bacteria) to cause a chemical change resulting in a gaseous or liquid fuel.

BIOMASS FUELS

Biomass conversion yields several types of fuel:

1. Biogas (also called synthetic natural gas) which is chiefly methane.
2. Ethanol and methanol alcohol.
3. Petroleum.
4. Solid fuels (char and chips).

Fig. 11-4. Animals reverse the process of photosynthesis. At low heat levels, their bodies use oxygen and enzymes to break down plant life into food for energy. (Cascade Pioneer Advertiser)

THERMOCHEMICAL SYSTEMS

Thermochemical processes include:

1. Direct combustion.
2. Pyrolysis.
3. Gasification.
4. Liquefaction.

DIRECT COMBUSTION

This is the simplest and best developed of all the processes. Dry wastes and residues from forests, farms, and cities are burned directly to produce steam, electricity, or heat for industries, utilities, and homes, Fig. 11-5.

Fig. 11-5. Some communities collect garbage and use the organic waste as a boiler fuel. (Union Electric Co.)

The forest products industry, for many years, has used biomass successfully to produce process steam and electricity for lumbering operations. According to estimates, forest wastes contribute 1.1 quadrillion Btu or about 45 percent of the energy used by the industry. Other industries producing textiles and paper products can also use biomass to provide heat for their manufacturing processes.

Home heating is also possible with a wood-burning stove or furnace. In a well designed and well insulated home, wood may well provide all the heat needed.

Direct burning of biomass has been limited because of the problems it presents. As a solid fuel it is quite bulky and difficult to transport long distances. Further, it causes some air pollution because of the gases and solids given off during burning. Therefore, it is more desirable to convert it to combustible gases through other thermal processes.

Gas fuels are clean burning and easy to distribute. Gases also convert easily to liquids or chemicals.

BIOCONVERSION BY PYROLYSIS

Pyrolysis is a breakdown of matter such as biomass by heat. Burning takes place in the absence of oxygen, Fig. 11-6. This process was developed by a research and development company in California. First the organic waste is separated from inorganic materials (glass, metals, and other minerals). Then the organic materials are shredded, dried, and fed into an air-tight furnace. It is incinerated (burned) at a temperature of 932°F (500°C) to produce oil, "char" (charcoal), and a low grade gas fuel. Some of the char and the gas fuel are recycled to provide heating for continuing the pyrolysis. Each ton of refuse produces one barrel of oil.

Coal, as well as biomass, may be converted by pyrolysis. The char is known as COKE.

ECONOMICS OF PYROLYSIS

Use of pyrolysis will be possible in some locations but not in others. Some of the organic waste in the United States is produced in small quantities at many sites. Collecting it in large enough quantities to feed the processing plant would be difficult.

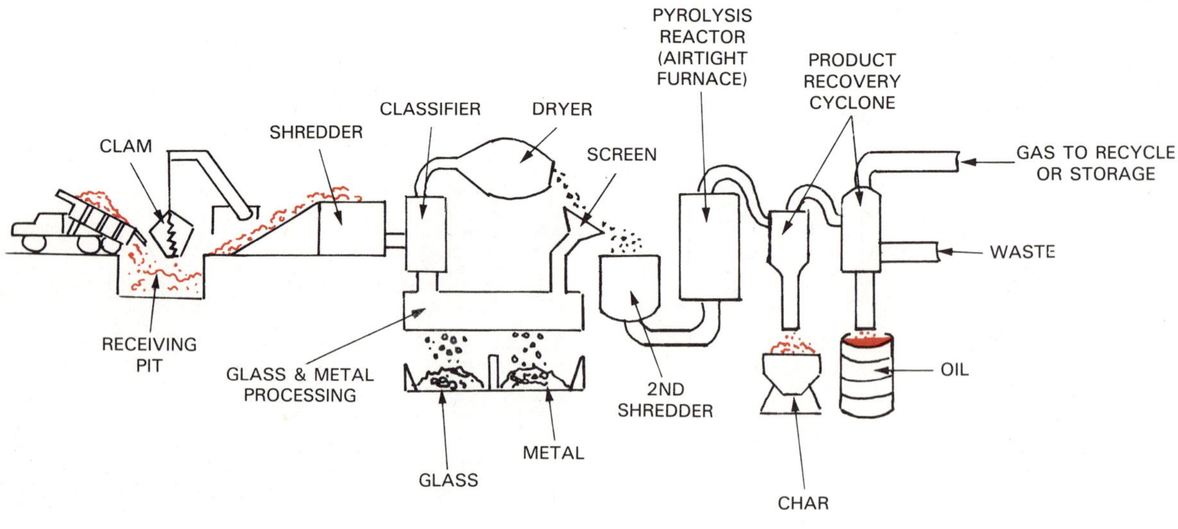

Fig. 11-6. Pyrolysis process. Organic waste is burned at a temperature of 932 °F in an airtight furnace. Oil and a combustible material, like charcoal, are the resulting energy products.

However, enough of it is accumulated at large factory operations, big cities, cattle feed lots, and sawmills to produce 170 million barrels of oil a year. It has been estimated that a 2000 ton-per-day plant could process refuse waste at $5 per ton and handle the waste output from a city of 50,000.

GASIFICATION

Pyrolysis is also used in gasification processes. The waste is transformed into fuel with heat plus air or oxygen. See Fig. 11-7 for gasification processes and their products.

Air gasification

A second type of gasification is known as air gasification. It uses small amounts of air and steam to convert char to gas. The process produces a low energy gas (LEG) which is diluted with nitrogen from the air. Though not suitable for pipeline distribution, it can be burned by boilers now using oil or natural gas. It is good enough also to power internal combustion engines in vehicles or in power stations.

Oxygen gasification

Like air gasification, oxygen gasification is a simple process. It is carried on in an oxygen atmosphere which produces a medium energy gas made up of carbon monoxide (CO) and hydrogen (H_2). It makes a good fuel and can be used as a feedstock to make methanol (a form of alcohol), ammonia, H_2, or gasoline. This type of gas is also known as "synthesis gas" or "syngas." Fig. 11-8

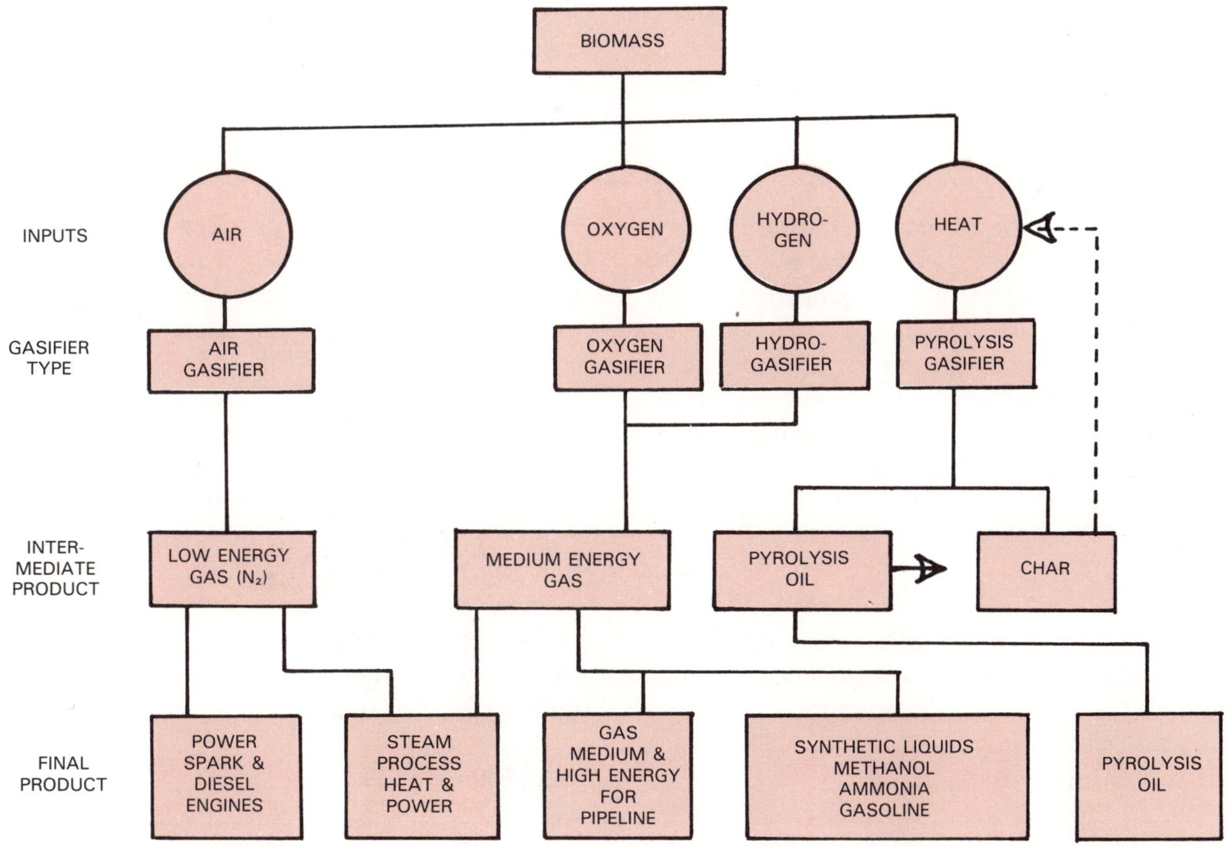

Fig. 11-7. Gasification of biomass has many intermediate and final by-products. (SERI)

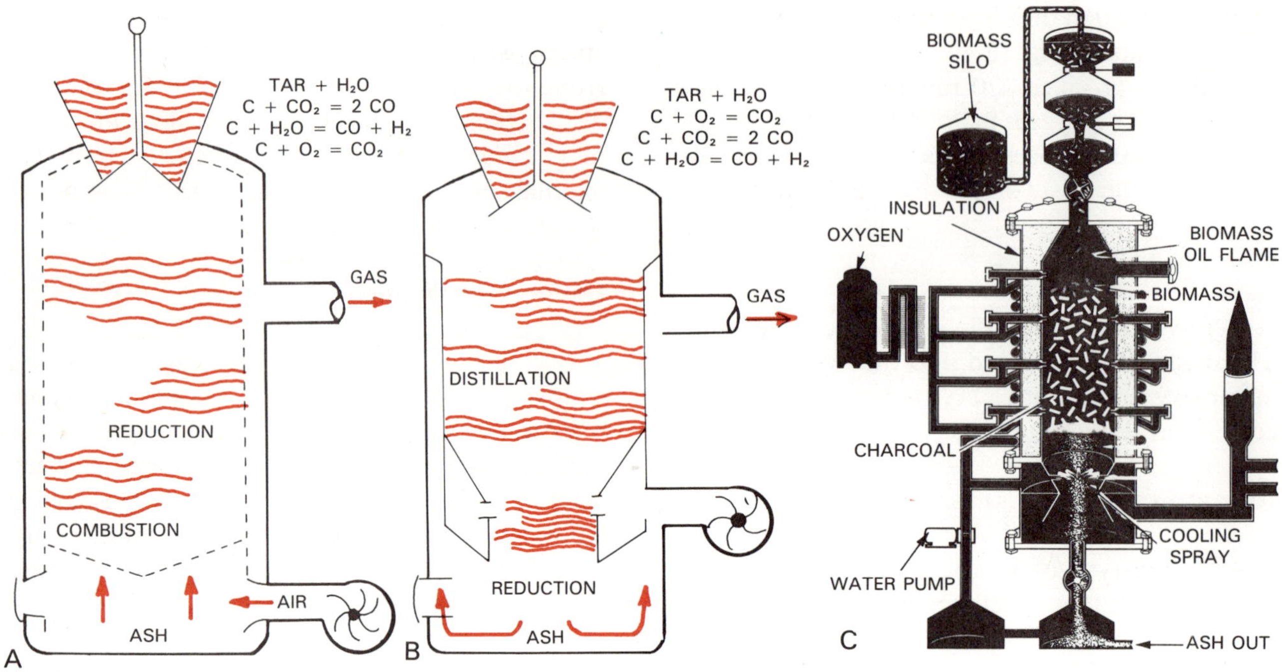

Fig. 11-8. Schematic diagram of retorts (burners) used for oxygen gasification of biomass. A—Updraft gasifier. B—Downdraft gasifier. C—Stratified downdraft gasifier.

shows three types of gasifiers.

Many industries produce wastes that would provide the biomass needed for gasification. Among them are sawmills, processors of animal feeds, food processors, and processors of raw materials for the textile industries.

Waste materials that can easily be gasified include: sawdust, corncobs, fruit pits, nut shells, and rice hulls. Cotton gins also produce waste which could be gasified.

LIQUEFACTION

Liquefaction is also called HYDROGENATION. It is a process used to produce a gaseous or liquid fuel. Generally, liquefaction takes place in reaction with hydrogen or a mixture of hydrogen and carbon monoxide.

Basically, the process has been used to make oil or gas out of coal. However, it has also been used successfully in experiments to make the same fuels from waste.

The waste is placed in a reaction vessel and heated to 716°F (380°C) in a mixture of carbon monoxide and steam under pressure. Within 20 minutes nearly half of the waste becomes fuel.

About 99 percent of the carbon content of the waste can be converted to fuel oil. A ton of dry waste will produce about 1.25 barrels of oil, Fig. 11-9. Animal wastes have a heat value of 15,000,000 Btu per ton. This is a little less than half the heat value of petroleum.

The fuel extracted by this process contains more oxygen than crude oil and needs much refining unless it is to be used only for direct combustion.

COGENERATION

Biomass conversion processes can be used by industry to produce electricity and process heat at

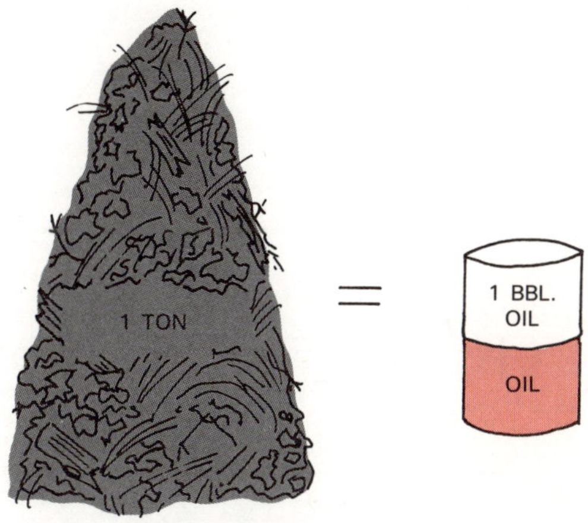

Fig. 11-9. A ton of dry waste will produce about a barrel of oil by hydrogenation process.

the same time, Fig. 11-10. (Process heat is the heat used by a manufacturing operation to process raw material into a product. For example, a sugar processing plant may use its own wastes to produce electricity for its plant machinery and steam for heating and evaporating processes.)

BIOLOGICAL CONVERSION

Biological conversion produces either liquid or gaseous fuels. It is accomplished through ANAEROBIC DIGESTION which is sometimes referred to as FERMENTATION.

ANAEROBIC DIGESTION

Anaerobic digestion is the controlled decay of organic matter without oxygen. Waste materials such as sewage, manure, paper, seaweed, and algae can be converted to produce methane gas.

The organisms which do the work are known as ANAEROBES. These cells are able to oxidize organic matter without using free oxygen. However, they do obtain oxygen by breaking up molecules containing oxygen. They then use this oxygen to produce chemical change in the oxidized matter.

The process is simple and cheap. Bacteria breaks down the waste at low temperatures—between 59°F and 122°F (15°C and 50°C) at atmospheric pressure. Two types of bacteria are involved. One breaks down the organic solids into organic acids, hydrogen, and carbon dioxide. The second feeds on the waste products of the first types and produces methane-rich biogas as its own by-product. See Fig. 11-11.

Fig. 11-10. Spent steam from the turbine of a power plant can be piped to a factory and used as process heat.

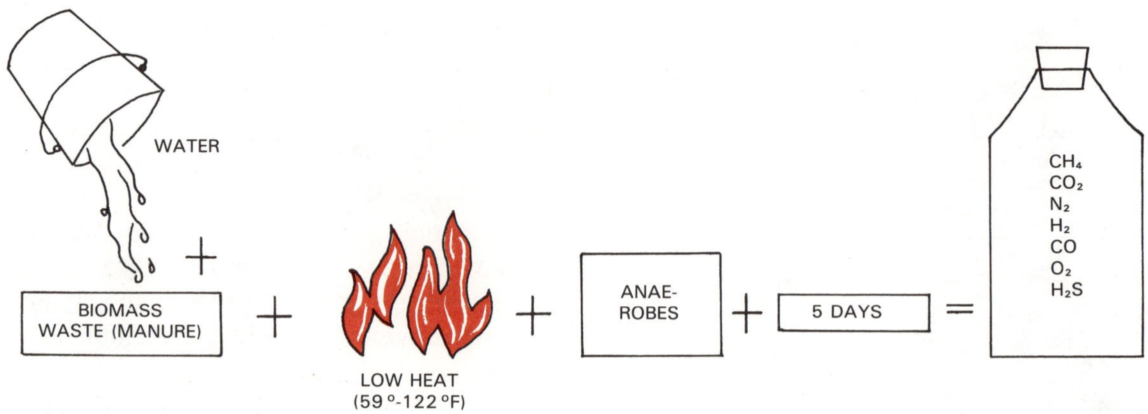

Fig. 11-11. In anaerobic digestion, low heat—between 59°F (15°C) and 122°F (50°C)—starts the process. In about five days it can produce methane-rich biogas from a waste slurry.

One of the foremost countries in development of anaerobic digestion is India. An Indian scientist, Ram Bux Singh, has built many units which he calls biogas plants. The residue of the process is a rich organic plant food which can be returned to the soil.

Rising costs of natural gas are making the construction of methane digesters at sewage disposal plants a more attractive alternative to purchasing gas. Small-scale digesters are being used on farms particularly in Europe and Asia.

Many U.S. farms have large enough biomass supplies to make small-scale anaerobic digestion possible. An estimated 20 percent of current U.S. natural gas requirements could be met with anaerobic digestion.

FERMENTATION

Fermentation is an anaerobic process for producing alcohol. The process depends on the action of enzymes on the starches and sugars in fruits, grains, and other biomass forms. It is a primitive form of respiration. The enzymes which carry on the process are secretions (yeasts) of microorganisms found in nature.

Alcohols are an important biofuel which can be blended with gasoline or burned alone. There are several types. The one produced by fermentation of grains, fruits, and other biomass is called ETHANOL or ethyl alcohol.

Fermentation produces a mixture of water and alcohol. Since water boils at 212°F (100°C) and alcohol around 173°F (78°C), the two are separated by distillation. The mixture is heated to a point where the alcohol boils off. The vapor is caught and condensed.

The formula for pure ethanol is CH_3CH_2OH. It is a colorless, sweet-smelling liquid that burns with a hot, smokeless, pale blue flame.

With the growing scarcity and cost of importing gasoline, alcohol has long been seriously considered as an alternate fuel. Already, some oil companies are blending it with gasoline at 10 parts of alcohol to 90 parts of gasoline.

ANAEROBIC DIGESTERS

A digester is basically an insulated, airtight container where gases from decaying organic wastes are trapped. Anaerobes break down the waste into a methane fuel.

The gas produced by anaerobic digestion is a

Makeup of Biogas

SYMBOL	GAS	PERCENTAGE
CH_4	Methane	54-70
CO_2	Carbon dioxide	27-45
N_2	Nitrogen	.5-3
H_2	Hydrogen	1-10
CO	Carbon monoxide	0.1
O_2	Oxygen	0.1
H_2S	Hydrogen sulfide	trace

Fig. 11-12. Biogas is made up of seven different gases. Usually the carbon dioxide is removed by other processes.

mixture of seven different gases. It is mainly methane and carbon dioxide, however. Usually it is called either BIOGAS or METHANE. See Fig. 11-12.

DIGESTERS

There are two basic types of digesters:

1. Batch load. This digester is filled with a soupy mix of waste called SLURRY. Then it is sealed. It will not be emptied and refilled until the raw materials have stopped producing gas. Fig. 11-13 shows a simple batch load design.
2. Continuous load. This type will accept a small amount of fresh slurry continuously. A small amount can be added daily. The digester will produce gas as long as it is being fed.

There are two types of continuous load digesters. The first type, shown in Fig. 11-14, is a vertical-mixing type. The materials are digested in vertical chambers. Spent slurry is lighter and will flow out over the top. Units like the one in Fig. 11-14, right, have two chambers. As the spent slurry overflows into the second chamber it continues to digest. Finally, it is drawn out the effluent pipe.

A second type designed for continuous load is known as a DISPLACEMENT digester. It is a long horizontal cylinder (often made from oil drums welded end-to-end). Fig. 11-15 is a simple sketch of a displacement digester.

By the time the spent slurry arrives at the opposite end of the digester it has separated into

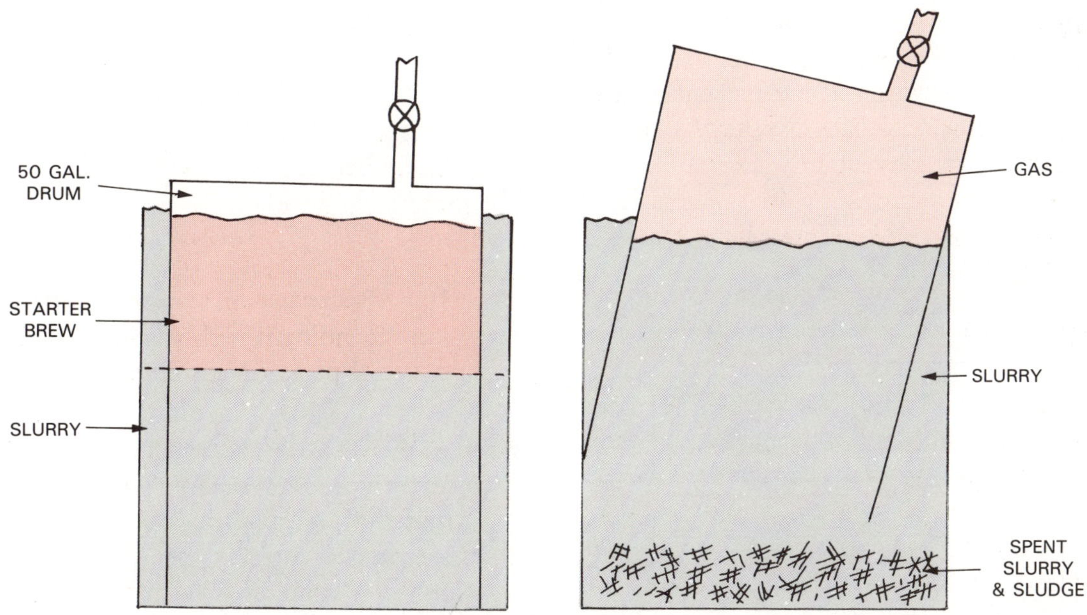

Fig. 11-13. A simple batch digester. When anaerobic action is finished the container must be dumped and refilled. Left. Freshly loaded digester. Right. Digester is working.

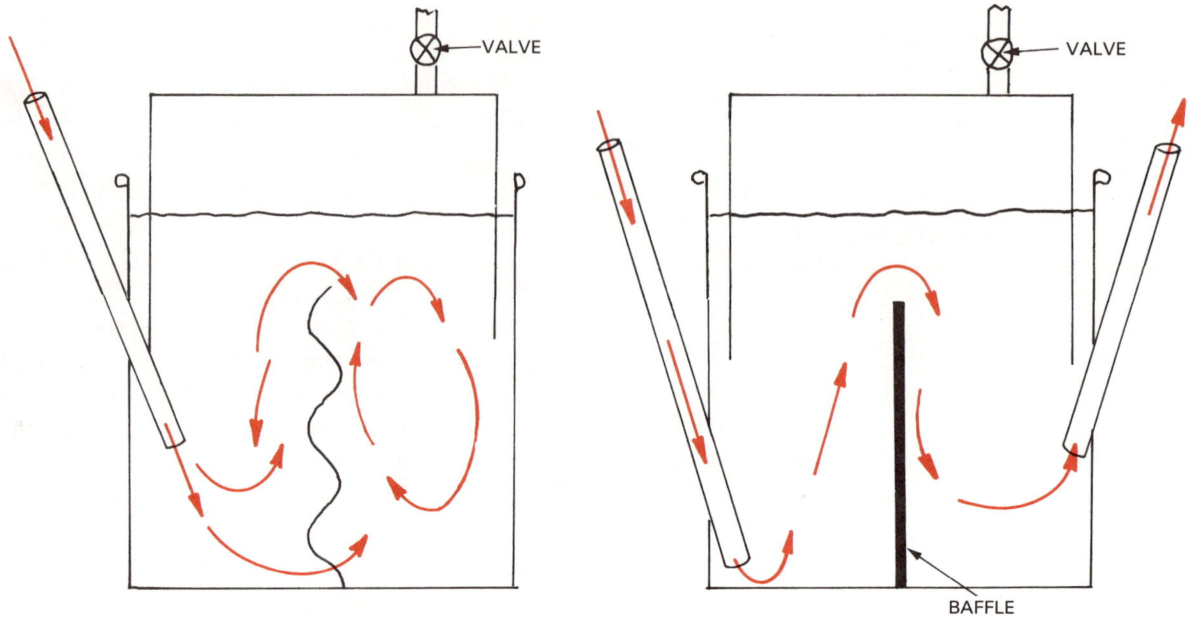

Fig. 11-14. Small amounts of slurry can be added daily to these continuous load digesters through pipes at left. Two types of vertical digesters are shown.

distinct layers. See Fig. 11-16.

Starting from the top, the working digester contains:

1. Biogas.
2. Scum. This is a mixture of fibers separated from the organic waste. If not present in large amounts, the scum insulates the slurry and helps the digestion process. However, scum is usually a problem in large amounts since it can cause the digester to stop working.
3. Supernatant. This is the spent liquid of the slurry. Its value as a fertilizer is equal to the sludge since it contains dissolved solids.
4. Sludge. These are solids contained in the original waste. They have been reduced to

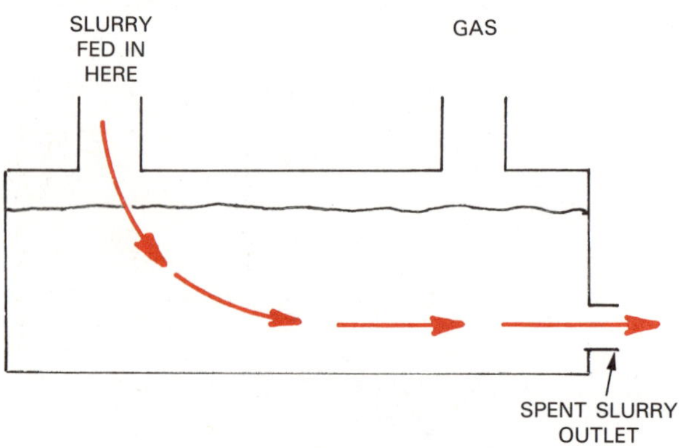

Fig. 11-15. A displacement digester. As waste is added it moves along, is digested, and moves to the outlet when it is spent.

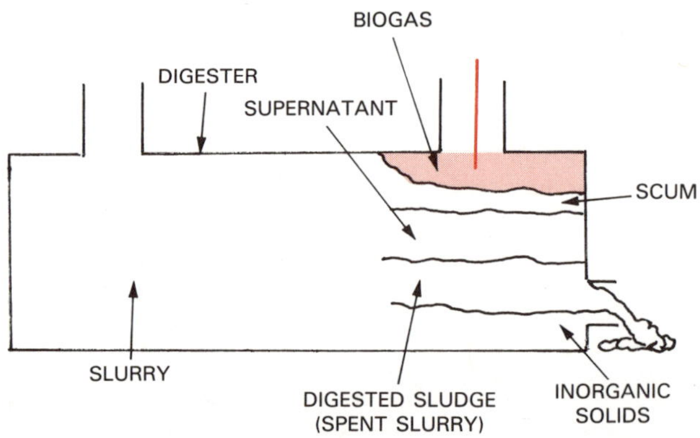

Fig. 11-16. Biogas and by-products of digestion separate into layers when the slurry is spent.

around 40 percent of their original volume. Sludge is used for fertilizer.
5. Sand and inorganic matter.

Anaerobic digesters are designed to be used on farms where large quantities of manure are available. The methane produced can be used as a fuel for stoves, agricultural dryers, or internal combustion engines.

BIOMASS ENERGY SYSTEMS

Many American cities may one day be able to reduce their trash disposal problems and reap energy rewards by burning their refuse. Municipal waste can be shredded, mixed with other fuel, such as coal or oil, and burned. The heat created will generate electricity or heat buildings. European cities have been harvesting the energy from trash and garbage for many years. It requires special incinerators equipped with moving grates. Some North American cities, such as Chicago, the Norfolk (Virginia) Naval Base, and several Canadian cities, already have programs.

ST. LOUIS WASTE BURNING PROGRAM

The city of St. Louis, Missouri, working with an electric utility company, built an experimental plant in the 1970s. The refuse, Fig. 11-17, was fed into a hammermill which shredded it. Massive metal blades of the mill, driven by a 1250 hp motor reduced the trash to particles less than 1 1/2 in. in diameter. See Fig. 11-18. Next, magnetic separators removed ferrous metals which were stored until they could be sold to a steel mill, Fig. 11-19.

Fig. 11-18. Shredded trash contains bits of paper and different fibrous materials.

Fig. 11-17. Garbage trucks brought trash to the processing plant for shredding and separating out metals. (Union Electric Co.)

Fig. 11-19. Separated metals were moved by conveyor and stored until they could be transported to a steel mill. (Union Electric Co.)

Fig. 11-20. Air classifier removes heavier particles such as glass, metals, and other material that would not burn.

Because the refuse still contained heavy, noncombustible (will not burn) materials, it was passed over heavy jets of air which separated light, burnable material from heavier materials, Fig. 11-20. Conveyors carried away the combustible material to the power plant boilers. Here it was mixed with coal and fed into the boiler through pneumatic tubes. Fig. 11-21 is a diagram showing the entire process in operation.

CROP RESIDUE FUELS GRAIN DRYING

Farmers have been experimenting with crop residues as a fuel for drying grain. A manufacturer of drying bins has been testing a biomass furnace. The furnace attaches to the drying bin and can burn nearly any form of combustible material, Fig. 11-22.

The furnace uses a heat exchanger (device which collects heat at one point and delivers it someplace

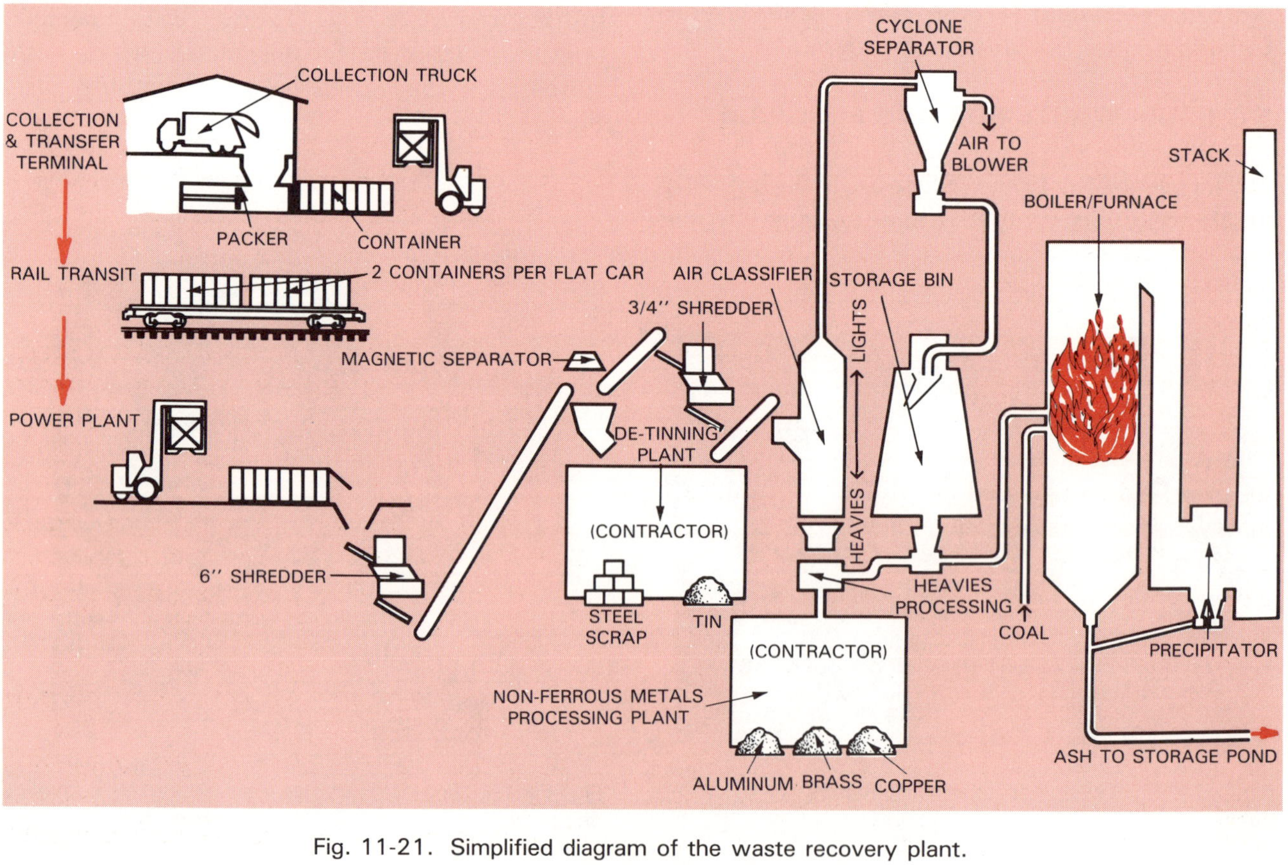

Fig. 11-21. Simplified diagram of the waste recovery plant.

Fig. 11-22. Agricultural biomass can be used for drying grain. Bales of dry straw are fed into a furnace like the one shown.

else). It heats the air blown into and through the grain but never comes in contact with the air it heats. According to the manufacturer, 95 bushels per acre of corn crop will yield 5500 lb. of cobs, leaves, and stalks per acre. This is equal to the energy of 320 gal. of propane.

DRYING OF SEED CORN

A seed corn company is using corncobs from its own field to dry seed crop. The cobs are fed into a gasifier chamber where they are burned with a limited oxygen supply. This creates combustible gases which fuel the seed dryer.

CORNCOB GASIFICATION RESEARCH

Iowa State University is also conducting research on corncob gasification. It is believed that gas engines will run on this fuel if carburetors are modified.

Compact corncob gasifiers are being designed for mounting on tractors. Crop residue would produce all the energy the tractor needs.

DENSIFYING BIOMASS

Because biomass is difficult to collect, store, ship, and use, efforts are being made to compress it. When so treated, it is called DENSIFIED BIOMASS FUEL or DBF.

Before it is converted, raw biomass must undergo some changes. First the noncombustibles must be separated. Then the remaining combustible materials are shredded and dried.

Five methods of biomass densification are now in use commercially:

1. Pelletizing. Pellets are formed by hard steel dies full of holes 1/8 to 1/2 in. (3 to 13 mm) in diameter, Fig. 11-23. The die rotates against

rollers forcing biomass through the holes under great force (10,000 psi or 7 kg/mm²). The extruded (pushed through dies) material comes out as pellets.

2. Cubing. This is a kind of pelletizing in which the dies create larger pellets or cubes 1 to 2 in. (25 to 50 mm) across. See Fig. 11-24. This

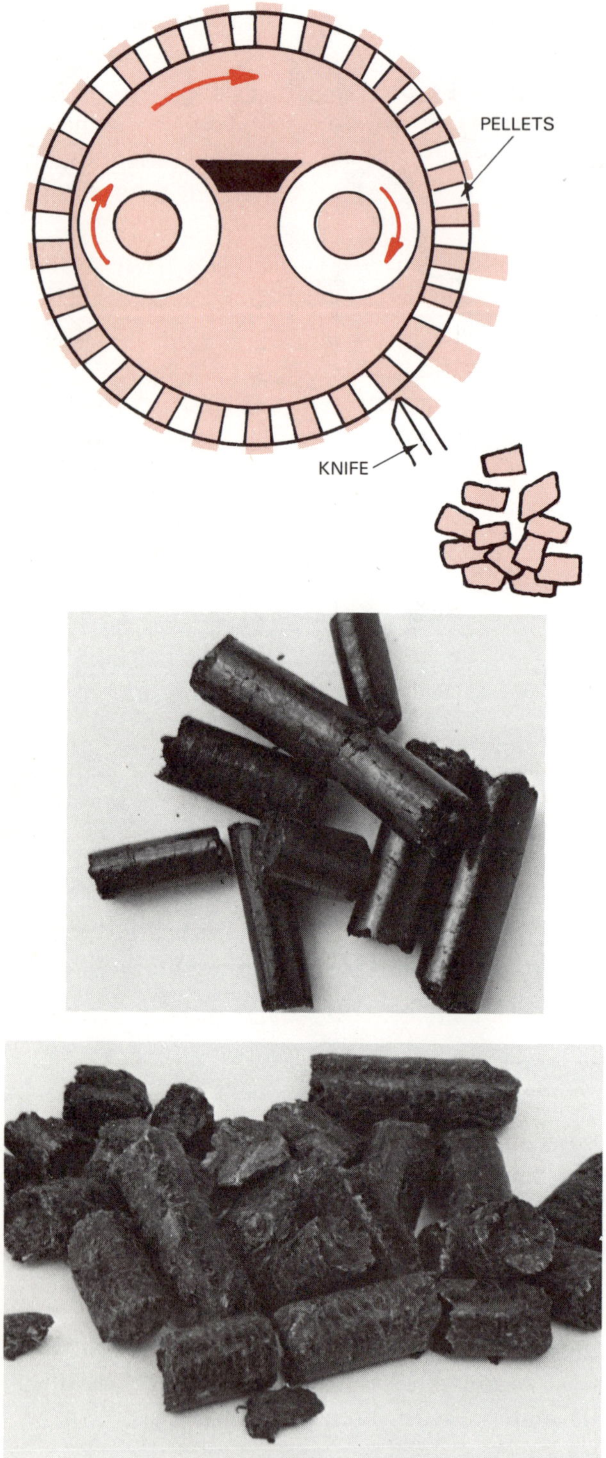

Fig. 11-23. Top. Type of die used in pelletizing biomass. The material is forced through the holes at great pressure. (SERI) Middle. Sample of pelletized wood/peat. Bottom. Pelletized wood particles. (Guaranty Fuels, Inc.)

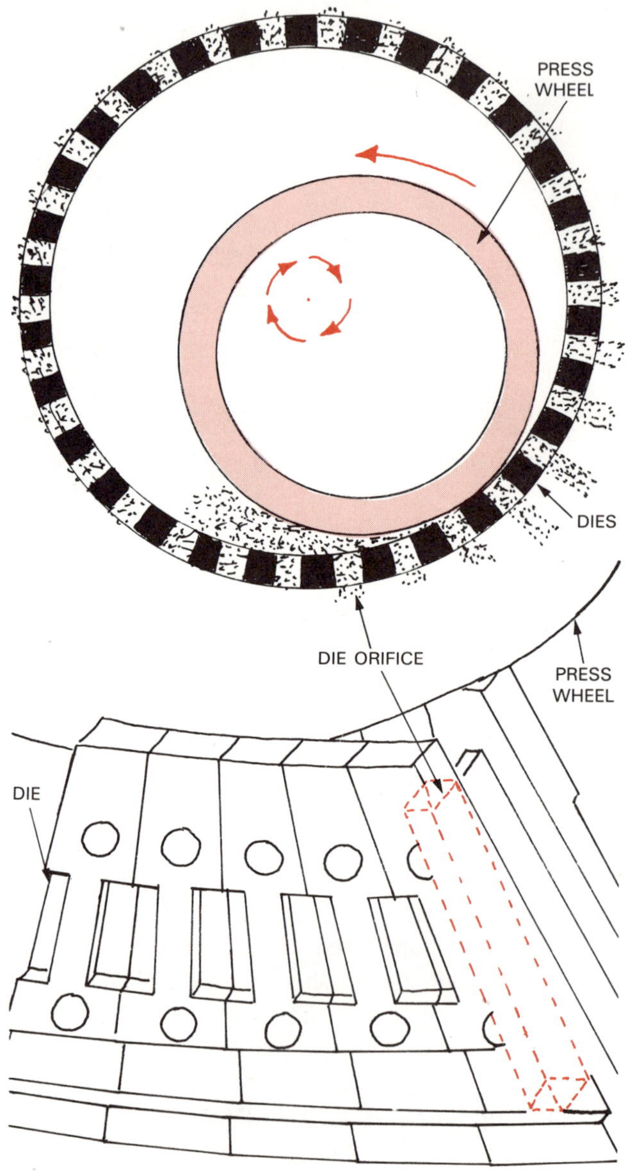

Fig. 11-24. Cubing process. Top. Press wheel and dies. Bottom. Enlarged section of die. Phantom lines show die opening through which biomass is pushed. (Papakube Corp.)

method is currently used with straw and paper. However, it has much wider potential.

3. Briquetting. The biomass is pressed between rollers which have depressions in them. They produce shapes like charcoal briquettes.
4. Extrusion. This method depends upon a screw such as is used in a food chopper to force the biomass through a large die. The particles are 1 to 4 in. (25 to 102 mm) in diameter. Pitch or paraffin is often added to the mix to bind the particles together and provide more heat content. Supermarkets sell them as fireplace logs. See Fig. 11-25.
5. Rolling-Compressing. In this process, fibrous material, such as straw or hay, is fed continuously into four rollers, Fig. 11-26. The rollers, being power driven and under heavy spring tension, squeeze and wrap the biomass into dense cylinders. As the rollers turn and compress, the rolled material is moving out of the rollers. This action produces a continuous cylinder of compacted material 5 to 7 in. (13 to 18 cm) in diameter. It is cut into cylinders 4 to 12 in. (10 to 30.5 cm) long, Fig. 11-27. This process is already used in farming to harvest hay and straw, Fig. 11-28. It could easily be

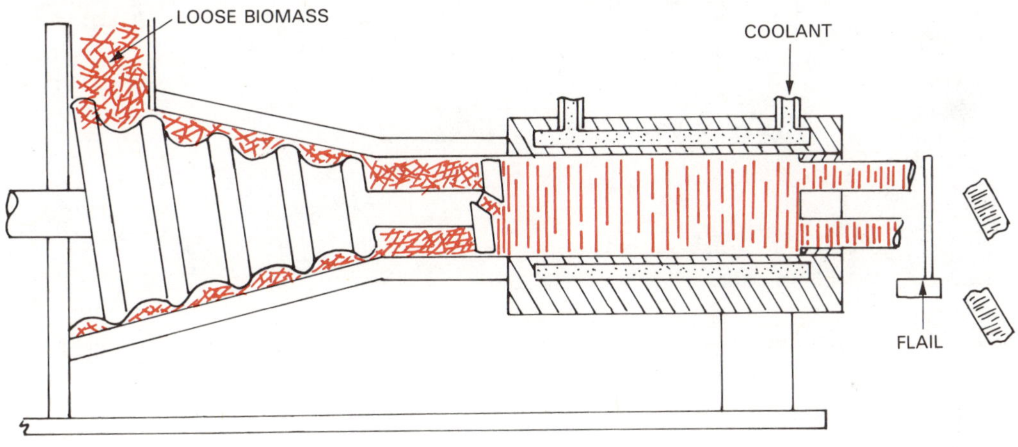

Fig. 11-25. Taiga extrusion process. Screw mechanism compresses biomass into large-diameter cylinders like small logs.

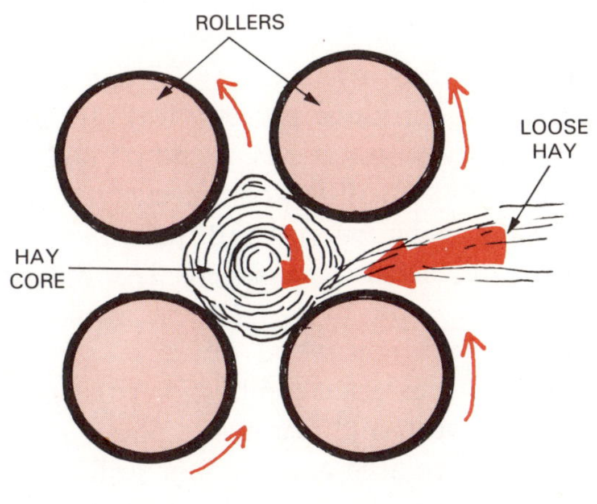

Fig. 11-26. Method of rolling and compressing straw or hay. Material is fed through the rollers and is wrapped in a cylinder. (Agropack)

Fig. 11-27. Hay sample wrapped into a densely packed cylinder.

Fig. 11-28. Compressor roller used to wrap hay and other fibrous crops. (Agropack)

adapted to biomass compaction.

Solid waste densification is an attractive process for city refuse disposal departments. It produces easily stored energy and solves waste disposal problems. The State of Maryland Environmental Service and Baltimore County are operating a pelletizing plant. The pelletized solids are sold to a paper mill in Pennsylvania. The mill mixes the pellets with bark, grinds the mix in a mill, and uses it as boiler fuel.

POWDERED BIOMASS FUEL

In another process presently under development, a dense powder, called Ecofuel II, is produced out of waste. The waste is shredded and milled under hydrochloric or sulfuric acid. The resulting powder is fed into suspension-fired boilers.

USES OF DBF

Densified biomass fuel (DBF) does not produce sulfur pollution like coal. The ash left after burning can be used as a fertilizer. DBF can be used in factories to produce process heat. It can be burned directly in modified gas and oil boilers or it can be fed into a gasifier for conversion to a fuel gas. Other possible uses of DBF are:

1. Fueling of residential, commercial, or industrial furnaces.
2. Fueling of airtight wood stoves.
3. As a fuel for external combustion engines such as the Brayton and Stirling engines.
4. Fuel for fireplaces and outdoor grills.
5. Feedstock for pyrolysis oil and high density charcoal.
6. Feedstock for gasification plants.

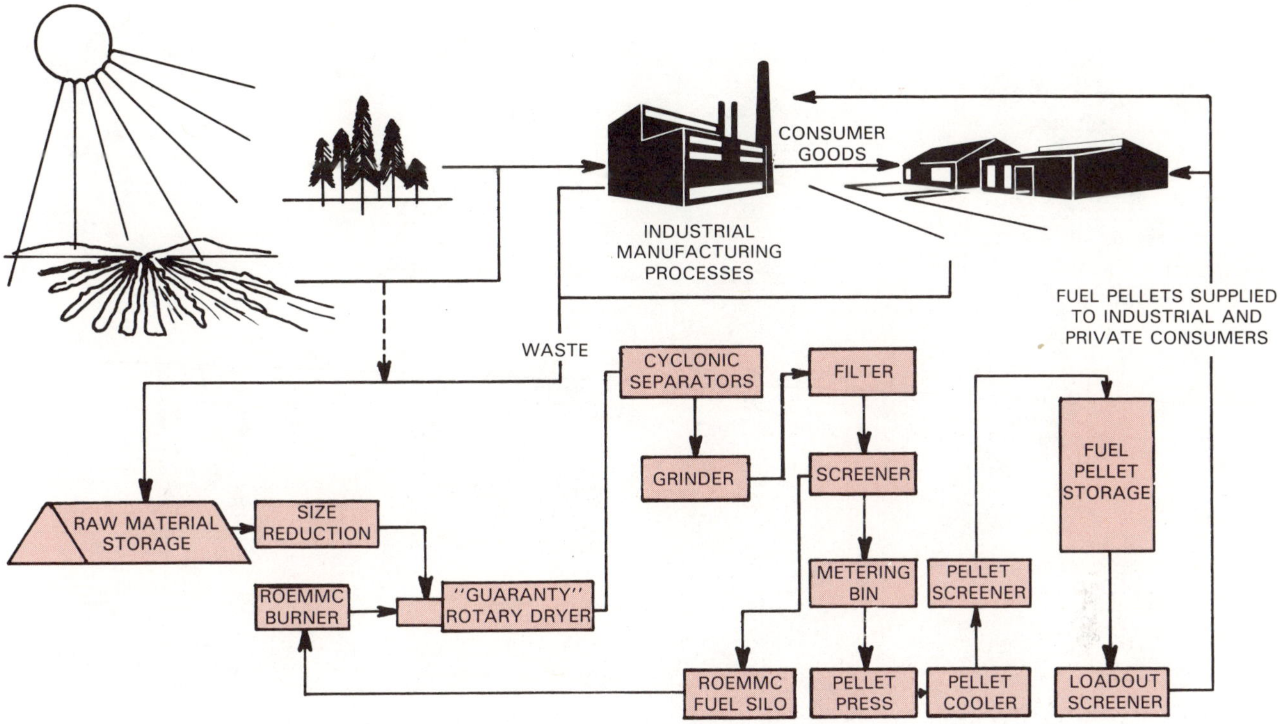

Fig. 11-29. Flow chart of typical biomass cycle from source to final use. (Guaranty Fuels, Inc.)

Fig. 11-29 shows a flow chart of biomass from its source, through processing, and final use. A typical biomass compaction plant is pictured in Fig. 11-30.

BIOMASS SOURCES

Organic matter making up the biomass could come from different sources. Included in these sources are:

1. Wastes and residues from cities and factories.
2. Wastes and residues from forests and farms.
3. Energy farming to grow plants in soil and fresh water.
4. Energy farming in salt water.

Wastes and residues are a valuable resource which, today, are too often thrown away. The forest and food processing industries have tons of waste which could be recycled to produce energy. The same is true of American cities which generate so much garbage and other solid wastes that their disposal is a national problem. See Fig. 11-31 for a partial list of wastes and residues.

TO GROW OR TO SALVAGE

The sources just listed represent two very different approaches to supplying biomass for energy. The last two sources would require a new industry which would grow plant life solely for the purpose of producing energy. The other two would simply redirect the disposal of often unused materials. It would require the setting up of new systems for collecting and converting the wastes and residues.

Either approach has its own set of problems. These will be discussed later.

BIOMASS SUPPLIES

Annually the U.S. generates about a billion tons of wastes from all sources. See Fig. 11-32. Not all of this is useful as fuel. Some of it is not organic

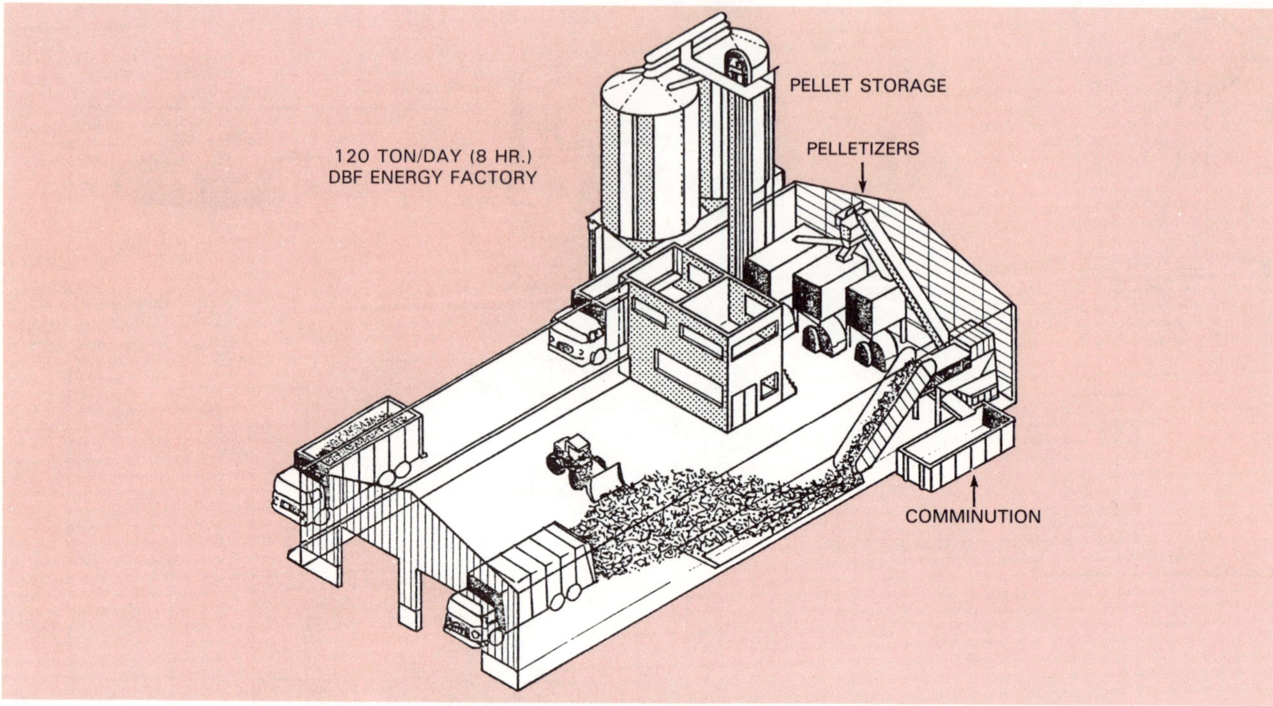

Fig. 11-30. Biomass compaction plant turns biomass into pellets which are easy to transport. (Papakube Corp.)

WASTES	RESIDUES
BY-PRODUCTS OF MANUFACTURING PROCESS	UNUSABLE "SCRAP" FROM HARVEST OF RAW MATERIALS FROM FORESTS AND FARMS
FOREST Sawdust Bark Paper pulp Wood shavings Scrap lumber Wood dust Paper	FOREST Noncommerical timber Diseased trees Slash
AGRICULTURAL & FOOD PROCESSING Corncobs Fruit pits Manures Rice hulls Sugarcane bagasse Nut shells	AGRICULTURAL Corn stalks Oat straw Wheat straw Barley straw Rice straw
MUNICIPAL Sewage Trash Garbage	

Fig. 11-31. Biomass materials such as these are wasted much of the time. Proper methods would turn them into useful energy.

Amounts of Organic Waste (and Residues) by Source

DRY WEIGHT IN MILLIONS OF TONS			
Source	**Readily Collectible**	**1971 Total Amount Generated**	**1980 Total Amount Estimated**
Urban refuse	55.0	100	222
Manure	26.0	200	266
Logging & wood manufacturing	5.0	55	59
Agricultural crops & food wastes	12.3	390	390
Industrial wastes	5.2	44	50
Municipal sewage solids	1.5	12	14
Miscellaneous	5.0	50	60
	110.0	851	1061

Fig. 11-32. Volume of organic waste generated in 1980 as compared to volume in 1971. Note that urban refuse more than doubled in nine years. (U.S. Dept. of Energy)

and will not burn. Much of it is too widely scattered to be worth collecting.

The inorganic materials found in municipal waste include:

1. Minerals. Among these are iron, aluminum, brass, copper, and tin.
2. Glass, stone, concrete, and bricks.

Many of these materials can be separated and recycled into the economy. Others can be used as clean landfill or building aggregate.

Recycling of metals represents energy saving also. For example, only 5 percent as much energy is needed to recycle aluminum as to produce it from raw ores.

Cities of the United States create huge quantities of organic waste in the form of trash and garbage, Fig. 11-33. Until now, little of this material has been recycled. Consequently it has posed a serious problem of disposal. Reports are seen with growing frequency that communities are running out of landfill sites for disposal of garbage and trash. The costs of trash collection and disposal are considerable. In 1964 it cost $26 per ton, but in 1984 is nearer to $50 per ton. These costs will continue to rise.

The energy that can be generated from municipal wastes would be roughly equal to 150 million barrels of oil per year. This would represent $7.2 billion in energy savings if oil were selling at $48 a barrel.

Annual Household Solid Waste and Recycle Energy Value

			ANNUAL HOUSEHOLD ENERGY SAVINGS EQUIVALENT	
Material	NATIONAL TOTALS (million tons)	AVERAGE HOUSEHOLD ACCUMULATIONS (pounds)	RECYCLING (gal. of gasoline)	BURNING (gal. of gasoline)
PAPER				
Newspaper	7.0	233	11.2 or	10.2
Other Separable	5.3	173	8.3 or	7.6
Nonseparable	10.9	370	—	16.3
GLASS	10.8	368	3.7	—
FERROUS METALS	6.3	210	10.1	—
ALUMINUM	.6	20	16.0	—
OTHER NONFERROUS	.2	7	—	—
PLASTICS				
Separable	1.0	33	4.9 or	2.9
Nonseparable	2.3	77	—	3.4
RUBBER & LEATHER	1.0	33	—	1.4
TEXTILES	1.0	33	—	1.4
WOOD	.6	20	—	.9
FOOD	17.8	593	—	26.1
MISC. INORGANIC	1.0	33	—	—
Subtotal	65.8	2203	54.2	70.2
BULKY MATERIALS	6.4	213	—	—
YARD WASTES	18.1	600	—	19.2
Total	90.3	3016	54.2	89.4

Fig. 11-33. Cities create mountains of trash and garbage.

ENERGY FARMS

Scientists have been experimenting with large-scale production of green plants for fuel since the early 1950s. Experiments have been underway to increase the crop yield of certain plants, Fig. 11-34.

An energy farm is operated solely to produce energy. It is one of the alternatives that have been considered to replace dwindling fossil fuels.

Such farms cannot be relied upon at present for several reasons:

1. Biomass farming would require large amounts of energy. This is true of most processes for converting biomass energy to other forms of fuel and chemical feedstocks.
2. Farmland of high quality is needed for biomass farming.
3. Energy needs would compete with rising food supply needs for a growing population.

There are 220 million acres of pasture and rangeland which are capable of growing biomass crops. In addition 160 million acres of forest lands might be used for biocrops. But, again, biomass farming of these lands would make them unavailable for growing food, feed, and fiber products.

AQUACULTURE

Algae, hyacinth, and kelp are three plants being studied for possible use in biomass aquaculture. While harvesting may be difficult, there is no problem of using precious water. In addition, yields are greater in water than on land.

For 20 years or more, algae has been the subject of experiments in fast-growing biomass farming. This single cell organism which is grown in either oceans or fresh water ponds is known all over the world.

A small-scale pilot plant was built more than 20 years ago to see if it was possible to produce the algae as a cheap source of biomass. Special nutrients and carbon dioxide were fed into pools set up on a rooftop to grow the algae. The test plant produced 75 to 80 lb. of dried algae. It proved that it is feasible to "grow" algae for energy. However, the cost of doing so is relatively high. Each pound of dried algae cost about $0.25 compared with the cost of $0.10 per pound for growing food.

In spite of its rapid growth rate, algae does not, at present, appear to be a suitable alternative for future energy needs.

BIOMASS TYPE	YIELD	
	(tons per acre)	(metric tons/km²)
Algae	12 to 30	0.044 to 0.108
Sunflower	10 to 20	0.036 to 0.072
Sugar Cane	12 to 50	0.044 to 0.108
Eucalyptus	8 to 30	0.029 to 0.108
Sorghum	8 to 30	0.029 to 0.108

Fig. 11-34. Some plants, such as those listed, produce large yields and may be grown in the future as biomass crops.

KELP FARMING

Dr. Wheeler J. North, a biological oceanographer at California Institute of Technology is testing kelp growing in the sea. Kelp is fast-growing (up to 2 ft. a day). Harvested kelp would be loaded into barges and allowed to decompose producing methane gas.

It is estimated that the kelp could supply the energy needs of 300 people at current U.S. levels. At the same time it could easily yield enough food for 3000 to 5000 people.

Since the oceans represent an area of more than 80 million square miles, kelp farming could support a population of from 20 to 200 billion people.

ENERGY FROM PHYTOPLANKTON

Phytoplankton is a name given to a huge number of plants and animals that drift near the surface in large bodies of water. Algae is one of them. They occur in both fresh and salt waters. Some are moved through the water by tides and currents. Others, called nekton, move about independently of the tides. An estimated 10 billion tons of plankton are produced in the sea every year making it the largest form of sea life and vegetation.

Dr. William Von Arx, noted marine scientist, speculates about the energy potential of plankton. He says that if the plankton were harvested for power alone, and converted to methane it would produce 10 times the present annual power needs.

COST OF BIOMASS FARMING

At present, producing biomass energy on energy farms would be expensive. The cost of such energy would be several times that of crude oil and coal. Sharp increases in the cost of fossil fuels and nuclear energy would not make biomass energy less costly by comparison since the production of biomass, its transport and processing requires a great deal of energy in itself. Two things would have to happen to make energy farming competitive.

1. The productivity of energy farming would have to increase greatly.
2. The efficiency of the conversion processes would have to be vastly improved.

The better the land on which the biomass is grown the better the yield. It is estimated that to produce 1 percent of U.S. energy needs would require 10 million acres of very good quality farmland. Poorer quality land would require more acres, possibly as much as 30 or 40 million.

ADVANTAGES OF BIOMASS ENERGY FARMING

There are a number of reasons for considering the use of biomass energy. For one thing, its use creates less pollution and health risk than either nuclear power or fossil fuel power. Unlike fossil fuels, biofuels are renewable and duplicate rapidly the whole process which formed the fossil fuels. Lower sulfur and ash content in biomass would reduce the environmental pollution from burning. The sulfur content of coal is 2500 percent higher than in biomass.

Biomass energy stores easily in organic matter. Other energy systems using solar energy require separate and expensive storage facilities.

DRAWBACKS OF BIOMASS ENERGY

Biomass, however, does have several disadvantages which make it less feasible for energy:

1. It is inefficient as a converter of solar energy. Photosynthesis generally converts less than 1 percent of the incoming solar energy to chemical energy. Even under the most favorable conditions efficiency does not exceed 4 percent.
2. Biomass fuels are spread out over a wide area and are expensive to collect for processing.
3. Being perishable and bulky, the raw materials present problems of transport and storage until they can be used.
4. The energy content of a ton of biomass is much lower than equal weight of fossil fuels or nuclear fuel. Dried biomass has less than half the heat content of oil, for example.

DO YOU KNOW THESE TERMS?

Algae, anaerobic, anaerobic digestion, biogas, biological conversion, biology, biomass, briquetting, char, cubing, densification, digester, enzyme, ethanol, extrusion, fermentation, gasification, hydrogenation, LEG, liquefaction, metabolic conversion, methanol, pelletizing, photosynthesis, pyrolysis, residues, synthesis gas.

SUGGESTED ACTIVITIES

1. Take part in a debate with other members of your class on one of the following topics:
   a. Should surplus grain be used for production of alternate fuels or should it be used to feed the hungry people of the world?
   b. Should large cities and large companies be required to separate the organic wastes they

generate and recycle them as energy?
   c. Should surplus crop acreages be turned to production of energy crops that can be converted to alcohol or methane gas?
2. Keep a record for a week of the organic wastes thrown out by one household. Report on:
   a. Your calculation of how much recyclable waste is disposed of in a year.
   b. Figure the amount of energy in this waste in terms of gallons of gasoline. Then, using the current cost of gasoline, determine the dollar value of the energy in the waste. Use the chart in Fig. 10-33 to make your calculations.
3. Construct a small methane digester and demonstrate it.

TEST YOUR KNOWLEDGE

1. What is biomass?
2. The process by which plants capture the energy of the sun and store it is called ____________.
3. List the three methods by which biomass energy is converted to other forms of energy.
4. List at least three of the biomass fuels.
5. Burning biomass to produce heat is called:
   a. Direct combustion.
   b. Pyrolysis.
   c. Liquefaction.
   d. Gasification.
6. A ton of dry waste will produce how much energy?
   a. Equivalent of one barrel of oil.
   b. Equivalent of 25 barrels of oil.
   c. Equivalent of 125 barrels of oil.
7. Anaerobic digestion is one of the forms of ____________ conversion.
8. Alcohol can be made from fruit. True or False?
9. List the two types of methane digester.
10. List three sources of biomass.

Firewood is the oldest form of biomass fuel.

12 THERMAL ENERGY

The information given in this chapter will enable you to:

- Describe the types of thermal energy which are found naturally on earth.
- Describe methods for tapping and using heat stored in the earth's crust.
- Discuss the amount of heat energy available in the earth.
- List and describe methods of tapping and using ocean thermal energy and geothermal energy.

Heat is stored naturally in the earth. One source of this heat, solar radiation, is collected and held in the top few feet of the earth's soil and in its oceans, Fig. 12-1. Another heat source is trapped deep within the earth. The sources are known as:

1. Ground thermal energy.
2. Ocean thermal energy.
3. Geothermal energy.

GROUND THERMAL ENERGY

Both ground thermal and geothermal energy come from under ground. They could both be con-

Fig. 12-1. The earth's surface—soil and water—absorbs and stores heat from solar energy.

sidered as part of the same source. However, the way in which the energy was created is different and the way in which it is captured and used is also much different.

Ground thermal energy is partly the result of solar radiation storage and partly the result of the heat locked in the earth when it was first formed. Some of the sun's rays are captured by the rock and soil and its temperature rises.

EARTH AS SHOCK ABSORBER

Unlike air, which loses its heat energy rapidly into the upper atmosphere, the earth gives off its heat energy much more slowly. At the same time, the ground warms up at a slower rate than does the air.

The large bulk of the earth acts as a temperature shock absorber. In one study, temperature was taken 9 ft. below the surface and at the surface over a period of a year. Purpose of the study was to compare temperature differences.

The coldest ground level temperature was recorded in February. The coldest temperature 9 ft. down, however, came in late May. The hottest ground level temperature was recorded in August. Below ground temperature reached its highest reading in November.

The temperature of the ground and ground water, except very near the surface, varies in most of the United States from 40°F (4°C) to 70°F (21°C). This is considerably higher than the winter air temperature in northern states.

Until recently, humankind made little use of the earth's heat. True, we did put down foundations for buildings deeply enough to get below the frost line. (This is the depth at which the ground does not get cold enough to freeze in cold weather.) However, use of this heat energy was not common until we began to construct earth homes and machines that could collect the heat energy and transfer it into dwellings.

EARTH PROTECTED HOMES

An earth protected home is one which uses the natural heat of the earth as protection against both excessive cold and excessive heat.

An earth home may be entirely in the ground, partially in the ground, or sheltered by earth. See Fig. 12-2. Because of the "shock absorbing" effect described earlier, the temperature of earth homes, even without artificial heating or cooling, is more even the year round.

HEAT PUMPS

Heat pumps, Fig. 12-3, are now often teamed up with sources of thermal energy to provide comfortable temperatures in living and working space. The working principles are explained in Chapter 13, Conservation of Energy.

Since the heat pump is designed to extract low-level heat from its surroundings, it is the perfect mechanism for drawing the heat from the ground. This is accomplished by placing the outdoor coil of the heat pump in the ground, down a well, or deep in a pond or lake. In cold climates these may be the only practical ways of using the heat pump. Even in the coldest climates, the temperature at 4 to 6 ft. down does not vary much from one season to the next. Fig. 12-4 is a schematic for a heat pump system with its outdoor coil located down a well.

OCEAN THERMAL ENERGY

Ocean thermal energy conversion (OTEC) uses ocean surfaces in and near the equator as a collector of heat energy. It converts the heat to electricity with a Rankine cycle heat engine that floats in the ocean.

The heat is used to drive a turbine which is connected to a generator. Whereas a fossil-fueled power plant uses steam to drive its turbines, OTEC turbines are driven by a liquid with a much lower boiling point than water.

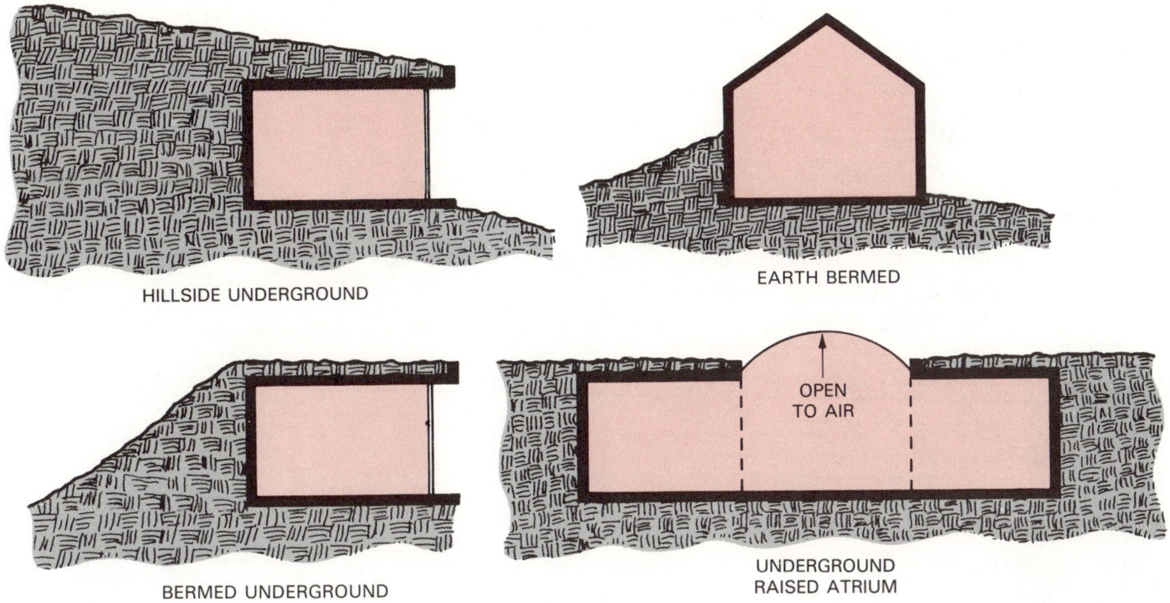

Fig. 12-2. Only a few feet down, earth temperatures fluctuate (move up and down) much less than the air. That is why earth protected homes have a more even temperature the year-round. Top. Types of earth homes and their names. Bottom. This home located in Mississippi is built into the side of a hill and has 18 in. of soil over the roof. The fireplace supplies the only additional heat needed during winter months.

HOW OTEC WORKS

The operation of the OTEC system can be explained simply. Warm water from the ocean surface is drawn into the power plant. As it passes through an evaporator it loses a quantity of its heat to a liquid such as ammonia, propane, freon, or some other type of refrigerant. The low-level heat causes the fluid to boil. This turns the fluid to a gas. As it becomes a gas it expands. This creates

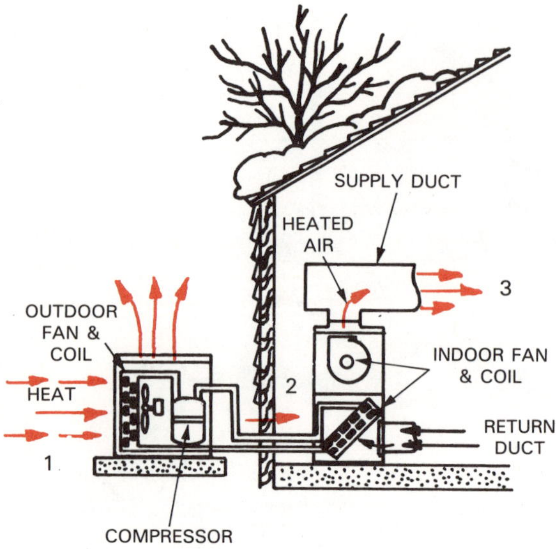

1—HEAT IS EXTRACTED FROM AIR BY OUTDOOR COIL.

2—REFRIGERANT GAS CARRIES HEAT TO INDOOR UNIT.

3—CIRCULATING INDOOR AIR PICKS UP HEAT AND CARRIES IT THROUGHOUT HOME.

Fig. 12-3. Heat pump extracts heat from one point and delivers it to another. In winter mode here, it pulls the heat from the outside air and brings it into the house. (Commonwealth Edison)

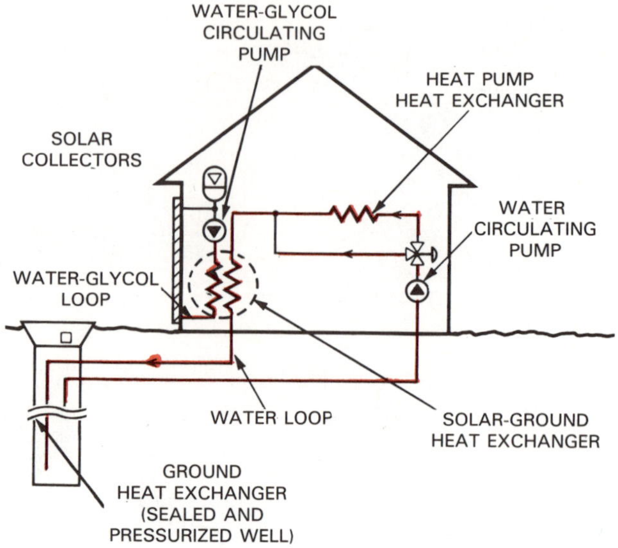

Fig. 12-4. Schematic of a heat pump which draws its heat from a well during winter months. In summer, heat is drawn from the house and delivered to the well.

pressure great enough to drive the turbine. After the gas has given up its energy to the turbine its pressure drops. Next, it passes into a condenser. Cold water from deep in the ocean passes around

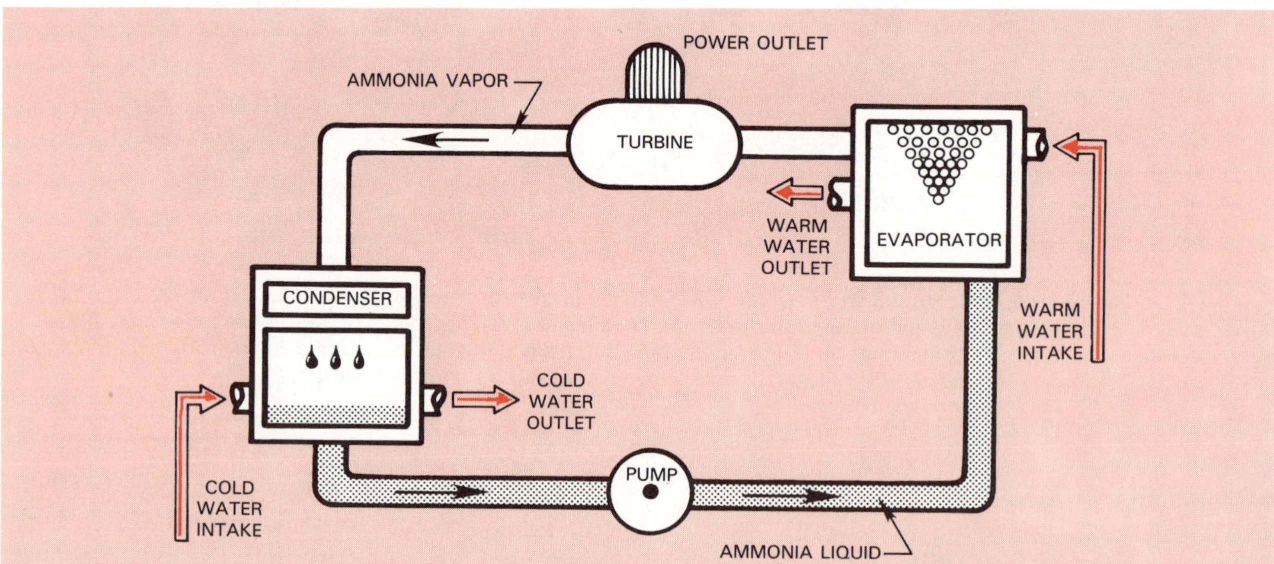

Fig. 12-5. Simplified drawing of how electricity is generated using the warmth of the ocean. Seawater heats up the ammonia. It turns to a gas and expands, driving a turbine. The turbine is connected to a generator. (SERI)

the condenser. The gas loses some of its remaining heat to the cold water and the water is returned to the ocean, Fig. 12-5.

The spent gas cools down and once more becomes a liquid. It is ready to repeat the cycle. Fig. 12-6 is a cutaway of a typical OTEC power station.

Generated electric power could be carried to shore through power cables and fed into an electrical power system. Excess electric power could be used to convert seawater to hydrogen. The hydrogen could be piped ashore or to a ship and stored.

Another use for the electric power generated is for the production of ammonia using hydrogen from the water and nitrogen from the air. At present, ammonia is obtained from natural gas which may soon be in short supply. Also, the ammonia can easily be changed back into hydrogen and

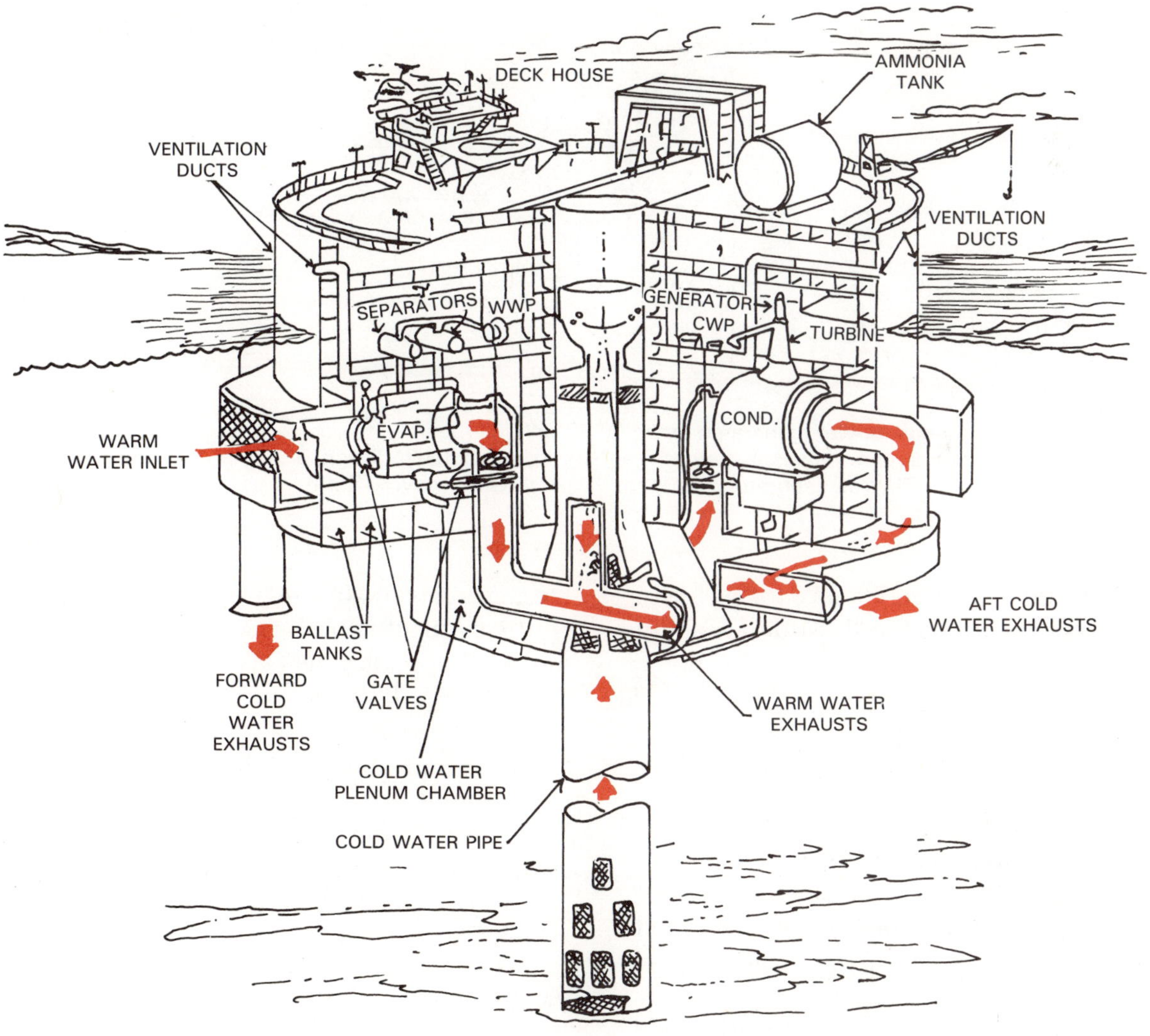

Fig. 12-6. Cutaway of an ocean thermal energy conversion plant. It floats on top of the ocean. (Lockheed Missiles and Space Co., Inc.)

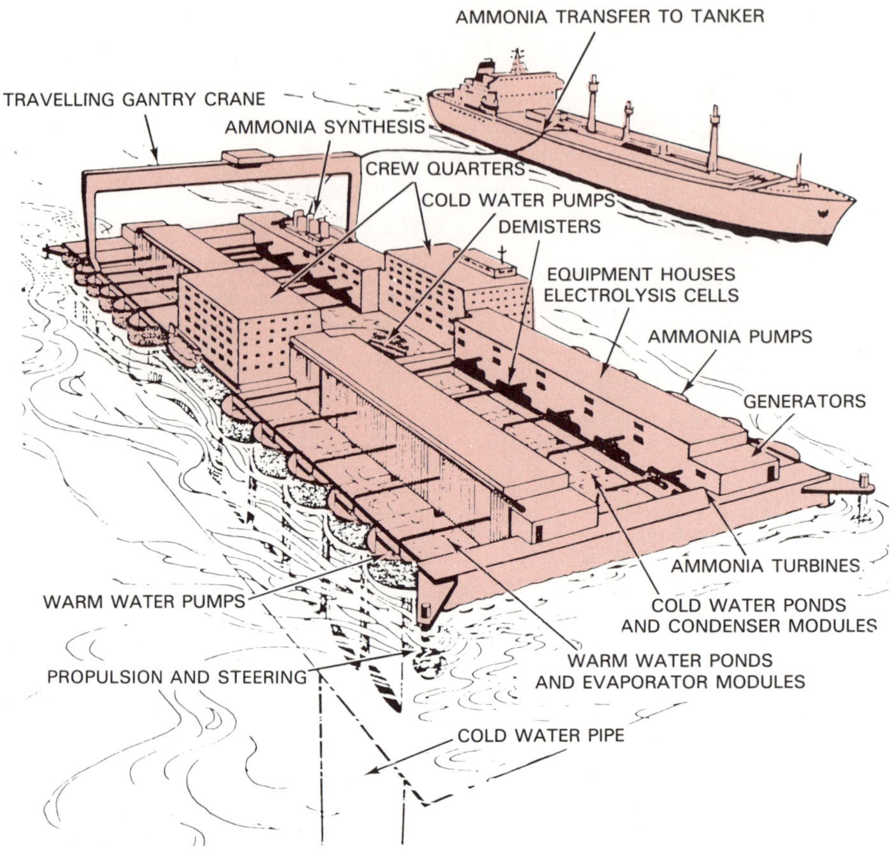

Fig. 12-7. Electric power, generated by an OTEC unit can be used to convert seawater to ammonia or hydrogen. This is a sketch of an ammonia plantship designed by the Applied Physics Laboratory at Johns Hopkins University. (SERI)

nitrogen and used as an energy source for fuel cells. See Fig. 12-7. (The fuel cell is discussed in Chapter 4.)

ENERGY SOURCE FOR OTEC

As we know, the sun bathes the earth each year with 18,000 times as much energy as humans can consume. Seventy percent of the earth's surface being water, a great deal of the energy is absorbed and stored in upper layers of the oceans.

A huge band of water 10 degrees on either side of the equator acts as a massive solar storage tank. Temperature of the surface water is a constant 80°F (27°C). Meanwhile, 3000 ft. below the surface, currents of 40°F (4°C) water flows from the North and South poles. The difference in temperature is what makes the OTEC power plant possible. See Fig. 12-8.

A great advantage of the OTEC plant is this ready source of stored heat.

1. It costs nothing.
2. It does not need to be transported.
3. It is constantly being replenished, and is non-polluting, generally speaking.

TESTING THE OTEC

No large OTEC plant has yet been built. However, the State of Hawaii and two private companies have entered a partnership agreement to

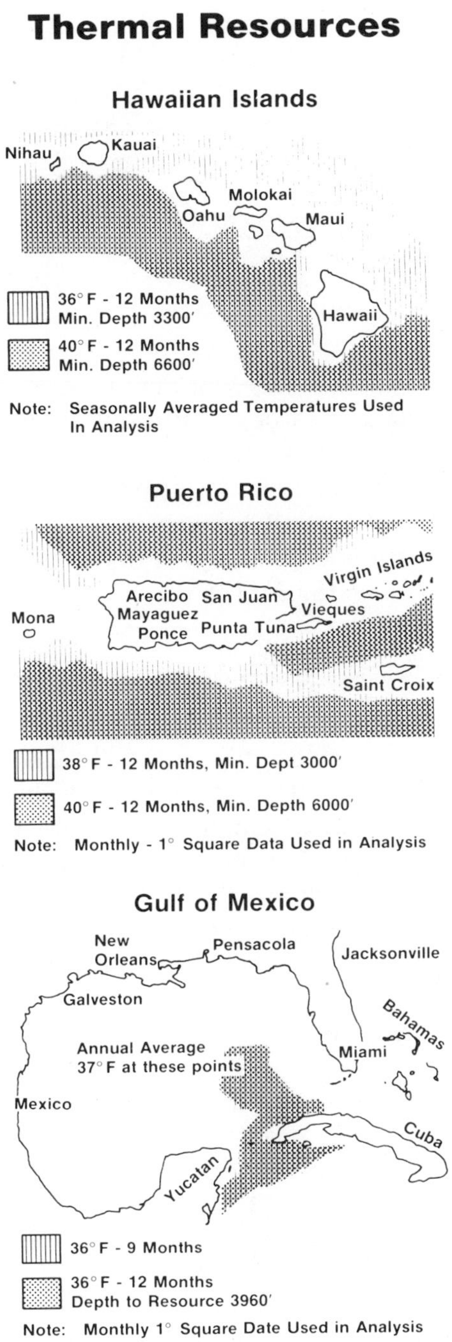

Fig. 12-8. Prime areas are shown for setting up OTEC power plants.

develop the first Rankine closed-cycle, self-sustaining OTEC system. It will be a miniature plant capable of generating 50,000 W of electric power, using ammonia as a working fluid.

The federal government is working on programs to develop and test OTEC components and subsystems. It is also planning storage and delivery systems. The first operating plant will generate about 25 MW of power.

At most sites, the cold water would be brought up from depths of 2000 to 3000 ft. (610 to 914 m) with a pipe diameter of 100 ft. (20 m). This size would serve a system with a generating capacity of 250 MW. The flow through such a pipe would equal the flow of the Missouri River past Omaha, Nebraska!

OTEC'S PROBLEMS

An OTEC power plant would have to be quite large to house both machinery and an operating crew. Dimensions would vary according to advances in design, but such a plant might be close to 400 ft. or about 122 m in diameter. Only living quarters and space for small equipment would be above the surface. The main structure, including buoyancy tanks would be underwater. Parts of the plant would extend downward 500 to 600 ft. (150 to 180 m).

Problems of such a structure might include:

1. Corrosion due to salt water.
2. Damage or instability due to stormy seas.
3. Clogging of inlets, ducts, and heat exchangers with algae and marine life.
4. Possible damage to environment.

Special materials would have to be developed to deal with salt water corrosion and special attention given to mooring systems that would secure the plant against turbulent seas. Discharge of cooling waters could endanger sea life accustomed to colder temperatures.

Some of the problems could be solved by a land-based OTEC plant. Several companies have received grants to develop 40 MW facilities in Hawaii. The facilities will either be based ashore or on an artificial island.

GEOTHERMAL ENERGY

Geothermal energy is heat energy which comes from deep within the earth. The earth is made up of a crust which varies in thickness. Under the land masses it is 20 to 25 miles (32 to 40 km) thick. Under the oceans it is only six to eight miles thick. Beneath this crust is a second layer called the MANTLE. This layer is probably solid rock and extends about 1800 miles (about 2900 km) downward. The temperature in the mantle is high enough for the rock to be molten if it were not for the tremendous pressure. (Molten rock is known as MAGMA.)

Less is known about the earth's core. It is thought that the outer shell is molten while the interior is solid. As in the mantle, the heat is tremendous.

ORIGIN OF EARTH'S HEAT

The origin of the earth's heat is not certain. According to one theory, the earth's core and most of its mantle contains much of the earth's original heat. It became trapped there after the cooling and solidification (hardening) of the crust and upper mantle. The heat of the crust and upper mantle could be the result of slow decay of radioactive elements. Some of these elements are plentiful in the crust. See Fig. 12-9.

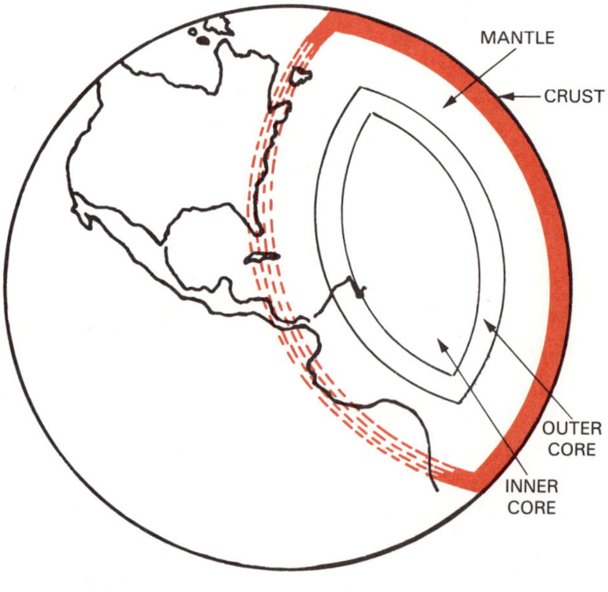

Fig. 12-9. Cutaway of earth showing source of geothermal energy. Temperature at core is estimated to be anywhere from 3632 °F to 18,032 °F.

TEMPERATURE LEVELS

From readings taken in mines and wells, it is known that the earth's temperature near the surface increases slightly less than 2°F for every 100 ft. At 400 miles down the temperature could be about 4532°F (2500°C). Temperatures at the center of the earth could be anywhere between 3632°F (2000°C) and 18,032°F (10,000°C).

GEOTHERMAL SOURCES

Reserves include all the geothermal energy known to exist and which is economically recoverable with present technology. Resources are the total amount of geothermal energy including estimates for deposits which have not yet been discovered. Ultimately the resources of geothermal energy may be 10 to 200 times the present reserves.

The actual amount of energy available and reachable from geothermal reservoirs under the earth is not certain. It has been said that if all the energy were to come from geothermal sources it would take 41 billion years to drop the earth's interior temperature 1°C.

One estimate places U.S. resources at 6×10^{24} calories. However, not all of this can be recovered. The heat must be concentrated in reservoirs where it can be stored over a long period of time.

Current estimates of reserves are 60,000 Mw of electrical power. This means that all of the recoverable energy from geothermal sources would generate 60,000 times 1 million watts of electric power annually. This is enough electricity to meet the needs of 60 million homes.

LOCATION OF U.S. RESOURCES

Most of the known U.S. geothermal reserves are in the western part of the country. See Fig. 12-10. California, Nevada, and Oregon have the most. Alaska, Idaho, Montana, New Mexico, Utah, and Washington also have large amounts. Probably, the greatest reserves will always be in the western regions.

The Electric Power Research Institute estimates that up to 6 million kilowatts of geothermal energy could be operating in the United States by the year 2000.

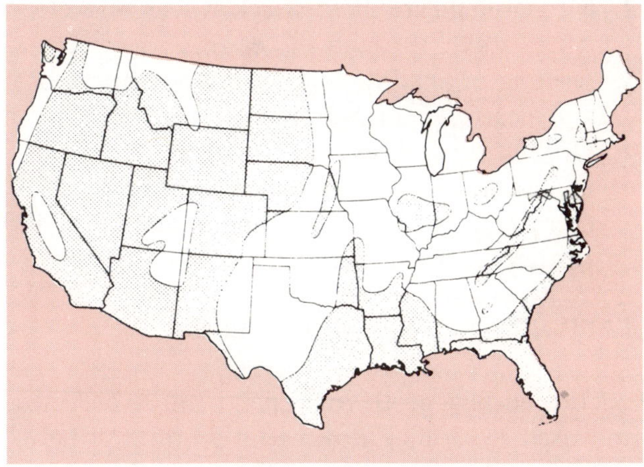

Fig. 12-10. Map of lower 48 states showing geothermal resources. Shaded areas are likely geothermal spots. In other areas, the source is too deep for drilling of wells to be practical. (U.S. Dept. of Energy)

FINDING GEOTHERMAL ENERGY

Often, geothermal sites are identified by the steam rising through cracks or from hot springs below ground. Without these signs the presence of geothermal energy is hard to detect. About 1000 hot springs have been mapped in the lower 48 states.

Many different techniques were borrowed from the oil and gas industry for finding "hot spots." These include:

1. Studying geologic maps and satellite photos which will show evidence of fault lines in the earth. Localities given to earthquakes and volcanoes are likely spots.

Fig. 12-11. Likely geothermal sources around the world. They are likely to be found where the earth's crust has shifted. Thus, volcanic activity is usually connected with geothermal activity. (Union Oil Co.)

2. Measurement of how shock waves and electricity travel through the earth.
3. Drilling of shallow holes to take temperature and heat flow measurements.
4. Drilling of deep exploratory wells in promising areas. These will indicate if the resource is there and if it contains enough heat to make further work worthwhile.

TAPPING GEOTHERMAL ENERGY

Theoretically, geothermal energy can be collected and used by drilling down to it from anywhere. However, in most parts of the earth this is not possible or practical. The hot interior mass lies too deep to reach. Even if we had the drilling technology, it would be too costly.

Fortunately, the magma sometimes escapes from its underground prison and works its way closer to the surface. This has occurred in a few spots around the world, Fig. 12-11. Geothermal sites are found in areas where the earth's crust has shifted. Volcanoes are a more violent form of geothermal energy working its way to the earth's surface.

TYPES OF GEOTHERMAL HEAT

Geothermal heat is found in three natural conditions:

1. Steam.
2. Hot water.
3. Dry rock.

As magma works its way through cracks in the earth, it heats up layers of rock above it. If the rocks are water soaked, hot water or steam is created. If the steam is close enough to the surface, hot spring, geysers, or fumaroles (wisps of steam) may occur.

GEOTHERMAL WELLS

Wells are drilled into geothermal fields. The steam or hot water is piped to its point of use, Fig. 12-12.

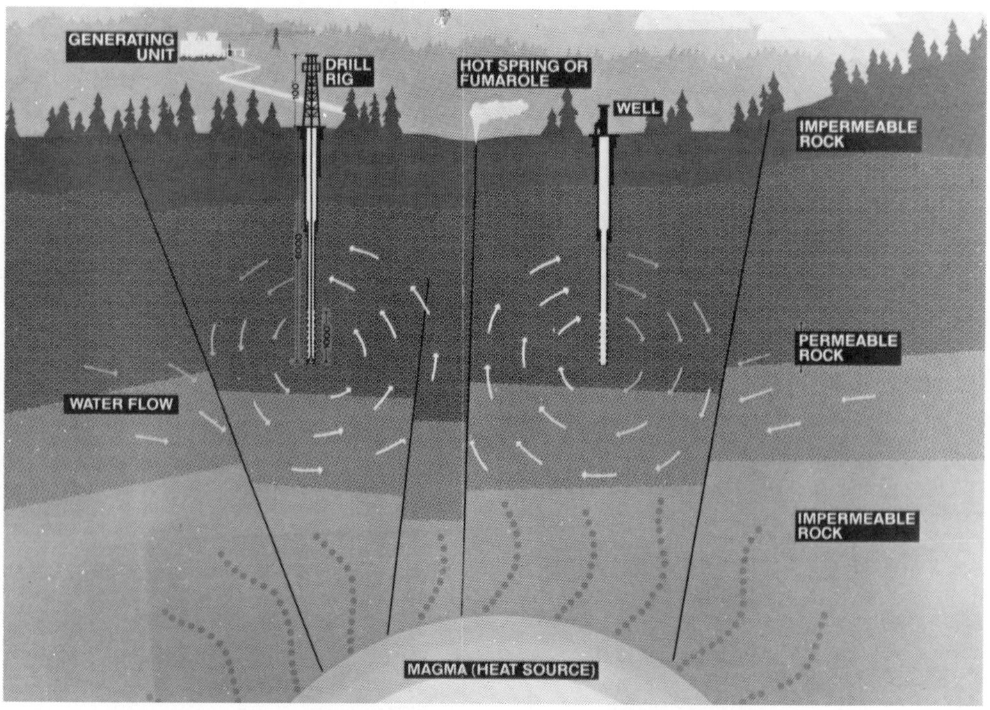

Fig. 12-12. Cutaway shows well drilling in a geothermal field.

PRODUCING ELECTRICITY

Steam must be "cleaned" first to remove impurities and dissolved minerals. Then it is piped to a power station where it turns the turbine blades of a generator.

Hot water cannot be used directly to power the turbine. Two methods are employed to use it:

1. Flash method. When the hot water is brought to the surface its pressure is reduced in special equipment. Some of the water "flashes" into steam. The remaining hot water is returned to the reservoir from which it came through an injection well.
2. Vapor-turbine method. The hot water is passed through a heat exchanger. Its heat is used to vaporize isobutane (a hydrocarbon liquid that vaporizes at low temperature). This is very similar to the exchangers used in ocean thermal conversion systems. The vaporized isobutane is under great pressure as it is piped away to drive the turbine.

Having given up its usable heat, the water is returned to the underground reservoir. See Fig. 12-13 and Fig. 12-14.

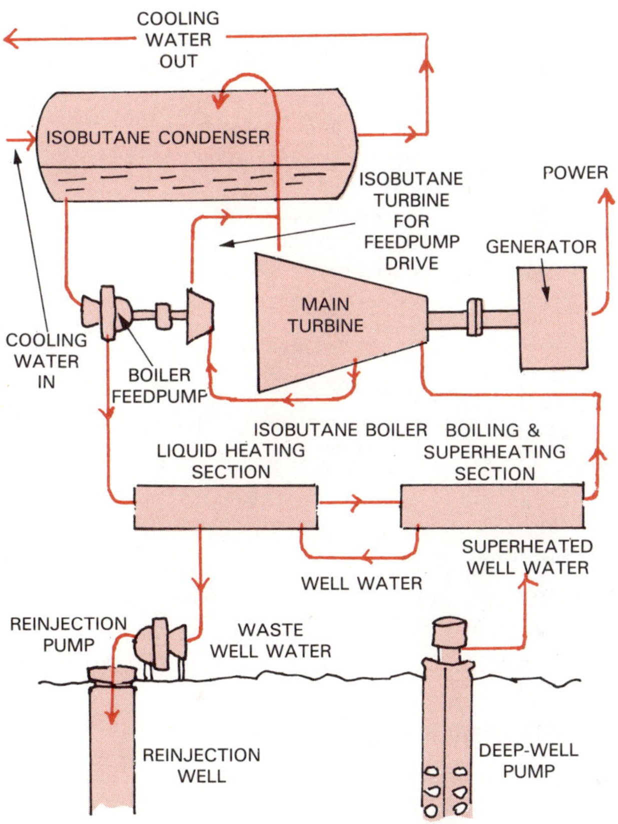

Fig. 12-13. Schematic of a type of heat engine that may be used to tap geothermal energy. It is called a vapor-turbine cycle because the refrigerant material, isobutane, becomes a gas at low temperature. It expands creating great force in the closed system. The gas is allowed to escape into the turbine where it creates mechanical energy to operate a generator.

Fig. 12-14. A geothermal power station. Steam is piped from the fumarole (steam vent) in the background to the power station in the foreground.

SPACE HEATING

It is often very efficient and economical to use geothermal energy as a direct source of heat. Some localities have long been using it to heat homes. Examples are Klamath Falls, Oregon and Boise, Idaho. For 80 years, Boise has been using geothermal hot water to heat houses. The system is now being expanded to include city and state public buildings. Klamath Falls is installing a similar system.

The city of Rykjavick, Iceland, heats over 70 percent of the community's homes with a geothermal system. An additional 20 percent are being added.

OTHER USES OF GEOTHERMAL ENERGY

Besides generation of electricity and heating of homes there are other uses for geothermal heat:

1. Agriculture and horticulture.
   a. Heating of greenhouses.
   b. Hay drying.
   c. Animal husbandry.
   d. Fish farming.
   e. Hatcheries.
2. Process heat.
   a. Paper mills.
   b. Drying and seasoning of lumber.
   c. Salt recovery from seawater.
   d. Brewing and distillation.

Fig. 12-15 is a chart of various uses and the temperatures required.

Temperature Needed for Various Uses of Geothermal

CELSIUS DEGREES	USE
180	Evaporation of concentrated solutions Refrigeration by ammonia absorption Digestion in pulp manufacturing
170	Heavy water by Hydrogen Sulfide Process Drying diatomaceous earth
160	Drying of fish meal Drying of timber
140	Drying farm products Canning of food
130	Evaporation in sugar refining Extraction of salt by evaporation
120	Distilling water Refrigeration by medium temperatures
110	Drying & curing light aggregate cement slabs
100	Drying organic materials (seaweed & grass) and vegetables Washing and drying wood
90	Drying of stock fish
80	Space heating Greenhouse space heat
70	Low temperature refrigeration
60	Heating buildings, housing animals Heating greenhouses
50	Mushroom growing Therapeutic baths
30	Swimming pools, biodegradation, fermentation
20	Fish farming

Fig. 12-15. Geothermal energy has various uses.

ARTIFICIAL GEOTHERMAL SYSTEMS

Hot dry rock contains vast quantities of geothermal heat that has not been used up to now. It has the potential of being the largest source of geothermal energy. One method, hydrofracturing, borrowed from the petroleum industry, shows promise for capturing great quantities of heat.

In hydrofracturing, water, under high pressure, is forced down a well into the dense rock, over an area of high temperature. The water fractures the dense rock. More pressurized water is forced through the cracks to extract the heat. See Fig. 12-16.

Another method still in the experimental stage is the creating of large artificial cracks by setting off underground explosions. Little is known about the possible effects of the blast waves on surface structures or the costs of explosive fracturing.

ENVIRONMENTAL EFFECTS

Geothermal energy has many advantages from the standpoint of enviroment. For one thing, the production of the energy takes place right where it is used. There are no mining, processing, or storage activities involved. Moreover, it produces no solid wastes or atmospheric pollution of a significant nature.

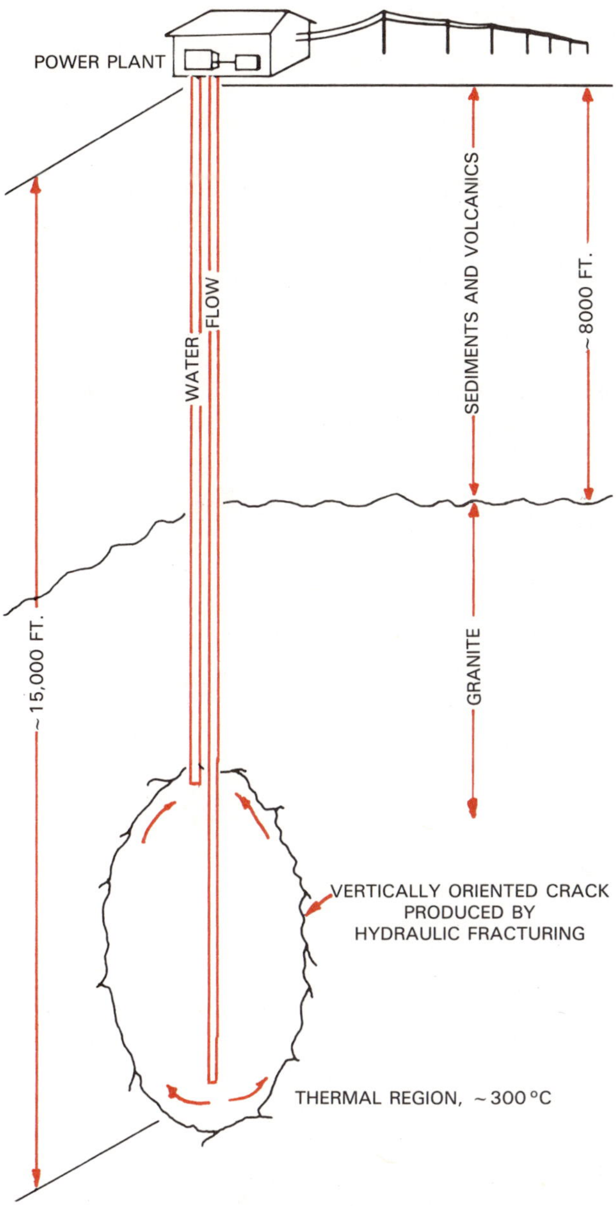

Fig. 12-16. Some geothermal reservoirs are located in dry rock. It is possible to fracture the rock and pump water into it from the surface to collect and "harvest" the heat energy. (Union Oil Co.)

This is not to say that there are no environmental problems, however. Within the immediate area of the power plant gaseous discharge such as hydrogen sulfides can be disagreeable. Other problems are subsidence (sinking) of soil in the well area, noise, and disposal of waste water which can contain minerals that affect plant life in the immediate area.

Even so, the problems caused by geothermal are not as serious as those connected with power plants using fossil or nuclear fuel. Most of the problems can be controlled or eliminated.

FUTURE OF GEOTHERMAL ENERGY

Geothermal energy will never contribute great amounts of energy to satisfy our need of electric power. Somewhat less than 1 percent of the total electrical power generated by the year 2000 would be all that can be expected. Its largest contribution will be in the South and Southwest for heating of homes and factories.

DO YOU KNOW THESE TERMS?

Compressor, condenser, evaporator, flash method, fuel cell, geothermal, heat pump, hydrofracturing, magma, mantle, ocean thermal energy, OTEC, vapor-turbine method.

SUGGESTED ACTIVITIES

1. Using two thermometers, compare the temperature a few inches under the ground with air temperature on any given day. Take the temperature readings at different times of the day. For example, first reading, 7 a.m.; second reading, 12 noon; third reading, one-half hour after sundown. Suggestion: use a darkroom or pool thermometer to take the ground temperature. Dig a hole with a spade, insert the thermometer and carefully replace the dirt around the thermometer. Discuss your temperature findings with your class.
2. Construct a cutaway of a geothermal well. Refer to Fig. 12-12 and Fig. 12-16.
3. Research construction of earth sheltered homes and build a scale model of one. Arrange with your instructor to display it at school.

TEST YOUR KNOWLEDGE

1. ____________ energy is the result of heat locked in the earth when it was first formed.
2. Why is the earth referred to as a "shock absorber?"
3. A ____________ ____________ is a device that will draw the heat from the earth and use it to heat buildings.
4. What does OTEC stand for?
5. A big advantage of the OTEC system is that the fuel costs nothing. True or False?
6. Geothermal heat comes from (pick best answer):
   a. Heat of the sun stored in the earth.
   b. Tremendous pressure of rocks in the earth's mantle.
   c. Molten matter in the mantle and core of the earth.
7. At a depth of 400 miles the temperature of the earth could be:
   a. About 2500°F.
   b. About 3600°F.
   c. About 4500°F.
   d. 10,000°F.
8. Since geothermal hot water cannot be used directly to power turbines, two methods must be employed to use it. Name them.

13 CONSERVATION OF ENERGY

The information given in this chapter will enable you to:

- Define energy conservation and understand why it is important as a part of a national energy program.
- List the various sectors of society which use energy and indicate how energy might be saved in each.
- Suggest the importance of each individual as an active conserver of energy.

Conservation of energy has two meanings:

1. In the study of physics, it means that the amount of energy in a system remains unchanged even though there are changes within the energy system.
2. In our everyday use of energy, it means to save any energy source by using it efficiently and frugally (not wasting it).

In Chapter 1 you learned that you cannot destroy matter. The total amount of matter in the world always remains the same. However, as it changes from one form to another it eventually enters a state where is it not usable by our technology. For example, when a gallon of gasoline

Fig. 13-1. Only about 28 percent of the energy in gasoline is used to move an automobile. The rest is absorbed by the atmosphere as low grade heat which can do no work.

is burned in an engine, only about 28 percent of the heat energy is changed to mechanical energy that drives a load. The rest is converted to random heat energy, Fig. 13-1. It is no longer useful as an energy source. A small amount can be used to heat the passenger compartment of the automobile, but most is lost to the atmosphere.

WHY CONSERVE ENERGY?

Not until the early 1970s did our society come to realize and appreciate that fossil fuel energy would soon be in short supply. Before energy costs began to rise, it was not always economical to preserve energy sources. For example, it was cheaper to buy fuel to heat homes than it was to install heavier insulation, Fig. 13-2.

Today, the situation has changed. We must find ways to save energy. We know that eventually there will be less and less fossil fuels. At the same time their cost will become higher and higher. The purpose of conservation, then, becomes twofold:

1. To reduce the cost of energy to individuals and to industry.
2. To save precious supplies of fossil fuels so they will be available to future generations. A further reason for saving lies in the fact that oil and gas are desirable feedstocks (raw materials) for medicines and health products. When they are gone, substitutes may be difficult to find.

A third purpose for conservation could be added: To reduce a growing dependence upon imported oil. The oil embargo of 1973 demonstrated the unreliability of imported energy in times of war or global unrest.

If the foregoing reasons are not enough to move us to conserve energy resources, then we should consider one more fact. We may well be visited by severe energy shortages in our own lifetime if we continue to use up energy at the rate we have in the past.

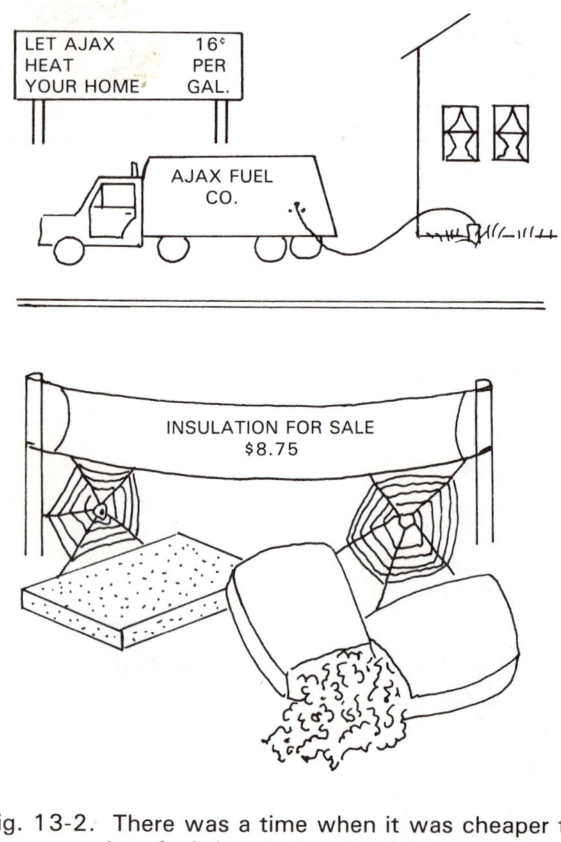

Fig. 13-2. There was a time when it was cheaper to buy fuel than to buy insulation.

HOW ENERGY IS LOST

It is an unfortunate fact that 50 percent of all energy available is lost as we convert it to useful forms. As we know from Chapter 1, some of this loss cannot be helped. It is a natural result of the laws of conversion. Still, there are ways to prevent some of the loss through correcting wasteful practices and through improving the efficiency of our conversion methods. For example, furnaces manufactured just a few years ago are able to use about 30 or 50 percent of the heat energy in fuel. Now, modern design improvements produce furnaces that are 80-96 percent efficient.

Improvement of design is a job for engineers and scientists, of course. However, elimination of waste requires the help and cooperation of everyone.

In 1977 the President of the United States emphasized the need for conservation. Some of the specific goals he called for included the following:

1. Reducing the annual growth in energy demand.
2. Reducing gasoline consumption.
3. Better insulation of homes. This was to involve upgrading the insulation of existing homes and higher insulation standards for new construction.

ENERGY-USING SECTORS

In Chapter 7 we defined the energy-using sectors in our society. Let us review them once more:

1. Residential/commercial. These include houses, apartment buildings, schools, office buildings, and stores. These buildings consume gas, oil, and coal to warm them in winter and cool them in summer. They also contain lighting, appliances, and equipment which consume electricity.
2. Transportation. In this sector are all types of vehicles—automobiles, trucks, buses, trains, airplanes—which move people and goods. Transportation also takes in pipelines for moving gas, petroleum, and water from one place to another.
3. Industrial sector. This sector takes in all businesses organized for producing goods and materials. Essentially these involve mining and manufacturing. Such enterprises use fuels and electricity for space heat, process heat, lighting, and operation of tools and machines. Examples would be:
   a. A steel mill which uses vast quantities of heat to melt ores and scrap until they are molten.
   b. A factory which produces clothing.

Total consumption of energy in all sectors of our society in 1979 amounted to 77.99 quadrillion Btu, according to the U.S. Department of Energy. See Fig. 13-3.

Energy Consumption in U.S. for 1979

SECTOR	QUADRILLION BTU	PERCENT
Residential and Commercial	14.96	19
Industrial	18.95	24
Transportation	19.73	25
Electric utilities	24.35	31
Total	77.99	

Fig. 13-3. Altogether, our society is using more than 77 quads of energy every year.

Generation of electric power takes a large part of our available energy. This power is distributed among, and is used by, every one of the sectors, Fig. 13-4.

CONSERVATION MEASURES

There are ways of saving energy in each of these sectors. Many, if not all, of the methods require the help of the whole community if they are to be effective. Individual action is required in all of the following:

1. Choosing and purchasing energy-efficient housing, appliances, and automobiles.
2. Providing better maintenance to retain or improve energy efficiency of housing and machines.
3. Developing habits which will conserve energy. Individual habits can affect all sectors.
4. Becoming active in collecting recyclable materials.

CONSERVATION RESIDENTIAL/COMMERCIAL

A major portion of the energy used in buildings goes for heat. About 75 percent of the energy is

SECTOR	QUADRILLION BTU	PERCENT
Residential and Commercial	29.42	37
Industrial	28.82	36.9
Transportation	19.76	25

Fig. 13-4. Use of energy by sectors with allocation for electric power. (Percentages do not add up to 100 percent because of rounding.)

spent on heating space and water. The rest is for lighting, air conditioning (cooling), and operation of a variety of appliances.

INEFFICIENT HOME ENERGY USE

Homes may not be efficient users of energy for many reasons:

1. Structures are not tight enough. Cracks around windows, doors, and foundations let air enter or exit, Fig. 13-5.

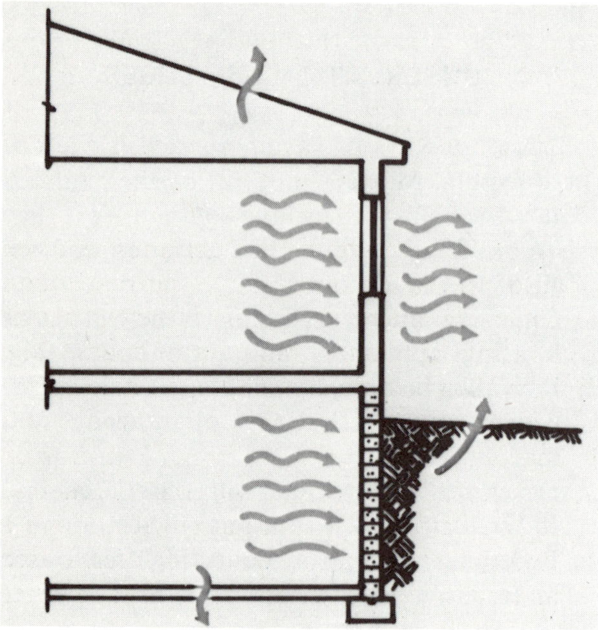

Fig. 13-5. Cracks around doors, windows, foundations, and elsewhere allow warm air to escape to the atmosphere. Cold winds can enter the building through the same cracks. (Portland Cement Co.)

2. Too little insulation. Heat escapes or enters through the walls, ceilings, and windows., Fig. 13-6. Furnaces must work harder and, thus, burn more fuel.
3. Electrical appliances use electricity inefficiently.
4. Uninsulated hot water pipes allow heat to escape.
5. Older furnaces do not deliver enough of the heat value in their fuel. Some units have an efficiency rating of less than 35 percent.

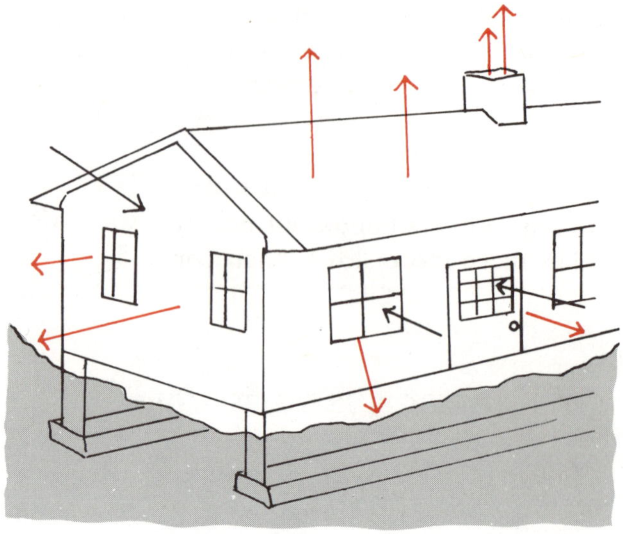

Fig. 13-6. Without adequate insulation heat can be conducted through foundations, walls, and ceilings.

6. Dwellings are often larger than they need to be. Too much energy is used to heat them.
7. Living quarters are often kept too warm.

It has been suggested that remedying these problems would save as much as 50 percent of our residential energy/commercial energy.

HOW HEAT IS LOST

We know from our study of the nature of heat (Chapter 1) that it will always move from a warmer substance or warmer surroundings to those that are cooler. It tends to do this until all the surroundings have the same temperature.

For example, a building which is heated to 70°F (21°C) will lose heat to the outside when the temperature outside is below 70°F. It will gain heat from the outside if the outside temperature is above the inside temperature. This will happen in spite of doors, windows, walls, and roofs.

Buildings gain or lose heat primarily through:

1. Conduction.

2. Convection which can be either exfiltration (moving out) or infiltration (moving in) of air currents.

When heat is conducted it moves through solid materials. Some materials are good conductors of heat. Others conduct it poorly. Conduction takes place more slowly than convection.

Convection losses are the result of air in motion. Currents of air can carry heat through cracks and other openings in the dwelling. It can be exhausted by such mechanisms as fans. Convection also carries heat up chimneys and through window or door cracks. Most of these same openings can also carry unwanted heat into a building.

Conserving energy in housing is closely related to the nature of heat. Conservation starts with the use of construction materials that will make movement of heat in or out of the building more difficult.

There are several methods of doing this:

1. Construct the building of materials which resist passage of heat through it. (You can tell the difference in materials on a cold day by simply touching a wood door and then the glass in a window.) Glass does not resist the passage of heat as well as the wood. A material which is good at reducing heat passage is called INSULATION. You will generally find that good insulators are able to trap air easily. Also, thicker material is generally a better insulator than thin material.

   Generally speaking, building materials must be strong to support loads. The very qualities which make them good at supporting loads make them poor insulators. Therefore, builders use strong materials to build a frame. It will support insulation, siding, wall coverings and more.
2. Seal up cracks in the building's frame or outside covering. Then hot or cold air cannot move in and out.

MATERIAL	RESISTANCE (R)
Still surface air	0.68
Air space	0.97
Face brick	0.39
Wood siding	0.98
1 in. wood floor	0.98
Concrete block	1.11
1/2 in. wood siding	0.85
Gypsum wallboard, 3/8''	0.32
Urethane insulation	9.0
Building paper	0.06
Linoleum or tile	0.05
Asphalt shingles	0.95
Plywood	0.95
Single pane, glass	0.9
Double glass, 1/2 in. space	1.54
Triple glass, 1/2 in. space	2.79
Storm windows	1.79

Fig. 13-7. R values of common building materials. These figures can be used in computing resistance values of different parts of the building.

R AND U RATINGS

To recognize energy efficiency in building, you should understand the U and R ratings of construction materials. An R rating is the value given a material's resistance to passage of heat. A high R value means the material is a good insulator.

The letter U is the symbol for heat conductance. A high U factor means that the material conducts heat easily. A material with a high U rating is not a good insulator. As you can see, U and R values are the opposite of each other.

Until the 1970s, when energy costs began to rise rapidly, little attention was paid to R or U values and insulation. Energy was so cheap. Insulation was considered expensive. However, with higher energy costs, it is important to be familiar with insulating standards for housing and insulating qualities of different construction materials, Fig. 13-7. How much insulation will be needed in a dwelling depends on where you live, Fig. 13-8.

DEGREE DAY

Recommendations for R values are set by the FHA Minimum Property Standards. They are

Normal Number of Degree-Days Per year

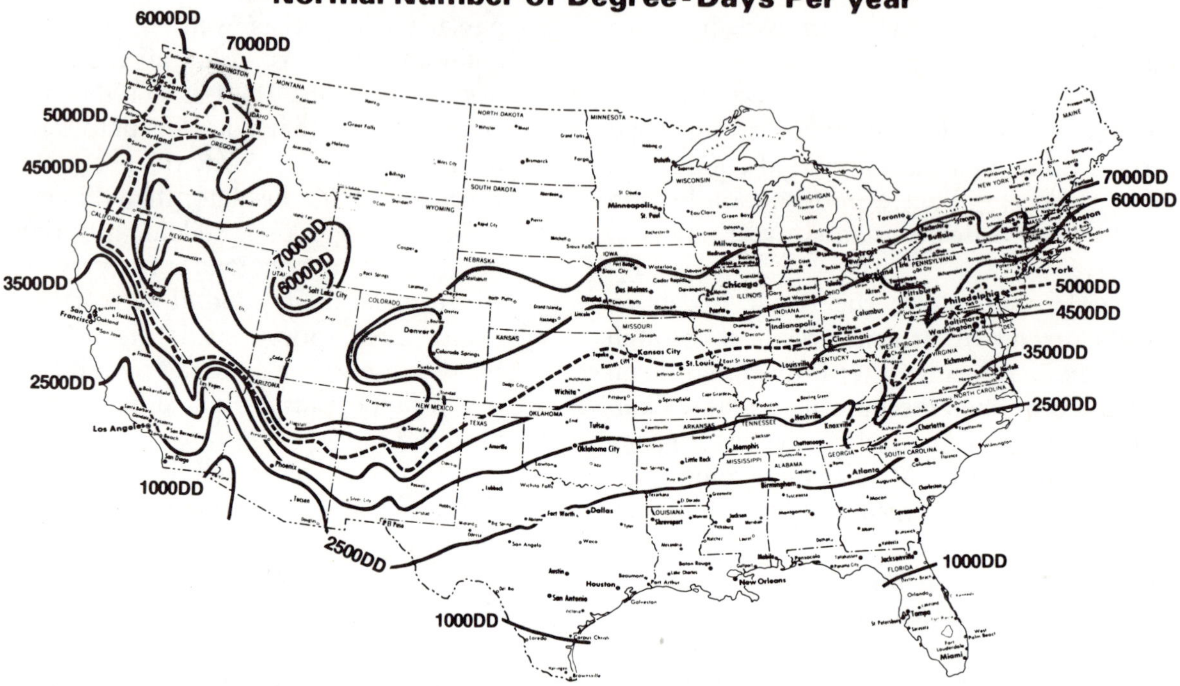

This map is reasonably accurate for most parts of the United States but is necessarily highly generalized, and consequently not too accurate in mountainous regions, particularly in the Rockies.

RECOMMENDED INSULATION R VALUES

DEGREE DAYS	CEILINGS	WALLS	FLOORS
Above 7000	38	17	19
6001-7000	38 (30F)	17 (12F)	19 (11F)
4501-6000	30	17 (12F)	19 (11F)
3501-4500	30	12 (17E)	11 (19E)
2501-3000	22 (30E)	11 (17E)	0 (11E)
1001-2500	19 (22E)	11 (12E)	0
1000 and under	19	11	0

Fossil fuel (gas & oil) = F
Electric resistance heat = E
Heat pumps with electric heat = H
R Values apply to all unless otherwise shown.

These insulation "R" values are recommended to meet the FHA Minimum Property Standards for thermal resistance. The R value shown for walls can be met with any combination of sheathing and cavity wall insulation, or blanket insulation alone. For additional information, refer to Owens-Corning Pub. No. 5-BL-9293-A.

Fig. 13-8. Insulation requirements for housing will be different depending on temperatures in the area where you live. (Owens-Corning Fiberglas Corp.)

based on the severity of the winters in a particular locality. This severity is measured in degree days.

There are heating degree days and cooling degree days. A degree day is measured against a standard temperature of 65°F (18°C). It is assumed that at this temperature neither heating nor cooling is needed.

This is how degree days are found:

1. Average the high and low temperature for a 24-hour period.
2. Subtract this temperature from 65°F. The answer is the number of degree days. For example, suppose that the average temperature for a 24-hour period is 40°F (5°C). Subtracting 40 from 65 leaves 25. There are 25 heating degree days for that period.
3. Total up all the degree days for a heating or cooling season to get the total degree days for

RESISTANCE OF VARIOUS INSULATING MATERIALS

MATERIAL	R VALUE PER IN.	INCHES NEEDED FOR VALUE INDICATED				
		R-11	R-19	R-22	R-34	R-38
Batts/Blankets						
Fiberglass	3.14	3.5	6	7	11	12.5
Rock Wool	3.14	3.5	6	7	11	12.5
Loose/Machine Blown						
Glass Fiber	2.25	5	8.5	10	15.5	17
Rock Wool	3.125	3.5	6	7	11	12.5
Cellulose	3.75	3	5.5	6	9.5	10.5
Hand Poured Loose Fill						
Cellulose	3.7	3	5.5	6	9.5	10.5
Rock Wool	3.125	3.5	6	7	11	12.5
Glass Fiber	2.25	5	8.5	10	15.5	17
Vermiculite	2.1	5.5	9	10.5	16.5	18
Rigid Board						
Polystyrene Beadboard	3.6	3	5.5	6.5	9.5	10.5
Extruded Polystyrene	5.41	3	5	5.5	8.5	9.5
Urethane	6.2	2	3	3.5	5.5	6.5
Glass Fiber	4	3	5	5.5	8.5	9.5

Fig. 13-9. R value of certain insulating materials.

the region's climate.

If the average is above 65°F, there are no degree days. No heat is required.

Insulation is manufactured from different materials and in different forms. Their R values are charted in Fig. 13-9.

INSTALLING INSULATION

Loose fill is used in ceilings or in space you cannot reach with batts or blanket insulation, Fig. 13-10. Rigid insulation is not normally used for floors or ceilings except in homes with cathedral type ceilings or in mobile homes. It is the preferred type, however, when insulating basement walls. The sheets can be placed between thin furring strips.

Batt and blanket insulation are suited for use between standard spaced joists and rafters. On new construction it is also used between wall

Fig. 13-10. Loose insulation requires special equipment but is preferred in areas that are difficult to reach. On older homes it is the only way to insulate walls. (Owens-Corning Fiberglas Corp.)

Fig. 13-11. Blanket insulation with vapor barrier.

studs. Batts are sold in precut sections or in long rolls, Fig. 13-11. Vapor barrier is always faced toward the heated space. Fig. 13-12 shows insulation being applied between ceiling joists. When adding additional batt or blanket insulation to ceilings, lay them across the joists as shown in Fig. 13-13. Loose insulation can also be applied on top of batts, Fig. 13-14. Keep insulation at least three inches away from ceiling fixtures, Fig. 13-15.

Fig. 13-12. Batts 12 in. thick should be placed in ceilings in cold climates. (Owens-Corning Fiberglas Corp.)

Fig. 13-13. Second layer of insulation should be laid across ceiling joists. (Certainteed)

Fig. 13-14. Loose insulation can also be added over batts. (Owens-Corning Fiberglas Corp.)

Fig. 13-15. Keep insulation away from light fixtures. They may become overheated. (Certainteed)

INSULATING FOUNDATIONS

R values of foundation walls are relatively low. This means that warm air travels through the concrete too easily. Fig. 13-16 shows how walls, sills, and headers should be insulated for most effective control of heat loss.

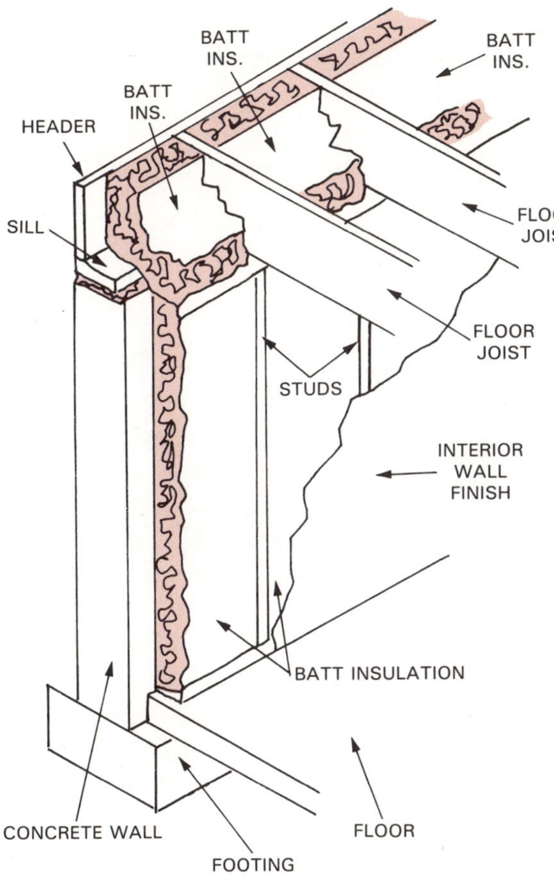

Fig. 13-16. In new construction, foundation walls can be insulated from the inside with stud walls and batt insulation.

CUTTING ENERGY LOSSES THROUGH GLASS

Since a single pane of glass has an R rating of 0.9, heat is easily transmitted. Thus, large areas of glass can cause significant energy loss during either heating or cooling of buildings.

Double or triple glazing are the most effective methods of increasing insulating value of windows, Fig. 13-17. Other effective ways of increasing the R value of window areas are:

1. Installation of storm windows. Besides the insulating value of an additional pane of glass, the air space between the storm window and the regular window has an R value about equal to that of the glass.

Fig. 13-17. Double and triple glazing add needed insulation to window areas. (Rolscreen Co.)

   Clear plastic film can take the place of storm windows. It should be taped to the inside of the window frame, Fig. 13-18.
2. Installation of insulated shades.
3. Installation of interior shutters which can be closed to reduce heat loss through the window during cold weather. Like the insulated shade, they could also be closed during daylight hours to block solar radiation in the hot summer months.
4. Closing shades and/or drapes.

Fig. 13-18. Plastic film is easy to install and keeps out drafts from ill-fitting windows.

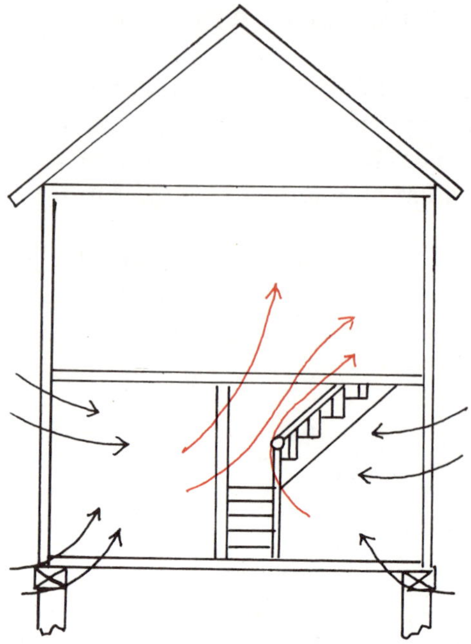

Fig. 13-19. Chimney effect robs rooms of heat. Cold air infiltrates through cracks and exits through openings at a higher level or through poorly insulated ceiling.

STOPPING AIR LEAKAGE

No building can or should be made entirely airtight. Good ventilation within the building requires one complete change of air every hour. Outside air leaks in; inside air leaks out. However, most buildings have too many leaks.

CHIMNEY EFFECT

One problem involving air leaks is the "chimney effect." Warm air rises and leaks out through cracks or openings at upper levels or near ceilings. Cold air leaks in at lower levels, Fig. 13-19. The cracks act like a chimney drawing in cold air at a low level and letting it out at an upper level.

FINDING THE LEAKS

Air leaks are usually found:

1. Where sills rest on foundation walls, Fig. 13-20.
2. Around doors and windows.
3. Siding joints at corners, windows, or doors, Fig. 13-21.
4. Around switches and electrical outlets and around doors and windows where insulation does not fill cavities.

It is easy to detect these leaks on a cold and

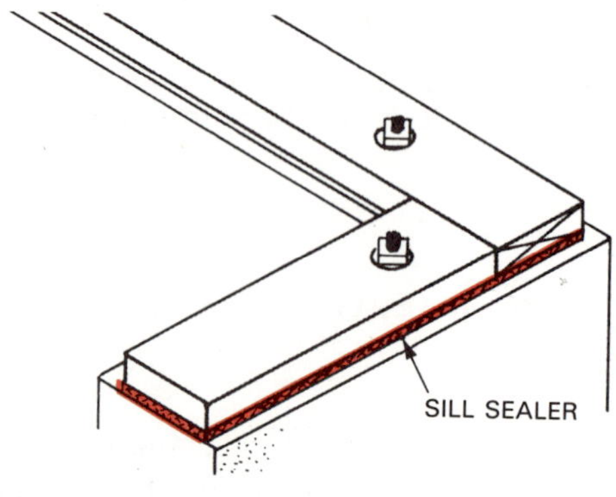

Fig. 13-20. Sealing up the joint between foundation wall and sills.

Fig. 13-21. Seal cracks around window frames with caulking.

windy day. Simply hold a lighted candle near a suspected leak. The movement of the air will cause the flame to flicker or even go out. More sophisticated methods are smoke testing and infrared photography.

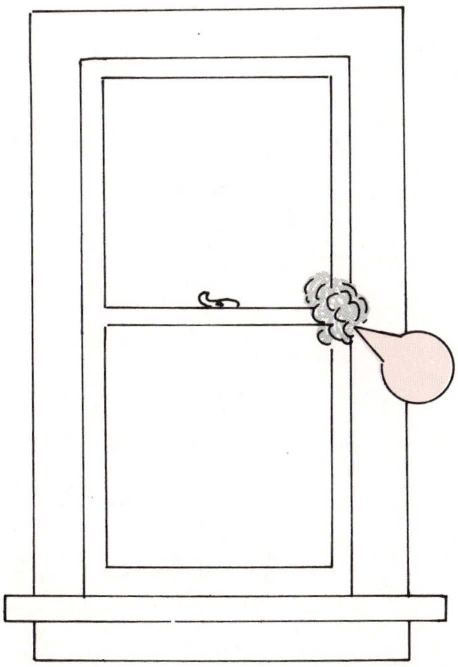

Fig. 13-22. Doing a smoke test. If the room or the house can be slightly pressurized, smoke will slip out indicating where the cracks are.

SMOKE TEST

In smoke testing the air pressure inside the house is raised slightly. Then smoke is released around areas where leakage is suspected. Air pressure in the house will force the smoke out, pinpointing the source of the air leak. See Fig. 13-22.

INFRARED PHOTOGRAPHY

Infrared photographs taken outside of the building will also indicate location and extent of heat loss through walls and roof. A chemical in the film emulsion is sensitive to infrared waves (heat). Poorly insulated areas of the walls and roof heat the air and the heat will appear bright red on the photograph.

SEALING CRACKS

Buildings which are not properly sealed against air infiltration and exfiltration can waste up to 40 percent of the energy used for heating and cooling.

Several commercial products are useful in sealing cracks. Included are caulking compounds, oakum, and weather stripping.

Caulk is a soft, putty-like material usually put up in tubes. The tube fits into a gun for application. It may also be sold in rope-like coils.

Oakum is a tarred hemp product made into a light rope-like material. It is used to seal large cracks and is sometimes called caulk rope. The rope is peeled off the roll and pressed into the cracks.

Weather stripping can be made of thin strips of metal, rubber, vinyl, spring bronze, foam rubber, or felt. The strips can be tacked or glued around doors and windows.

To seal air leaks:

1. In exterior walls. Inspect all joints where

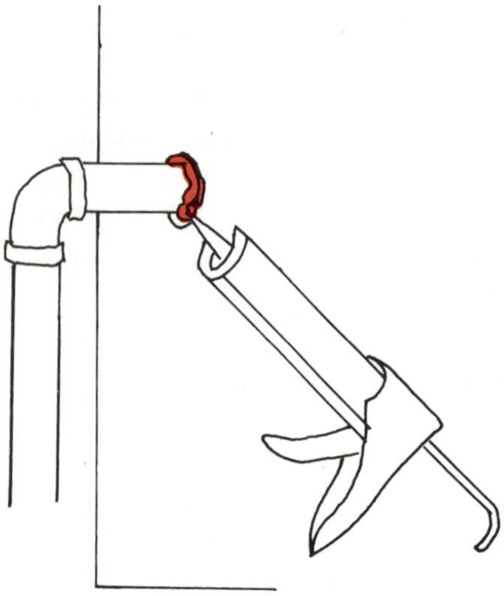

Fig. 13-23. Caulk around openings where ducts or pipe go through a wall.

Fig. 13-24. Simple remedy like foam pad under light switches helps eliminate air infiltration (drafts).

siding meets corners, window frames, and door frames. Also check points where siding meets the foundation. Seal joints with a thin bead of caulk. If old caulk has dried out and cracked, remove it and recaulk. Caulk openings where pipes or ducts come through exterior walls, Fig. 13-23. Foam rubber gaskets are made to slip under switch plates, Fig. 13-24.

2. Doors and windows. Weather strip edges where air leakage is occurring. Reglaze sashes

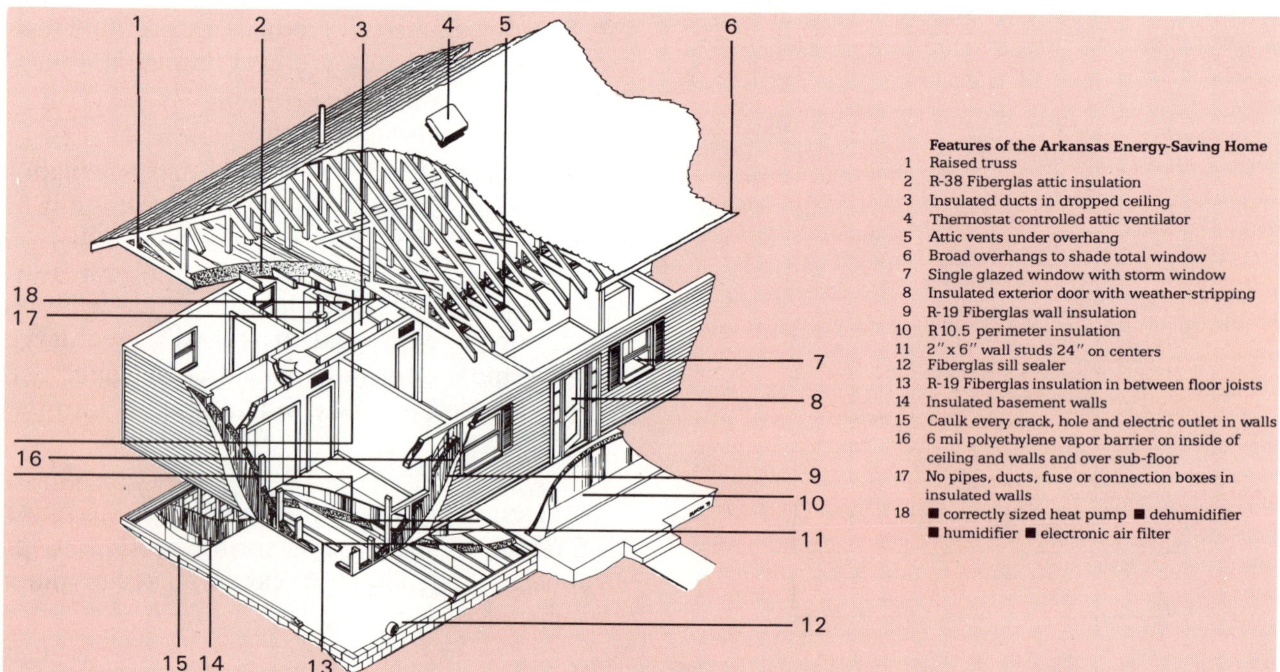

Fig. 13-25. Energy-saving home has 18 special features. (Owens-Corning Fiberglas Corp.)

where glazing has cracked and fallen out. On old windows with sash weights, stuff small pieces of sponge rubber into openings around pulleys. Weather strip doors. Special gaskets are available to seal gaps between the bottom edge of the door and the threshold.

NEW CONSTRUCTION

New homes can be designed to reduce the amount of energy needed to heat and cool them. The savings come about in two ways:

1. If the home is designed to use less energy, a smaller, less costly, heating/cooling system can be used.
2. Tighter, well insulated homes and a smaller heating/cooling system will consume less energy to keep the home comfortable.

One house, known as the Arkansas Energy-Saving Home, was part of a study which compared nine energy-saving homes with nine similar homes which were built to the specifications of the 1974 Minimum Property Standards (MPS) of the FHA. The energy-saving homes used 50 percent less energy over a three-year period. Fig. 13-25 is an exploded view of one home design and its energy saving features. Fig. 13-26 is a list of construction practices to avoid.

DON'TS

1. Avoid anything that will interfere with the insulated envelope, upset the humidity control, or create a need to readjust the thermostat controls once they have been balanced.
2. Avoid using sliding glass doors. They are hard to keep satisfactorily sealed, and the large glass areas present large hot and cold surfaces that can cause discomfort leading to higher- or lower-than-necessary thermostat settings. Substitute French doors that can be weather stripped to close tightly.
3. Don't use recessed lighting in the attic. Fire codes require a 3-inch uninsulated space around them, which leads to undesirable heat losses into the attic. Tracked lighting on the ceiling surface can provide similar aesthetic effects without the heat loss.
4. Avoid a fireplace unless the combustion chamber is supplied directly with outside air and has a tight-fitting glass screen and damper. (The exterior of the fireplace must be as well insulated as the other exterior walls.)
5. Don't allow anything in exterior walls, such as pipes, ducts, electrical distribution, or medicine cabinets, that will reduce insulating efficiency.

Building an energy-saving home is no longer something for the future. If you are planning to build or buy a home, make sure it is energy efficient. These energy-saving homes are an economical and comfortable way to fight rising fuel costs. Remember, in Arkansas, owners of energy-saving homes are saving nearly $500 a year.

Fig. 13-26. Avoid these five situations when building.

EFFICIENT APPLIANCES

Appliances account for about 25 to 28 percent of the energy used in the average household. Fig. 13-27 lists estimated electrical power consumed annually by 49 different appliances.

Appliances can be designed to do their job using less energy. Since 1976, the appliance industry has been efficiency labeling their different models. This information is displayed on most new appliances and is known as an "energy efficiency rating" (EER).

EER RATINGS

Following the energy crisis of 1973, the National Bureau of Standards set up an appliance labeling section. Its job is to sponsor a program of efficiency rating of household appliances. Appliances covered by the rating program include: room air conditioners, refrigerators, freezers, water heaters, clothes washers and dryers, ranges, ovens, furnaces, and central air conditioners.

EER labels are attached to the appliances and contain:

1. The energy efficiency rating of the appliance. (The higher the rating, the more efficient the appliance in its use of electricity, gas, or fuel oil.)
2. The range of efficiency or energy consumption of other similar products on the market.
3. An estimate of how much the appliance will cost to operate during the year at various levels of usage.

A typical label is shown in Fig. 13-28. Usually, they are referred to as "Energy Guide" labels.

ANNUAL ENERGY REQUIREMENTS OF ELECTRIC HOUSEHOLD APPLIANCES*

*Source: Edison Electric Institute

	EST. kWh
Major Appliances	
Air-conditioner (room) (Based on 1000 hours of operation per year. This figure will vary widely dependong on geogrpahic area and specific size of unit)	860
Clothes Dryer	993
Dishwasher including energy used to heat water	2100
Dishwasher only	363
Freezer (16 cu. ft.)	1190
Freezer—frostless (16.5 cu. ft.)	1820
Range with oven	700
with self-cleaning oven	730
Refrigerator (12 cu. ft.)	728
Refrigerator—frostless (12 cu. ft.)	1217
Refrigerator/Freezer (12.5 cu. ft.)	1500
Refrigerator/Freezer—frostless (17.5 cu. ft.)	2250
Washing Machine—automatic (including energy, used to heat water)	2500
washing machine only	103
Washing Machine—non-automatic (including energy to heat water)	2497
washing machine only	76
Water Heater	4811

	EST. kWh
Kitchen Appliances	
Blender	15
Broiler	100
Carving Knife	8
Coffee Maker	140
Deep Fryer	83
Egg Cooker	14
Frying Pan	186
Hot Plate	90
Mixer	13
Oven, Microwave (only)	190
Roaster	205
Sandwich Grill	33
Toaster	39
Trash Compactor	50
Waffle Iron	22
Waste Disposer	30
Heating and Cooling	
Air Cleaner	216
Electric Blanket	147
Dehumidifier	377
Fan (attic)	291
Fan (circulating)	43
Fan (rollaway)	138
Fan (window)	170
Heater(portable)	176

	EST. kWh
Heating Pad	10
Humidifier	163
Laundry	
Iron (hand)	144
Health & Beauty	
Germicidal Lamp	141
Hair Dryer	14
Heat Lamp (infrared)	13
Shaver	1.8
Sun Lamp	16
Toothbrush	.5
Vibrator	2
Home Entertainment	
Radio	86
Radio/Record Player	109
Television	
Black & White	
Tube type	350
Solid state	120
Color	
Tube type	660
Solid state	440
Housewares	
Clock	17
Floor Polisher	15
Sewing Machine	11
Vacuum Cleaner	46

Note:
When using these figures for projections, such factors as the size of the specific appliance, the geographic area of use, and individual use should be taken into consideration.

Fig. 13-27. This chart will help you to determine the annual energy costs in your household. If you had to, which ones could you do without?

FIGURING EER

The EER of an electrical appliance is calculated by dividing the energy output (in Btu per hour) by the electrical energy used (in watt/hr). For example, if a window air conditioning unit has a capacity of 7000 Btu/hr with an input of 840 watts of electricity per hour the EER is equal to $\frac{7000}{840}$ or 8.4. This information can be found on

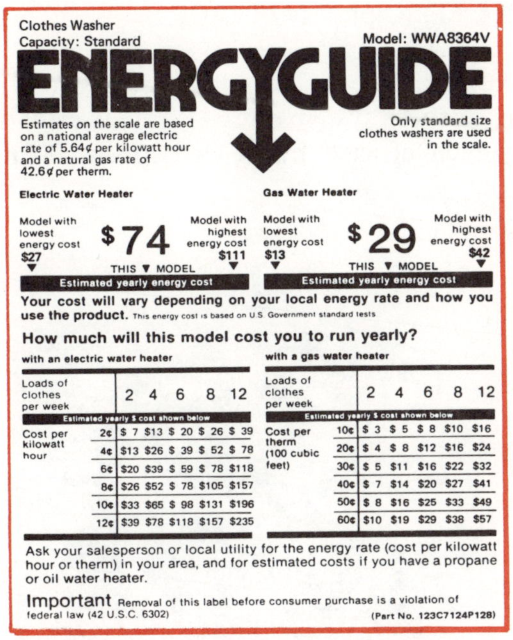

Clothes Washer
Capacity: Standard

Model: WWA8364V

ENERGYGUIDE

Estimates on the scale are based on a national average electric rate of 5.64¢ per kilowatt hour and a natural gas rate of 42.6¢ per therm.

Only standard size clothes washers are used in the scale.

Electric Water Heater

Model with lowest energy cost $27 — $74 THIS MODEL — Model with highest energy cost $111

Estimated yearly energy cost

Gas Water Heater

Model with lowest energy cost $13 — $29 THIS MODEL — Model with highest energy cost $42

Estimated yearly energy cost

Your cost will vary depending on your local energy rate and how you use the product. This energy cost is based on U.S. Government standard tests

How much will this model cost you to run yearly?

with an electric water heater

Loads of clothes per week		2	4	6	8	12
Estimated yearly $ cost shown below						
Cost per kilowatt hour	2¢	$ 7	$13	$ 20	$ 26	$ 39
	4¢	$13	$26	$ 39	$ 52	$ 78
	6¢	$20	$39	$ 59	$ 78	$118
	8¢	$26	$52	$ 78	$105	$157
	10¢	$33	$65	$ 98	$131	$196
	12¢	$39	$78	$118	$157	$235

with a gas water heater

Loads of clothes per week		2	4	6	8	12
Estimated yearly $ cost shown below						
Cost per therm (100 cubic feet)	10¢	$ 3	$ 5	$ 8	$10	$16
	20¢	$ 4	$ 8	$12	$16	$24
	30¢	$ 5	$11	$16	$22	$32
	40¢	$ 7	$14	$20	$27	$41
	50¢	$ 8	$16	$25	$33	$49
	60¢	$10	$19	$29	$38	$57

Ask your salesperson or local utility for the energy rate (cost per kilowatt hour or therm) in your area, and for estimated costs if you have a propane or oil water heater.

Important Removal of this label before consumer purchase is a violation of federal law (42 U.S.C. 6302)

(Part No. 123C7124P128)

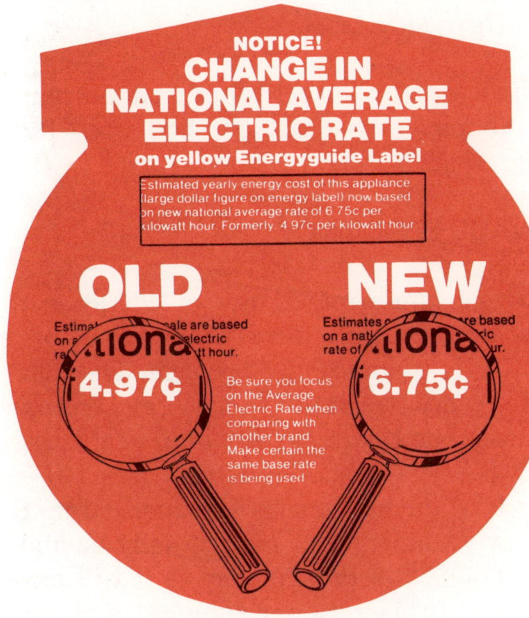

Fig. 13-28. Left. An Energy Guide label is found on many appliances. By comparing different models, shoppers can select models that will do the job using less energy. Right. Annual cost of operation may rise as costs of energy rise. (G.E. and Irv Wolfson Co.)

About 8 percent of all the energy used in the United States goes into running electrical home appliances, so appliance use and selection can make a considerable difference in home utility costs. Buying an energy-efficient appliance may cost a bit more initially but that expense is more than made up by reduced operating costs over the lifetime of the appliance.

Energy efficiency may vary considerably though models seem similar. In the next few years it will be easier to judge the energy efficiency of appliances with the Government's appliance labeling program. (See page 31 for details.) In the meantime, wise selection requires a degree of time and effort.

You will find a number of tips on how to save energy when buying or using appliances in other sections of this booklet, but here are a few general ideas to consider.

- **Don't leave your appliances running when they're not in use.** It's a total waste of energy. Remember to turn off your radio, TV, or record player when you leave the room.
- **Keep appliances in good working order** so they will last longer, work more efficiently, and use less energy.

The list shows the estimated annual energy use of some household appliances. With this information, you should be able to figure your approximate energy use and cost for each item listed. You also should get a good idea of which appliances in your home use the most energy and where energy conservation practices will be the most effective in cutting utility costs.

- **Use appliances wisely;** use the one that takes the least amount of energy for the job. For example: toasting bread in the oven uses three times more energy than toasting it in a toaster.
- **Don't use energy-consuming special features on your appliances if you have an alternative.** For example, don't use the "instant-on" feature of your TV set. "Instant-on" sets, especially the tube types, use energy even when the screen is dark. Use the "vacation switch," if you have one, to eliminate this waste; plug the set into an outlet that is controlled by a wall switch; or have your TV service man install an additional on-off switch on the set itself or in the cord to the wall outlet.

Fig. 13-29. Wisely used appliances are more efficient. (U.S. Dept. of Energy)

the energy guide label.

Wise use of appliances is just as important as wise shopping for the most efficient ones. Refer to Fig. 13-29.

FURNACE EFFICIENCY

Furnaces are a major user of energy in the home. Let us assume that the efficiency of an older unit is around 50 to 60 percent. This means that for every 100 Btu of fuel energy consumed the furnace delivers 50 to 60 Btu of heat energy.

Much of the heat energy is exhausted up the chimney. Temperatures must usually be above 300°F (149°C) in the chimney so it will draw. When the furnace is not running, additional warm room air is lost as it rises up the open chimney.

Automatic vent dampers are available which close off chimney flues when furnaces and other gas burning appliances are not in operation. Fig. 13-30 shows a vent damper in both open and closed modes.

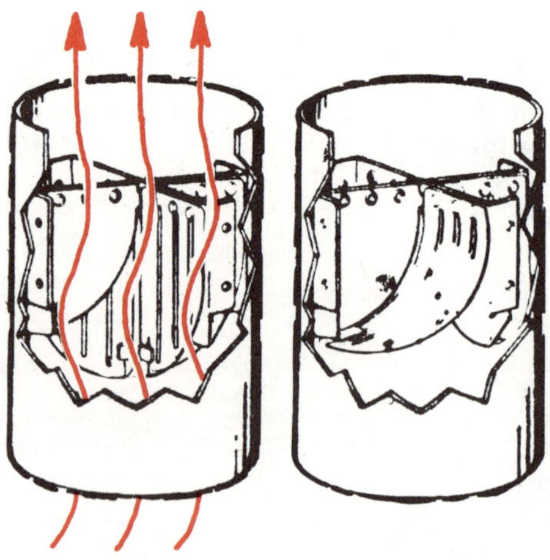

Fig. 13-30. Typical automatic damper. It opens when heat from fuel-fired appliance strikes it. When the appliance shuts down the damper closes. (Ameri-Therm)

FUEL EFFICIENT FURNACE

New furnace designs make better use of fuel. One of these systems, Fig. 13-31, is rated 91 to 96 percent efficient by the Department of Energy.

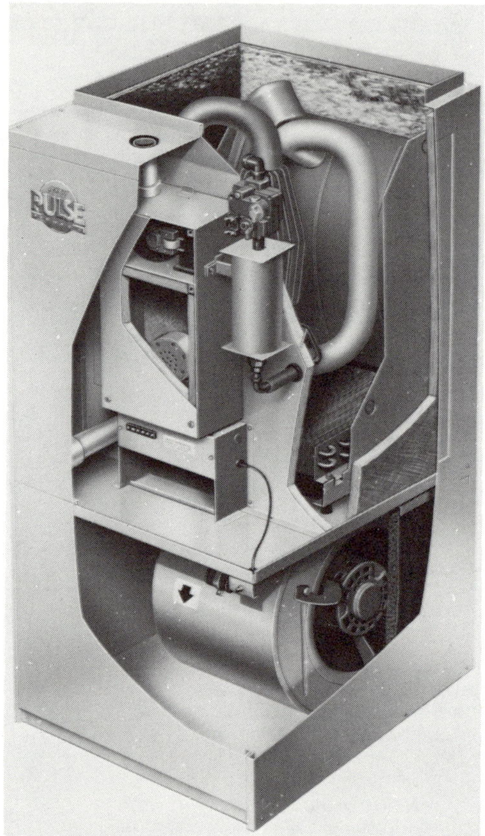

Fig. 13-31. New furnace designs are more efficient than older units found in residential use. Annual Fuel Utilization Efficiency (AFUE) ratings are provided by the manufacturers. (Lennox Industries Inc.)

The system is designed with heat exchangers which extract most of the heat of combustion. Exhaust gases are in the 100°F (38°C) range. Therefore, the furnace does not require a chimney. Spent gases can be exhausted through a small plastic pipe. The fuel charge is ignited by a spark plug. Fuel is burned in a series of small explosions called "pulses." The combustion cycle is shown in Fig. 13-32.

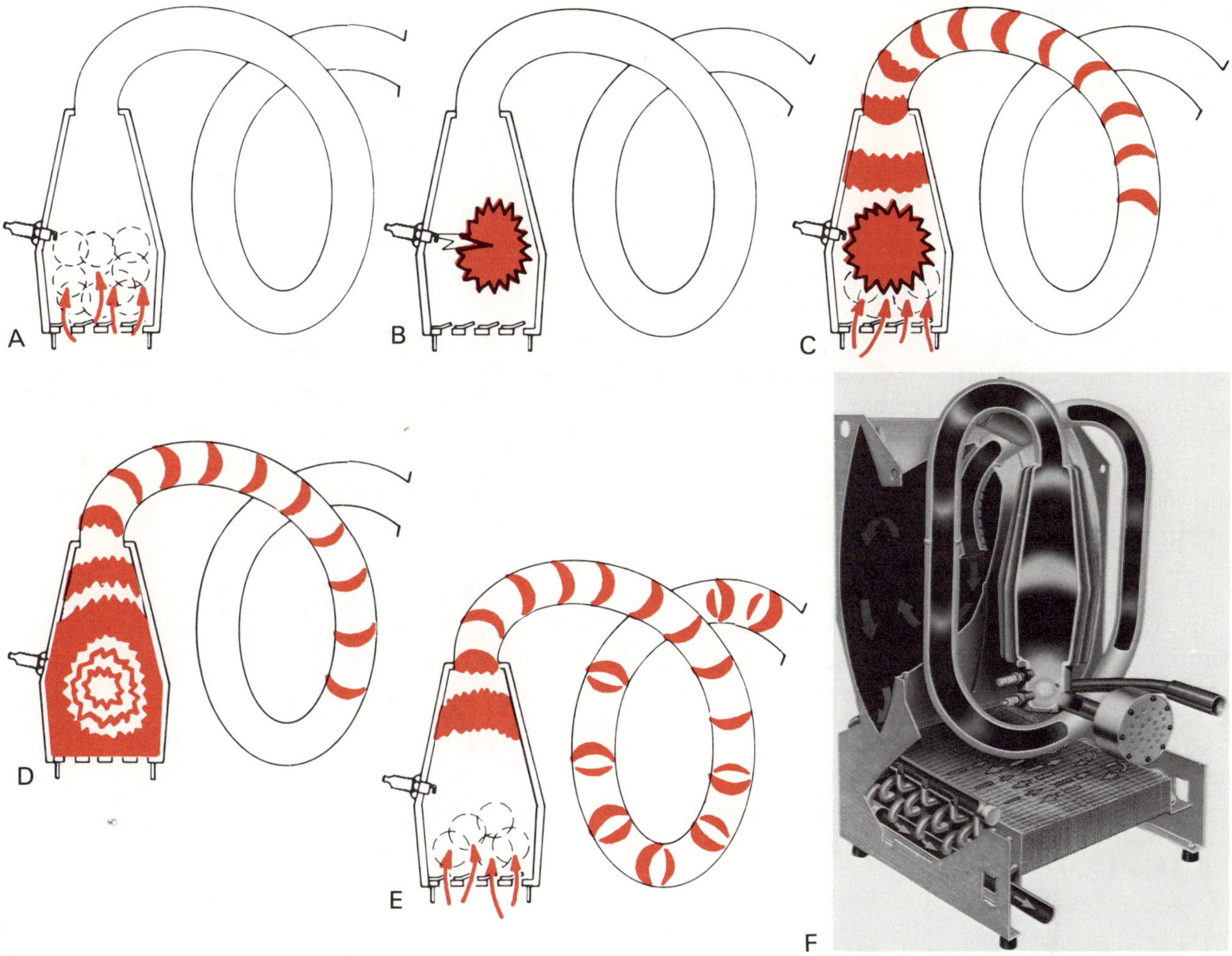

Fig. 13-32. Operation of pulse furnace. A — Gas and air enter combustion chamber at flapper valves. B — Spark plug ignites mixture, causing first pulse. C — Flapper valves close and heated gases are forced into tailpipe. D — First pulse reaches end of tailpipe and bounces back. Meanwhile, flapper valves open, admitting more gas and air. E — Returning gases ignite new charge and cause pulse process to begin over. F — Cutaway of combustion chamber. Fin and tube condensing coil at bottom draws heat from the system. (Lennox Industries Inc.)

HEAT PUMPS

An appliance which may become an important energy saver in the future is the heat pump. It is basically a refrigerator which is capable of either heating or cooling. In winter it pumps heat indoors; in summer it pumps it outdoors.

The pump, which uses 60 percent less electricity than an electric heater, makes use of three well-known principles of heat energy:

1. Heat energy will always move from a warm area to a colder one.
2. A gas will heat up if it is compressed (squeezed together).
3. When allowed to expand the gas will become colder.

These are also known as THERMODYNAMIC principles (thermo for heat and dynamic for motion). In other words, they deal with movement of heat.

HEAT PUMP SYSTEM

Heat pumps are appliances which can heat or cool as seasons change. The basic system includes:

1. A pump for compressing a gas. This pump is usually called a compressor.
2. Two chambers or heat exchangers. One collects heat and is called an evaporator. It is where the gas expands. The other gives off heat and is called a condenser.
3. Pipes which connect the evaporator and condenser with the compressor.
4. A reversing valve which changes direction of the gas flow for heating or cooling.

HOW THE HEAT PUMP WORKS

The system is filled with a gas, also called a REFRIGERANT. When the reversing valve is in the "heat" position, Fig. 13-33, the chamber located outdoors is like an evaporator. The refrigerant expands into it. Being cooler, the refrigerant absorbs heat from the warmer air. Moved along by the compressor, the refrigerant flows to the chamber inside the house which is the condenser. Here the gas gives up its heat load to the indoors. When the valve is reversed, the heat pump reverses and acts as a refrigerator. Now it takes heat from the dwelling and moves it outdoors, Fig. 13-34.

HEAT PUMP EFFICIENCY

For the same energy use, a heat pump will deliver up to three times as much heat as an electrical resistance heater. Also, since it both heats and cools, it is cheaper than having both a furnace and an air conditioner.

In the heating mode, the heat pump becomes less efficient as outside temperatures drop. Below 20°F (−7°C) it usually will not be able to provide enough heat for a building. Other methods of

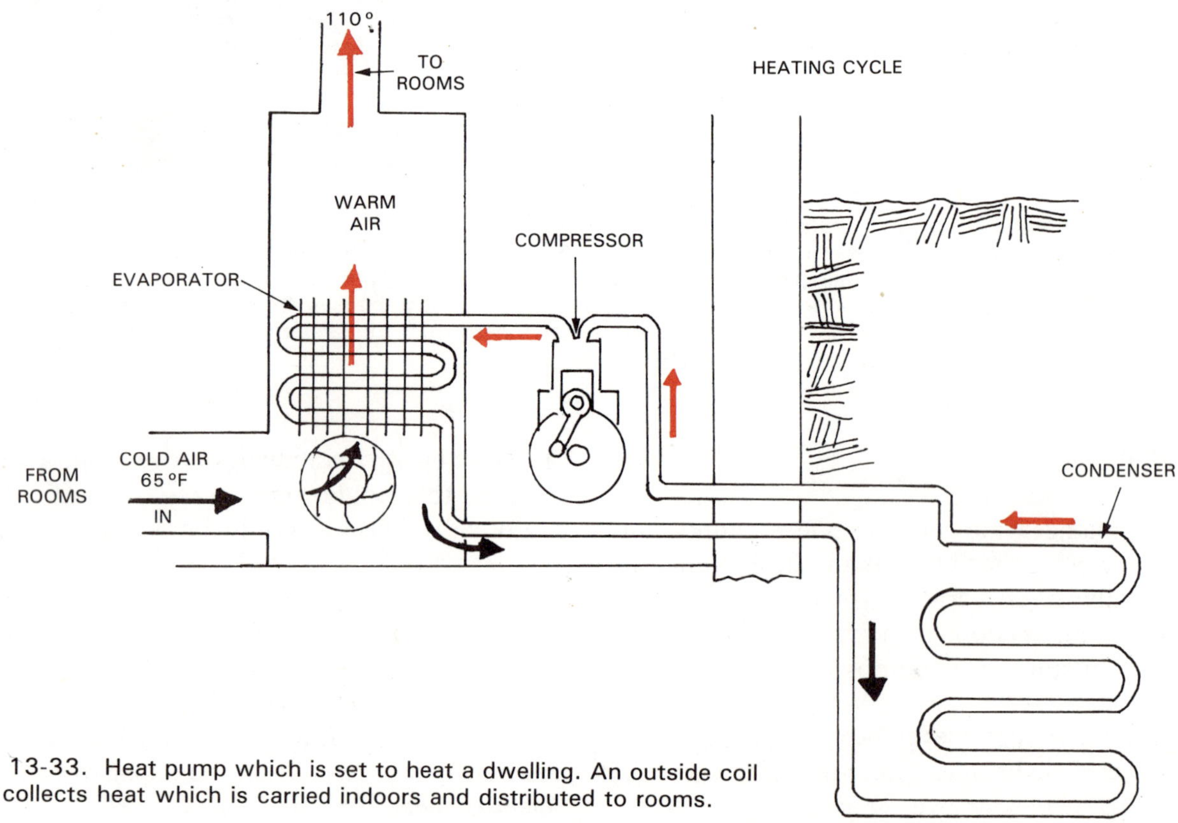

Fig. 13-33. Heat pump which is set to heat a dwelling. An outside coil collects heat which is carried indoors and distributed to rooms.

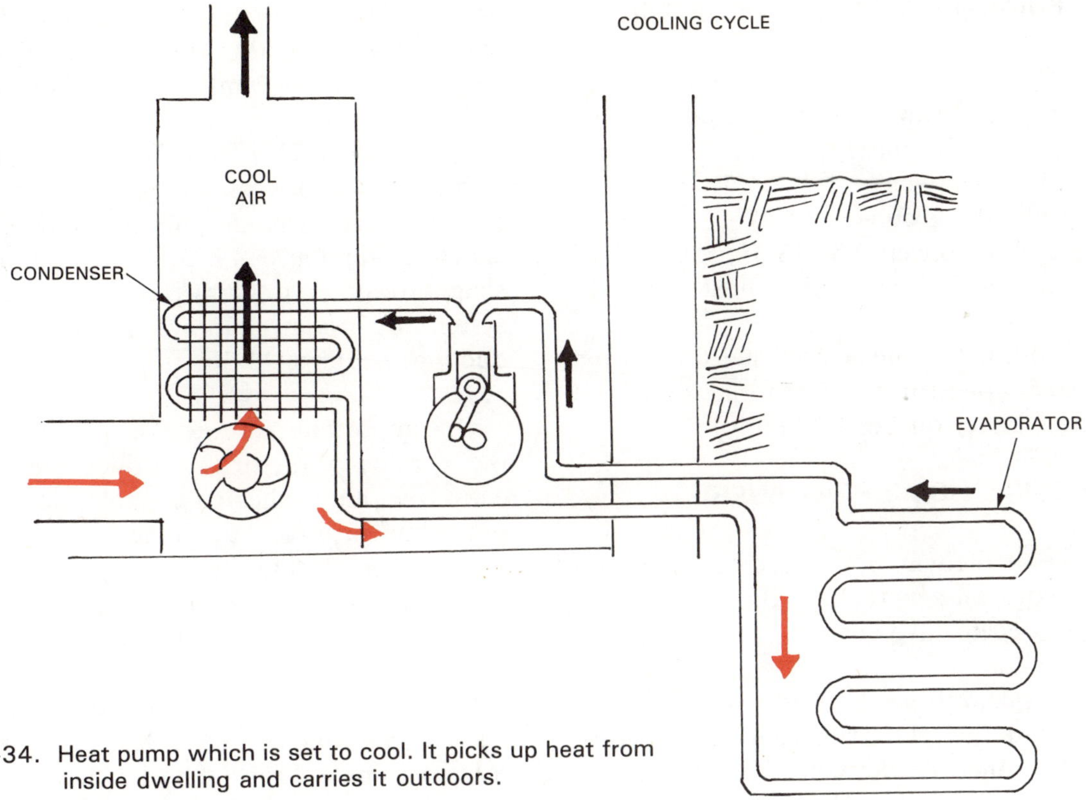

Fig. 13-34. Heat pump which is set to cool. It picks up heat from inside dwelling and carries it outdoors.

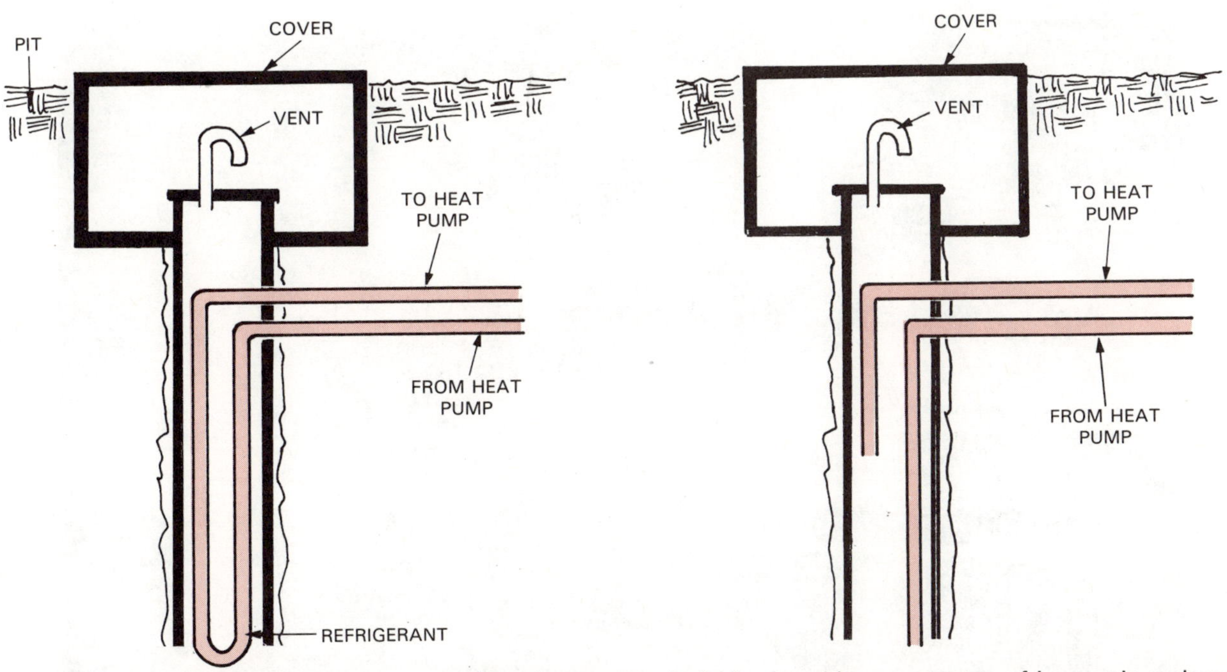

Fig. 13-35. Using a well as a heat source for a heat pump. Left. In closed loop system a refrigerant in a closed coil circulates through the pipe in the well. Right. System in which well water is brought to the heat pump to heat or cool refrigerant.

heating will then have to be used to assist the pump.

This problem can also be solved in cold climates placing the outdoor coil in a better heat source. For example:
1. Bury it in the ground.
2. Place it down a well, Fig. 13-35, or on a lake bottom where it can collect heat from the water.
3. Install one coil in the heat storage tank or bin of a solar collector. (Collectors are described in the chapter on Solar Energy.)

REDIRECTING HOME HEAT ENERGY

Because of the way homes are designed and constructed, heat generated in the house is not generally used to save energy.

For example, refrigerators and ovens give off heat which, in the cold weather, add to the comfort of the home. At the same time, the added heat reduces the fuel consumption of furnaces and stoves. While this heat is beneficial in cold weather, it is not wanted during hot weather.

Similarly, much of the heat carried through uninsulated duct work and pipes is lost. In 1976 the federal government published a report on an experimental house in Hampton, Virginia which sought to conserve energy through an active solar system, by insulating its systems, and by proper ducting. See Fig. 13-36.

One important feature of the model home was the location of ducting. All ducts were located entirely within the heated or cooled space. At the same time, heat from appliances could be ducted out of doors during the cooling season. During the heating season the appliance-generated heat could be distributed throughout the home.

Fig. 13-37 is a schematic for the heating system for a home that was remodeled in Canada. It serves as a model for the Saskatchewan Conser-

Fig. 13-36. This experimental energy-saving home was a project of the U.S. Department of Energy.

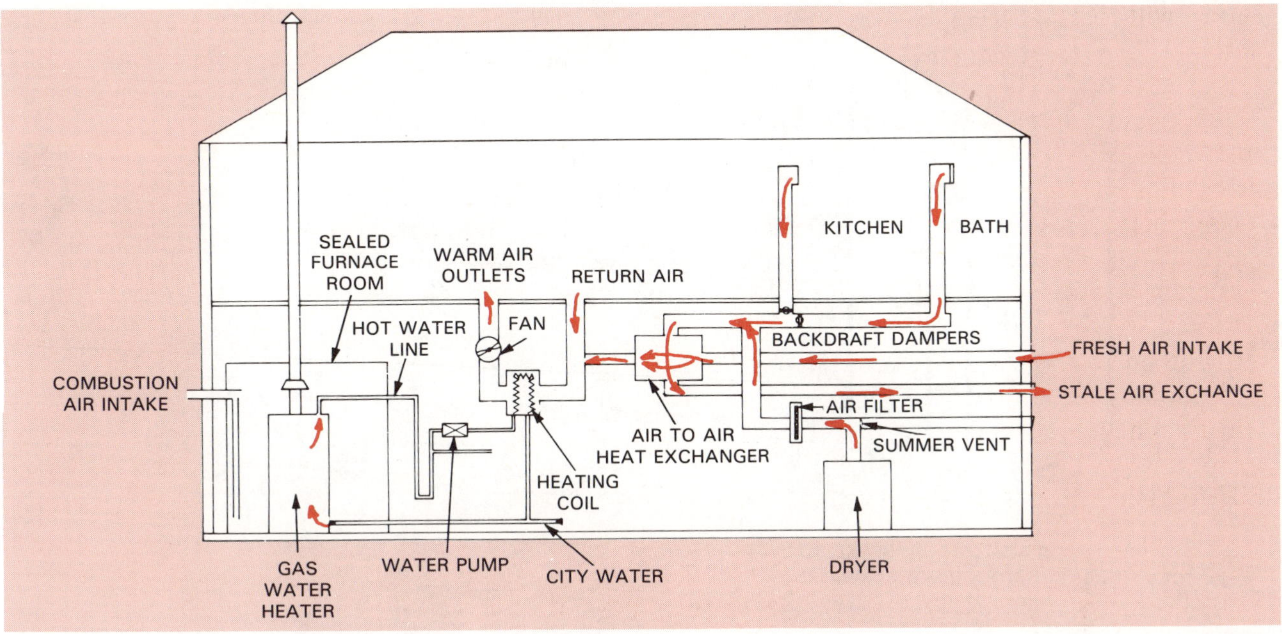

Fig. 13-37. Diagram of an energy-saving heating system for a home that has been made energy efficient. There is no furnace. Heat is supplied by the gas-fired water heater. Hot exhaust from the dryer is also fed into the heating system. (Saskatchewan Energy and Mines)

vahome Program.

The home was made so weathertight and so well insulated that it did not need a furnace. In fact, it was estimated that the home's heating needs had been reduced by 78 percent. Even the smallest conventional furnace would have been too large and, therefore, inefficient. Instead, the home's water heater became the primary heating source. It heats the house as well as supply hot water needs.

Hot water from the heater is pumped through a coil in the warm air distribution system. Warm exhaust from the clothes dryer is also used. It, too, is fed into the heat distribution system. During hot months, the dryer is vented outside.

The hot water heater draws its combustion air from outside. Thus, it does not rob warm air from the house.

Hot water systems in any home are another source of heat loss. Pipes carry heated water from the water heater to faucets in the kitchen, bathroom, and laundry. Some of the heat is lost to the cooler air surrounding the pipe. Insulating the pipe will prevent this heat loss.

Insulating the water heater itself will reduce heat loss by about 400 kWh a year of electricity or 3600 cu. ft. of natural gas. See Fig. 13-38.

Other methods of saving energy for heating of water include:

1. Lowering of the temperature setting on water heaters to the lowest temperature setting acceptable to household needs.
2. Fixing leaking faucets.
3. Using the dishwasher only when fully loaded.
4. Installing flow restrictors in shower heads.
5. Laundering soiled clothing only in cold or warm water when it is practical to do so.
6. Loading washing machines fully or adjusting the water level control lever on the washer to the size of the load.

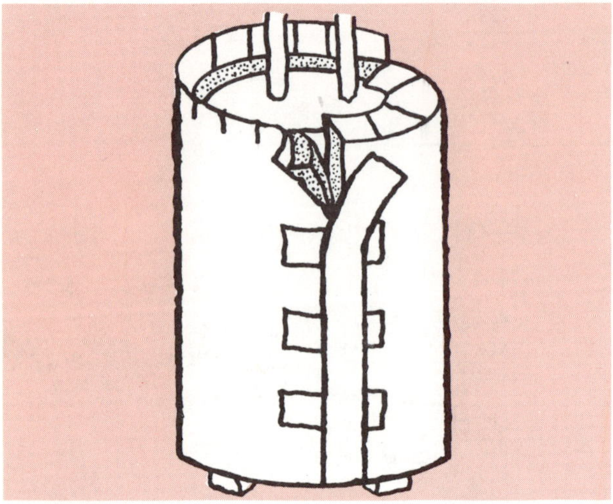

Fig. 13-38. Insulating blankets are designed especially for installation around water heaters.

LIGHTING

According to the U.S. Department of Energy[1], 20 percent of all electricity generated in the United States is used for lighting in homes and businesses. Saving of energy for lighting is possible in three respects:

1. Selection and purchase of more efficient bulbs.
2. Proper fixtures or placement of light.
3. Changing of personal habits regarding use.

Light from all sources is measured in LUMENS. A lumen is the amount of light which falls on a surface. A light source such as an electric light bulb uses energy, which is measured in WATTS. A watt is the measure of electrical power. One watt equals a current of 1 ampere at a pressure of 1 volt.

The efficiency of a light source is determined by how much light (lumens) is produced by a watt of electrical energy. It is like miles per gallon for an automobile.

When you purchase light bulbs, the package will indicate the wattage of the bulb. This is the indication of the amount of energy the bulb will use in a given amount of time. Usually, the package will also indicate the lumens the bulb will deliver. The more lumens delivered per watt the more efficient the bulb. Fig. 13-39 compares the efficiency of various types of bulbs.

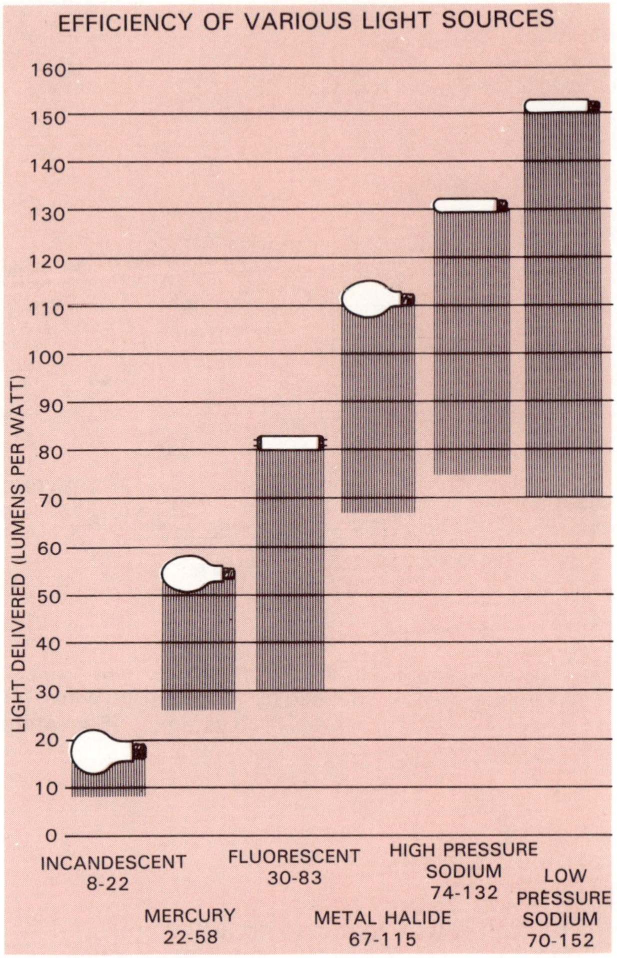

Fig. 13-39. Types of artificial lighting vary greatly in their efficiency. Least efficient is the incandescent since it must first convert the electrical energy into heat.

As you can see from this chart, the incandescent bulb, Fig. 13-40, is the least efficient. It delivers from 8 to 22 lumens per watt. Incandescent bulbs waste so much electricity because they produce light through heat. The heat is wasted because it cannot be used for anything else.

[1] "Lighting", U.S. Dept. of Energy., July, 1980., Tech. Info. Center., Oak Ridge, TN.

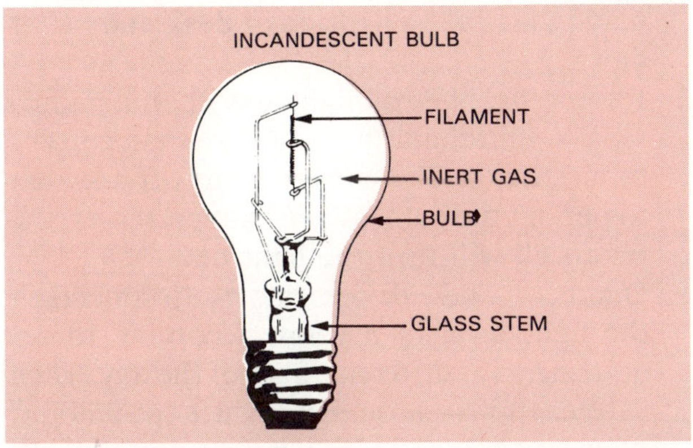

Fig. 13-40. The incandescent bulb. Efficiency increases as the wattage increases. Thus, one 100 watt bulb produces about the same amount of light as two 60 watt bulbs and uses less electricity. (U.S. Dept. of Energy)

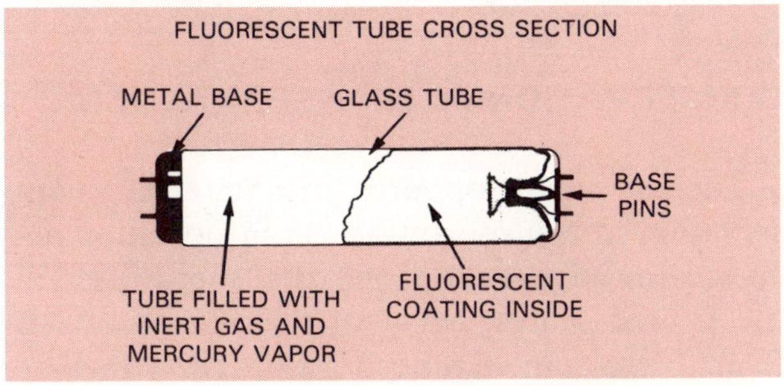

Fig. 13-41. Cross section of fluorescent tube.

FLUORESCENT LIGHTING

Fluorescent lights, Fig. 13-41, do not depend upon heat buildup to produce light. They use an electric charge to stir up atoms of a gas trapped inside the tube. A small transformer called a BALLAST, Fig. 13-42, causes the electric charge or spark and keeps a current moving through the tube. The motion of the gas affects the phosphor coating on the inside of the tube and makes it glow. Fluorescent lights are five times more efficient than incandescent bulbs. Fig. 13-43 is a guide to selection of fluorescent lighting.

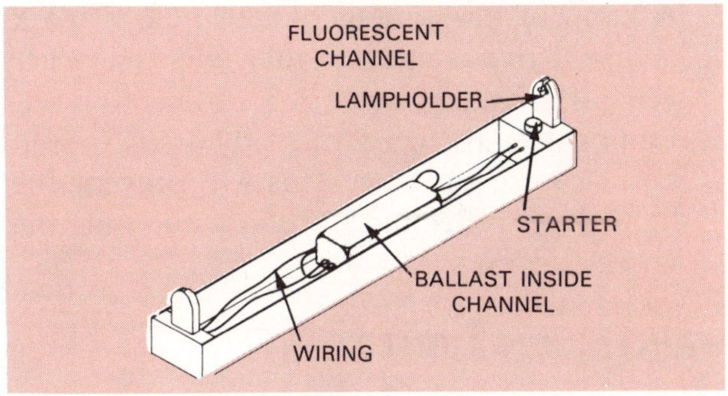

Fig. 13-42. Fixture required for fluorescent light is more expensive than for incandescent lighting but fluorescent bulbs are more efficient and longer lasting.

PROPER PLACEMENT

Placement of light fixtures in relation to surroundings affects their efficiency. The following tips will help increase the amount of light you can get from lighting:

1. Light-colored walls and bright surfaces reflect more light than dark surfaces.

	EFFECT ON "ATMOSPHERE"	COLORS STRENGTHENED	COLORS WEAKENED OR GRAYED	REMARKS
Cool white	Neutral to fairly cool	Orange, yellow, blue	Red	Blends with natural daylight
Deluxe cool white	Neutral to fairly cool	All nearly equal	None appreciably	Simulates natural daylight
Warm white	Warm	Orange, yellow	Red, blue, green	Blends with incandescent light
Deluxe warm white	Warm	Red, orange, yellow, green	Blue	Simulates incandescent light

Fig. 13-43. Fluorescent tubes are manufactured for different qualities of light. (U.S. Dept. of Energy)

2. A lamp of low wattage placed near a work area will give as much usable light as a larger general area light such as a ceiling fixture.
3. Lamps used for work or study areas should be placed to the side. This will prevent the light from reflecting off the work into the viewer's eyes.

MAINTAINING LIGHTS

Frequent cleaning of lights, particularly the reflectors, helps to keep them delivering the light levels for which they are designed. A lamp producing 20 lumens per watt when installed may produce only half the light after it is covered with dust. Fluorescent reflectors, especially, are known to collect dust. They need frequent cleaning.

EFFICIENT USE OF LIGHTING

Wise usage of lighting will produce more light for less wattage. The checklist in Fig. 13-44 suggests how to increase lighting efficiency.

AUTOMATIC CONTROLS

Lights can be controlled with automatic devices that will turn them off should they be forgotten. All of these devices, if used consistently, will reduce the amount of energy used. The savings more than offset their cost. Devices include:

1. Timers. These can be set to turn lights on and off at set times. For example, they can be made to turn the light on at dusk and off at dawn.
2. Photocells. Natural light acts upon a photosensitive mechanism causing a small current of electricity. The current turns off a switch to put out the light. When light decreases, the photocell will turn the light on.
3. Dimmers. These devices allow adjustments to the light level in a room according to requirements at different times of the day. They save energy because they reduce the amount of energy being used when light demands are low. There are two types: solid state and variable automatic transformers. A third device, a rheostat, is not an energy saver. Though it does cut down the light level, the excess energy is turned into heat and so it is wasted.

- Do not overlight an area. Try lower wattages and choose one that is adequate. Use daylight when you can.
- Larger bulbs use energy more efficiently. Try to replace two bulbs with a single bulb which delivers as much light as the two.
- Four-watt night lights have a clear finish and deliver almost as much light as 7 watt night lights. They use only about half as much energy.
- Put fixtures on individual switches. Use only as many lights as you need.
- Turn off decorative outdoor gaslights or replace them with low-wattage incandescent lamps.

Fig. 13-44. Checklist for more efficient use of lights.

CONSERVATION IN TRANSPORTATION

There are many opportunities for energy conservation in transportation. Transportation accounts for 26 percent of the total energy used in the United States. Most of this energy comes from petroleum. Transportation fuels include gasoline, diesel fuel, and alcohol.

The energy conversion machine is primarily the internal combustion engine. These include the gasoline engine, diesel engine, gas turbine, and jet engine. Electric motors are used in a few types of transportation units such as commuter trains.

Transportation has two purposes: to move people and things. Seventy percent of transportation energy is spent in moving people. By far the most used mode of transport is the automobile. Next most popular for passenger service is the airplane and then the train and the bus. Trucking is the preferred mode for moving freight. See Fig. 13-45.

When we compare the costs of using each type of vehicle for moving people we can readily see that we pay a high price for convenience and speed when we travel. The automobile and the

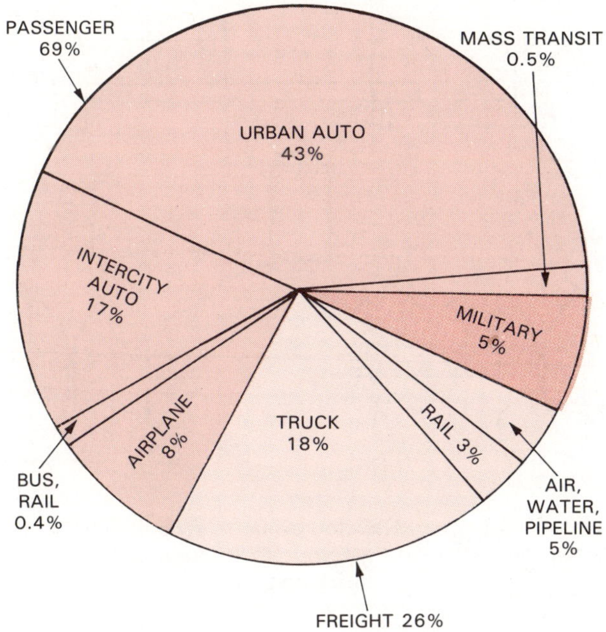

Fig. 13-45. How energy use is distributed among types of transportation. Percentages are approximate.

	Inter-City		
Mode	Average Passenger Capacity	Average Load Factor (percentage)	Energy Cost (BTU per (passenger mile)
Railroad	69/car	35%	2900
Bus	41	45%	1600
Automobile	5	48%	3500
Airplane	106	50%	8400
	Urban		
Bicycle	1	100%	200
Walking	—	—	300
Rail (Trolleys, Subways)	?	?	2300
Bus	55	20%	3800
Automobile	5	28%	8500

Fig. 13-46. Comparison of modes of transportation and energy cost per mile passenger is transported. What does it tell you about efficiency of the automobile?

airplane, though most used, cost more energy per passenger mile than other types of transportation, Fig. 13-46.

METHODS OF CONSERVATION

The automobile is not an efficient way to transport people in the city. However, we can use it more efficiently by joining a carpool to get to and from jobs. The Federal Highway Administration estimates that on a 30 mile round trip a driver can save from $280 to $650 per year by pooling.

Other methods of conserving fuel might include:

1. Planning shopping trips so that a number of chores could be accomplished at one time. For example, if your Saturday schedule includes trips to the supermarket, lumberyard, and post office, try to do it all in the same trip with the most efficient routing, Fig. 13-47.
2. Using other modes of transportation. Consider walking four blocks to pick up the newspaper. Use public transportation for getting to and from work.
3. Changing driving habits to get better gas mileage from the automobile. Consider the following:

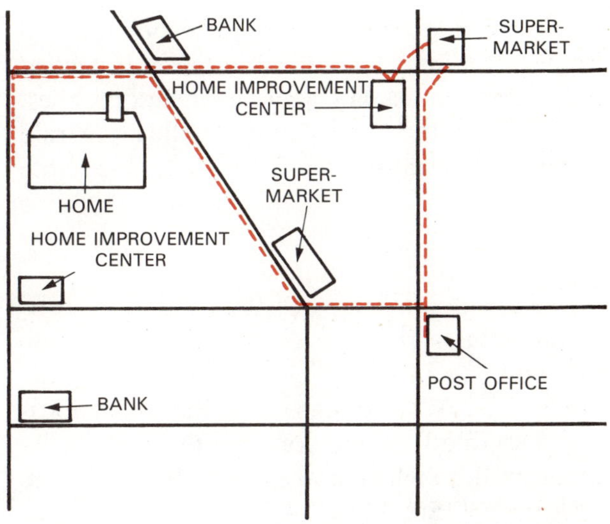

Fig. 13-47. Plan to take care of your shopping trips in the most direct and efficient way. Look at the proposed route. Is it the shortest?

a. Accelerate at a smooth, steady pace. It saves on engine wear as well as tires and fuel.
b. Keep up a steady pace if possible.
c. Anticipate slowdowns and stops; hold braking to a minimum.
d. Avoid long warmups in cold weather. A couple minutes should be enough.
e. Keep air filters clean. Dirty filters cause an overly rich fuel mixture which wastes fuel.
f. Keep tires properly inflated. When buying new tires consider radials; they reduce energy consumption.
g. Remove unnecessary loads. Extra weight reduces fuel efficiency.
h. Keep windows rolled up, if possible, it reduces wind drag and improves gas mileage.

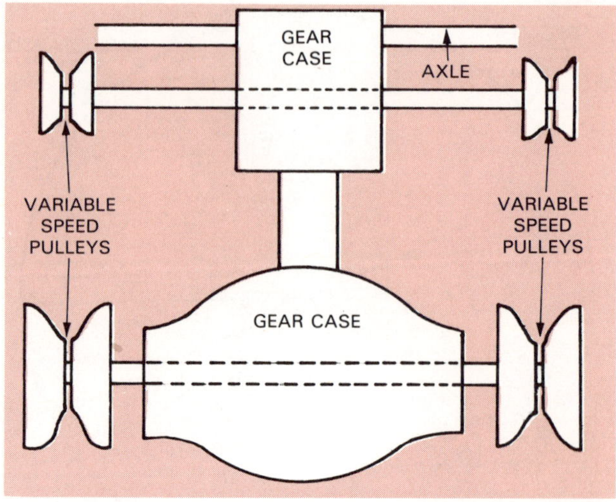

Fig. 13-48. New CVT (continuously variable transmission) design is one attempt to improve efficiency of automobile.

IMPROVING VEHICLE EFFICIENCY

Internal combustion engines convert from 22 to 38 percent of the energy they use to useful motion. The rest escapes as heat or is absorbed as friction. It is not possible to greatly improve this level of efficiency because of the law of thermodynamics. Still, some improvements are possible.

One improvement is to reduce the size and weight of the vehicle so that less power is needed and smaller engines can be placed in automobiles. Still another is to reduce slippage in the drive train when power is transmitted from the engine to the back wheels.

This is possible through:

1. Use of manual transmissions and more efficient gear ratios.
2. Producing more efficient automatic transmissions. A new design called the continuously variable transmission (CVT) uses v-belts and pulleys which can automatically change their size. See Fig. 13-48.

The principle of the CVT is shown in Fig. 13-49.

If you drive a large pulley with a small one, the shaft attached to the large pulley will turn more slowly than the shaft on the little pulley. It will also take less power to turn the small pulley. If a large pulley is driving a small pulley, the small pulley shaft will turn much faster than the shaft of the pulley from which the power is coming. At the same time, more power will be needed to turn the larger pulley.

The variable pulleys in the CVT can be made larger and smaller through the use of weights in the hubs of one set of pulleys from which the power is coming. Thus, as less power is needed the power pulleys become larger and the driven pulleys smaller.

REDUCING WIND DRAG

Auto and truck manufacturers are working on ways to reduce the wind drag on vehicles. Drag is due to the laws of aerodynamics.

Aerodynamics is the study of the motion of air and gases and the forces on bodies moving through such gases or air. Vehicle body designs are placed in wind tunnels where the effects of

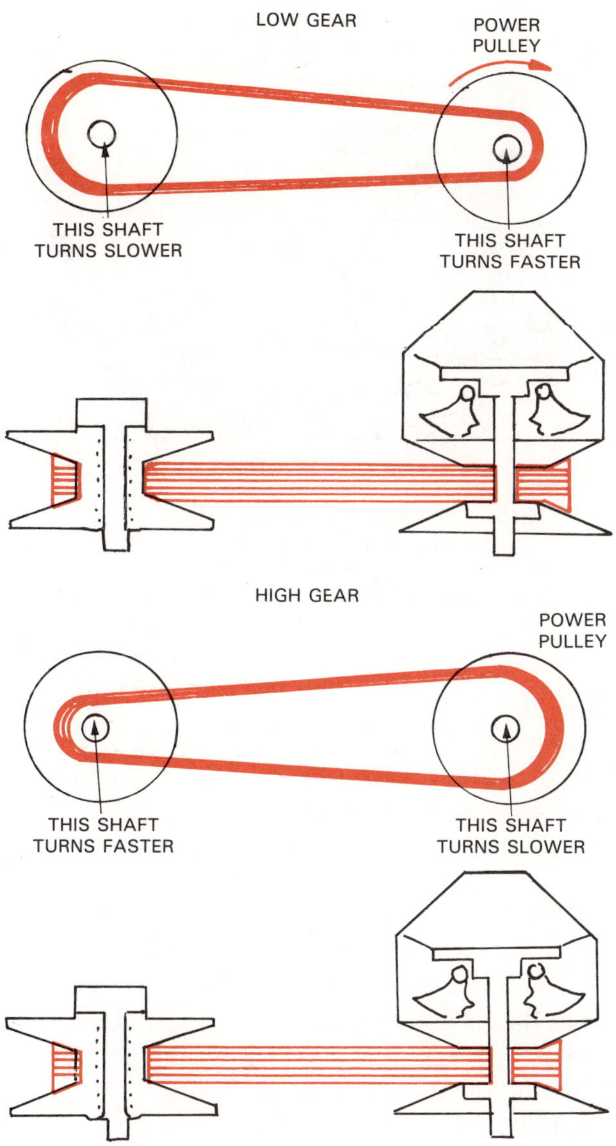

Fig. 13-49. How the continuously variable transmission works. Top. At low speeds the righthand pulley is small and lefthand pulley is larger and easier to turn. Bottom. As auto speed increases, righthand pulley is large while lefthand pulley is smaller and harder to turn.

Fig. 13-50. Three types of drag reduce the efficiency of automobile as its speed increases.

wind drag on different shapes can be studied. Usually, a model of the body is placed in the tunnel while air is blown past it.

There are three types of drag imposed by motion of the car body, Fig. 13-50.

1. Pressure drag. This is the type you experience when riding a bicycle. Air is forced against the front of the body and creates greater air pressure. Eddies and whirls form in the back of the body as the air flows past and absorbs energy. Streamline forms are shapes which reduce this drag. The classic streamline form is the teardrop. A cross section of an airplane wing takes this basic shape.
2. Induced drag. This type does not concern automobile body design much. It occurs mostly at the tips of airplane wings. It is created when air being compressed under the wing slips over the tip to the lower pressure on top of the wing. It creates eddies which drag on the wing.
3. Skin friction drag. This is the tendency of the air to "stick" to the surface of a moving object. This is somewhat like the effect of pulling

an object through thick molasses. Of course, the air is not as viscous, but at high speeds the effect is similar.

REFINEMENTS TO INTERNAL COMBUSTION ENGINES

Fuel consumption guidelines for the automobile have been established by the federal government. The guidelines set a fleet average of 27.5 mpg (miles per gallon) to be reached by 1985. (To establish a fleet average you would add up the averages of gas consumption of all the different models offered by a manufacturer and then divide by the number of models.)

Many improvements are being made on automotive engines to make them more fuel efficient. The most important of these advances currently in production are fuel injection and electronic ignition. At least one manufacturer is testing a new design in the combustion chamber known as stratified charge.

FUEL INJECTION

Fuel injection is a system of delivering air and fuel separately to the combustion chambers of the engine. It is much different than a carburetor fuel system.

In a carbureted system, the carburetor mixes air and fuel which then moves into the intake manifold, a group of passages leading to the cylinders. In the fuel injected system the air and fuel are mixed just before they enter the cylinders.

CONTROLLED COMBUSTION

Researchers for Texaco are developing an improved internal combustion design which has been giving more than 35 percent improvement in gas mileage during tests. Known as the Controlled Combustion System, Fig. 13-51, it is adapted to a conventional gasoline engine. The system carefully controls fuel injection, ignition (spark), and air swirl in the combustion chamber. See Fig. 13-52.

Fig. 13-51. Texaco Controlled Combustion System combines fuel injection (arrow) with carefully controlled ignition and air turbulence in the combustion chamber of a six cylinder engine. (Texaco, Inc.)

For a number of years, automotive manufacturers have been experimenting with different types of power units for automobiles.

STIRLING ENGINE

The Stirling cycle engine gets its energy from heat that is generated outside the cylinder. This is vastly different from the internal combustion engine where fuel must be burned inside the cylinder.

The cylinder of the Stirling engine is filled with trapped air or gas. The cool gas is heated by an outside source. The heat can be caused by any source, including sunlight.

The gas is compressed and then heated. The hot gas expands, driving the piston which turns a crankshaft. After the gas has expanded, an outside source cools it and it contracts again. The piston returns and the cycle is repeated. Greater efficiency is possible with the Stirling engine because it has a power stroke every revolution. The four-stroke cycle internal combustion engine

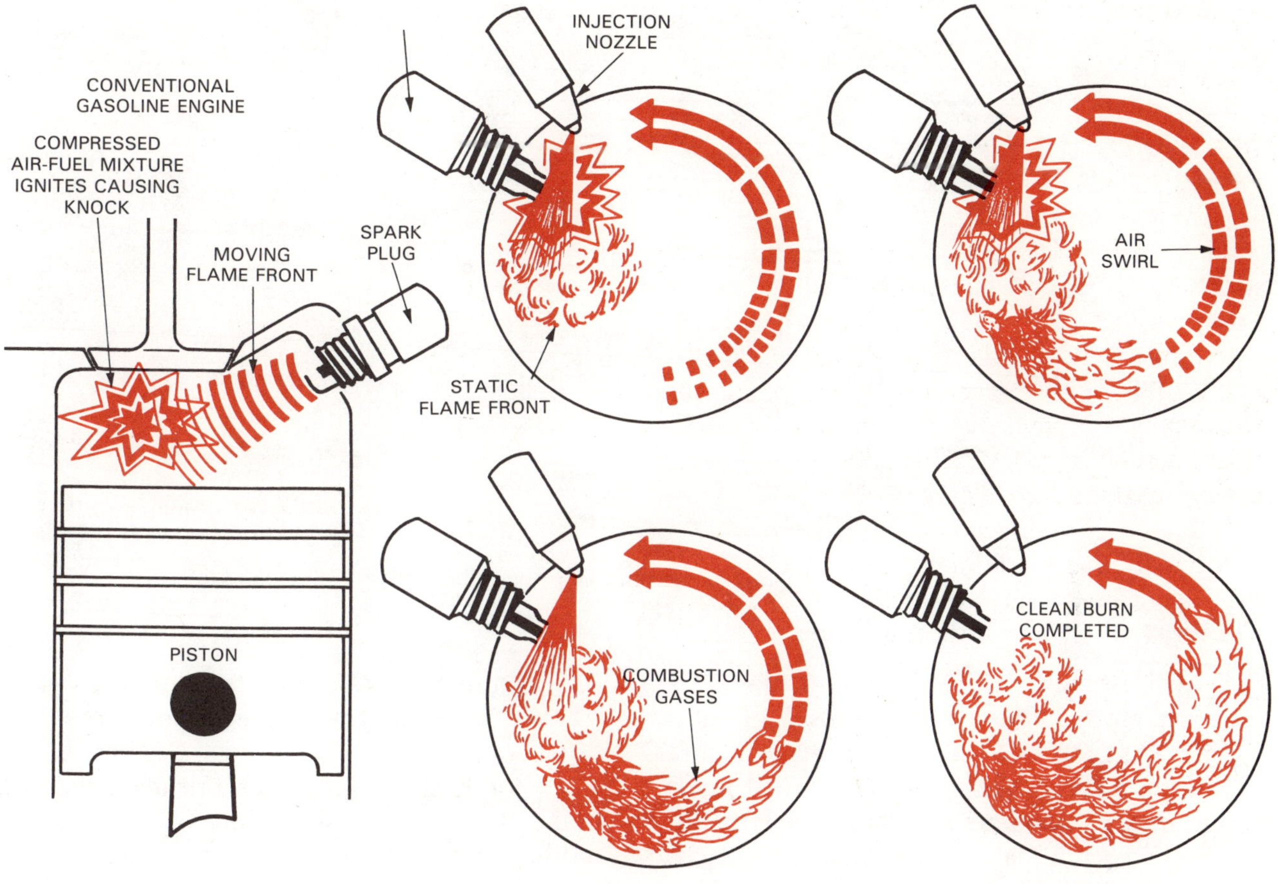

Fig. 13-52. How controlled combustion system works.

has one power stroke every two revolutions. Efficiency of the Stirling cycle engine is above 48 percent.

It has been found that hydrogen and helium are the best working gases for the Stirling engine. The gas must be contained at 200 atmospheres pressure.

The Stirling engine has several disadvantages. One is that it must dissipate (get rid of) great amounts of heat. Cooling is a problem when the engine is mounted in an automobile. A second problem involves the use of hydrogen. At high pressure, the hydrogen is hard to contain. A seal leak could be a safety hazard. However, engineers are working on ways of solving these problems.

GAS TURBINE

Gas turbines operate on the same principle as the water turbine. They have been used successfully to power airplanes, generators, helicopters, large trucks, and boats. See Fig. 13-53. Cool air is taken into the turbine and heated by compression and an exchanger to 1100°F (593°C). As hot air enters a combustion chamber fuel is sprayed into the chamber and ignited. Hot gases at 1700°F (927°C) exhaust into a turbine and cause it to rotate at high speed.

The turbine is more efficient than the internal combustion engine. Where the latter are 22 to 38 percent efficient, turbines are 42 to 48 percent efficient. Moveover, they have greater power for

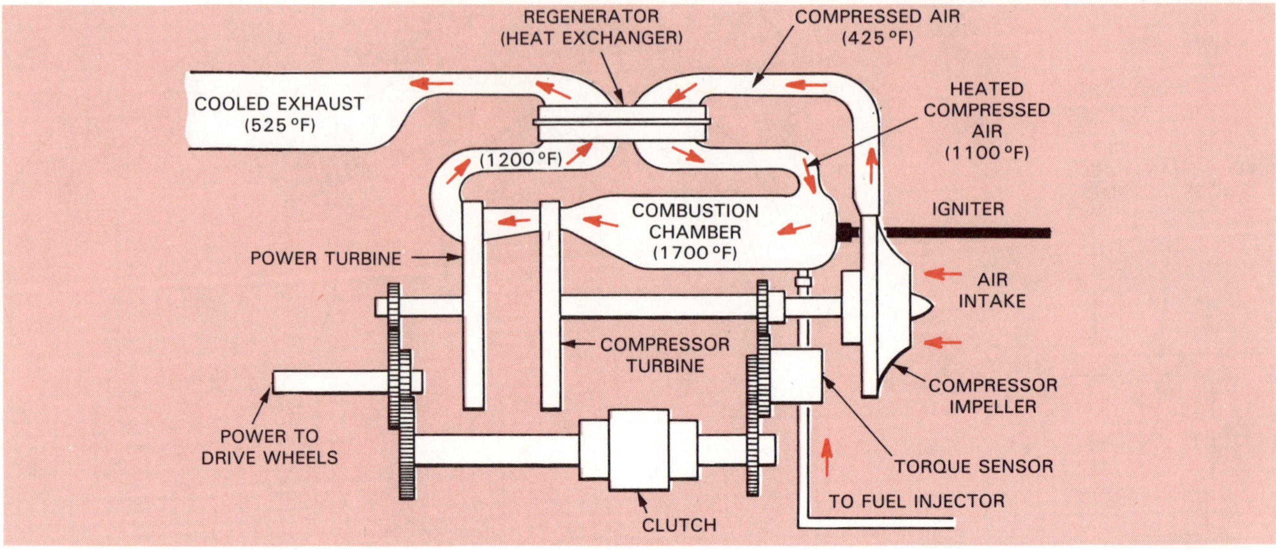

Fig. 13-53. Simple schematic of a gas turbine engine.

their size. Because of the expense of manufacturing them, they have not been used in automobiles. As fuel costs rise, price of manufacture will be less a factor. Another thing in their favor is that they will operate on almost any fuel including powdered coal, grain alcohol, and biogas.

ELECTRIC AUTOMOBILE

The electric motor has many advantages as an energy converter supplying power for an automobile. It is efficient, makes hardly any noise, does not use hydrocarbon fuels, and does not pollute the air. Electric vehicles have been in use since 1920 and are still manufactured today, Fig. 13-54.

That the electric automobile is not more popular is due to the problem of how to store the electrical energy the automobile uses. Today's lead acid batteries cannot store enough energy and they are too heavy. There are other disadvantages:

1. They cost over $600 for a set.
2. They do not last long. A set must be replaced every three years.
3. Constant charging and discharging shortens battery life.
4. Too much bulk. The batteries take up space that is needed for passengers or cargo.

Engineers are working on refinements that will make electrics more appealing. A new type of battery that is lighter and more efficient is one approach. Lithium sulphur batteries are one type being developed, Fig. 13-55. Other electric car technology under development includes:

1. Regenerative braking. This is simply using the motor for braking by running it as a generator to recharge the batteries.
2. Using a "super flywheel" to store electric power. The flywheel would be brought up to top speed by an electric motor plugged into house current at night. When the automobile is in use, the flywheel would drive a generator. The current would be fed to individual electric motors at each wheel. Again, regenerative braking would feed electricity back to the motor driving the flywheel.

LONG RANGE PLANNING

Looking into the future, it is considered possible that the need to conserve fuel will affect the planning of communities and change the modes of travel from what we know today.

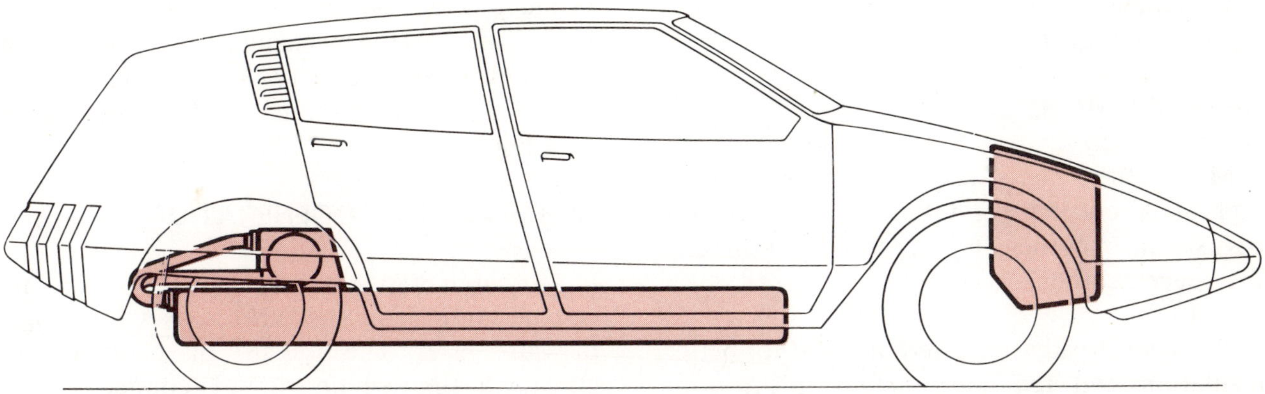

Fig. 13-54. Basic concept of an electric car. This design is for a four-passenger vehicle which would use zinc-chloride energy storage system. Long container under passenger compartment holds storage batteries. Forward storage is for battery electrolyte. (Energy Development Associates, A Gulf & Western Co.)

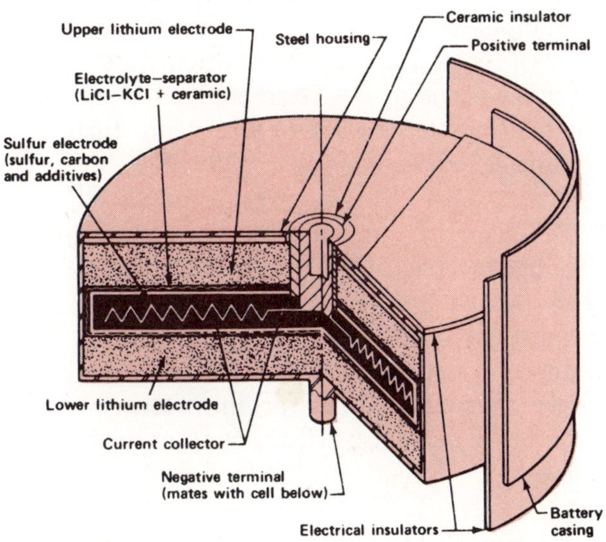

Fig. 13-55. Cutaway of a lithium sulphur battery. (ERDA)

It is possible that cities would be much smaller and laid out in a circle. Private vehicular traffic would be prohibited. Bicycles would become the most-used conveyance. Shopping and other necessary community services would be within walking distance. Small neighborhood stores would spring up. Truck traffic would be permitted to service stores and other businesses. Automobiles would be used for intercity traffic within 500 miles (805 km). Aircraft would be used for more distant travel.

CONSERVATION IN INDUSTRY

Industry—manufacturing, mining, agriculture, forestry, and fishing—consumes 40 percent of U.S. energy. Just six industries use half of this energy. They include: metals, chemicals, petroleum and coal products, stone, clay, glass, pulp and paper, and food processing.

ENERGY MANAGEMENT FOR INDUSTRIAL CONSERVATION

In some respects, conservation in industry is like residential conservation. One area of similarity is the confinement of heat so it is not wasted. Thermostats can be lowered. Offices and factories can be better insulated. Air infiltration can be controlled through sealing up cracks and installing doors and windows with better thermal insulation qualities.

A second area where industrial conservation is an entirely different problem is in improving machines and processes to get more product for the same input of energy. Control of heat and improvement of energy usage is called ENERGY MANAGEMENT.

HEAT RECOVERY

Much of the energy used by industry is spent on heat. About 39 percent is used to produce process steam. Another 26 percent is used directly for space heating.

We have already suggested the need for better insulation and reduction of air infiltration. Another important method of conservation is recovery of heat that is now being carried away by smoke stacks, exhaust systems, and water cooling systems. There are several systems for recovering this heat.

COGENERATION

This is the production and use of heat energy to give space or process heat and electrical power at the same time, Fig. 13-56. The steam used for powering turbines in electric power plants is typically released to the atmosphere. So, too, in industrial plants, steam is produced for manufacturing a product. The energy in the steam can also produce electric power. The savings in energy are high. Basically, the mechanical energy in the heat is used to create electricity. The remaining energy provides heat for buildings and manufacturing processes.

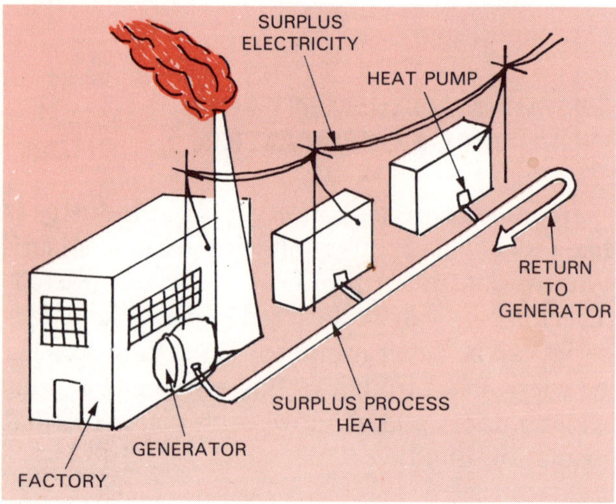

Fig. 13-56. Cogeneration system. Mechanical energy of heat generates power. At same time exhaust heat is used to heat buildings and to provide heat for manufacturing processes.

SYSTEMS FOR COGENERATION

Several different systems can be used for cogeneration. The gas turbine uses hot, high-pressure gases. The pressure of the expanding gases spin the turbine which is connected by a shaft to a generator. Spent exhaust gases from the turbine are directed into a boiler system. From 62 to 90 percent of the energy in the gases can be captured and used.

The steam turbine is driven by high-pressure steam from a boiler which burns fossil fuels or municipal waste. It, too, produces electricity. Low-pressure steam, from which most of the mechanical energy has been removed, then is used for heating. Efficiency of the steam turbine is around 85 percent.

Diesel engines may also drive a generator and produce electricity. Waste heat can be collected from the water cooling system and from the hot exhaust gases. This heat can be used to heat more water or even produce steam. Diesels usually have lower efficiencies than turbines. However, cogeneration can raise efficiency to as much as 90 percent.

A new type of conversion system uses fluidized bed combustion. This is simply a new method of burning coal that is more efficient and nonpolluting. Finely crushed coal is mixed with limestone. It rides into the firebox of the boiler on a stream of air. The coal burns at a lower temperature that reduces the pollution. At the same time the limestone absorbs the sulphur dioxide released by the burning coal.

While most systems for cogeneration now use fossil fuels, they are likely to be using more renewable energy sources in the future. A large

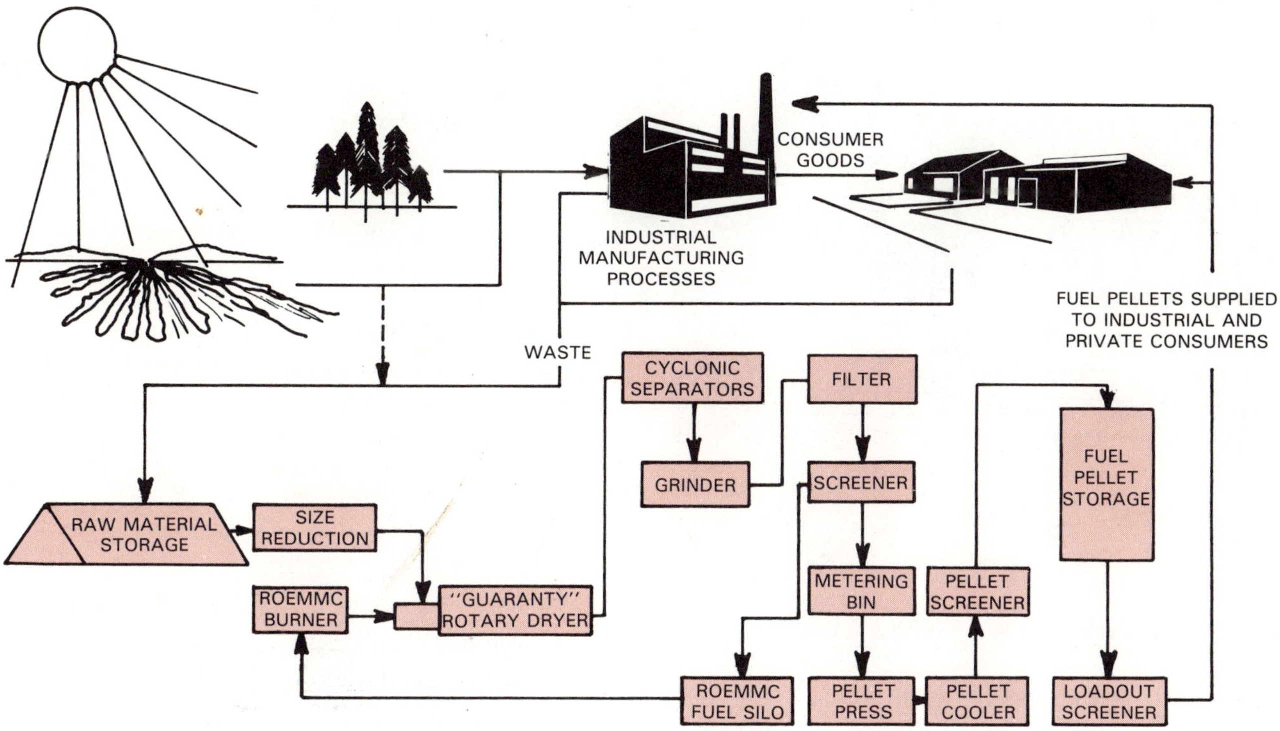

Fig. 13-57. Ideal biomass recycling system. Waste wood is used to provide energy for manufacturing. (Guaranty Fuels, Inc.)

walnut processing plant in California, for instance, feeds walnut shells into its turbine generator to produce electricity. The generating plant produces 32 million kilowatt hours of electricity. Fig. 13-57 is a schematic of an ideal biomass energy system.

WASTE MATERIAL RECOVERY

Recovery of metals and minerals from discarded and wornout products is called RECYCLING. Some efforts at recycling have been and are being made. These include collection of scrap paper and aluminum, old iron, and junked automobiles.

Efforts at widespread recycling are not very successful because of the amount of labor involved in collection. However, large-scale recycling of metals and glass could save energy in the manufacture of new products. For example, it requires only 700 kilowatt hours of electricity to produce a ton of steel from scrap. To produce the same amount of steel from ore would require the use of 2700 kW/hr of electricity. It takes four times as much energy to produce new bottles as to refill returnable bottles. Reprocessing of aluminum cans takes five times less energy than processing from raw bauxite from which aluminum is made.

INFRARED HEATING

Most heating is done through convection and conduction. Infrared heat moves from one place to another through radiation. Infrared heating in factories, warehouses, and public buildings is more efficient than conventional heating. The radiant heat travels through space and releases its heat energy when it strikes something solid. See Fig. 13-58.

Fig. 13-58. Infrared system is a more efficient form of heating.

DO YOU KNOW THESE TERMS?

Ballast, caulk, cogeneration, degree day, embargo, EER, exfiltration, heat pump, infrared heating, infrared photography, lumens, molten, oakum, recycling, regenerative braking, rheostat, R rating, U rating, wind drag.

SUGGESTED ACTIVITIES

1. Survey your community to see what your neighbors may have done to conserve energy with respect to any of the following:
   a. Their automobile.
   b. Their commuting arrangements.
   c. Their house or apartment.
   d. Their business.
2. Have a class discussion on how to plan a city which would save energy resources. Assume that you are designing a new community and can place recreational, business, residential, and industrial areas for maximum convenience.
3. Plan and build a device which will be useful in recycling some type of household organic waste or mineral waste.
4. Plan a new energy saving residence and produce a scale model to demonstrate your plan. Ask your instructor to recommend reference books on energy saving construction.
5. Contact a company which conducts energy audits of housing. Collect literature and interview them on how an energy audit is done. Report to the class.

TEST YOUR KNOWLEDGE

1. Conservation of energy means to ________ any energy source by using it ____________ and by __________ __________ ________.
2. List two purposes of conservation of energy.
3. Name three types of personal action you can take in energy conservation.
4. Dwellings may waste as much as ________ percent of the energy they use.
5. An R value is:
   a. Unit for conductance of heat through matter.
   b. Unit of resistance to transfer of heat through any matter.
   c. A symbol for insulation.
6. Severity of climate in a certain locality is measured in ____________ ____________.
7. If temperature averaged 32°F (0°C) over a 24 hr. period what would be the degree days for that period?
8. Warm air rising in a closed space to leak out of cracks or into unheated upper levels is called the chimney effect. True or False?
9. What is an EER rating and how is it calculated?
10. A ____________ ____________ is basically a refrigerator which is capable of either heating or cooling.
11. Why are fluorescent lights more efficient than incandescent lights?
12. Automotive internal combustion engines convert from __________ to __________ percent of the energy they consume to useful motion.
13. How does fuel injection differ from a carburetor in feeding fuel to the automotive engine?
14. ____________ is the use of heat energy to produce space or process heat and electrical power at the same time.

14 CAREERS IN THE ENERGY FIELD

The information given in this chapter will enable you to:

- Discuss energy jobs opportunities in different industries.
- List some types of jobs offered by energy industries today.
- Describe the kinds of work done by energy industry employees.
- Discuss the effects of conservation and pollution control efforts on workers in energy jobs.

Energy companies in the United States are one of the largest groups of employers. This is easy to understand since every year the United States alone uses a third of the energy consumed in the world today.

As you learned in Chapter 7, energy is consumed by various sectors of society. Some industries, such as steelmaking, drug manufacturing, and plastics fabrication, use petroleum, natural gas, or coal as feedstocks. (That is, these products are the raw materials from which other products are manufactured.)

Other fuel users include farms and factories which consume energy to operate machines and produce process heat. Each of us represents two other groups which are the heaviest users of gas and liquid fuel. Our residences, offices, schools, and automobiles consume the lion's share of these energy sources to help us live, ride, study, and work in relative comfort.

THE JOB MARKET

It is not easy to predict what the job market will be for energy workers in the years ahead. A study by Congress in 1979 predicted that 2.1 million jobs could be created in energy-related fields. But the study added that we would have to work hard at conservation and solar energy.

We cannot expect that energy jobs will continue to be the same as they have been. The energy industry is entering a period that will see many changes.

NEED FOR CONSERVATION

As mentioned before, there is a need to conserve scarce fossil fuels. We need also to develop renewable resources as well as the know-how to harness them for our energy needs.

There will continue to be jobs in traditional energy fields and coal mining will, without question, continue to increase. See Fig. 14-1. The need to conserve precious fossil fuel reserves will have an effect on every person working in the industry. Anyone entering the field today will need certain additional understandings and skills:

1. Why conservation is necessary.
2. How we can cut down our use of fossil fuels.
3. Technical know-how for conservation.
4. Why continued pollution of our environment is not acceptable.

Fig. 14-1. Mining of coal will offer long-term job opportunities in the fossil fuel energy field. This coal miner is operating a low-slung electric truck called a shuttle car. It hauls coal away from a continuous mining machine. (National Coal Assoc.)

5. Skills for developing and finding ways to use alternate energy sources.

FOSSIL FUEL MARKET

U.S. consumption of energy is much greater than that of any other country, Fig. 14-2. Our per capita (each person) usage is four times that of all western Europe and 20 times that of Communist Asia. Our autos alone use one out of every nine gallons of oil consumed in the world.

Most of the job opportunities in the energy field will continue to be in the fossil fuel industries. This will be true for at least the next decade. However, as we will see later in this chapter, new jobs are going to be opening up in alternate energy fields.

Coal, being our most plentiful fossil fuel, offers career opportunities on a long term. As oil supplies dwindle in the next century, it is likely that energy companies will turn to coal as a source, not only for solid fuels, but for liquid and gaseous fuels too. Several pilot projects on coal gasification are already in operation. Other plants are under construction.

FOSSIL FUEL CAREERS

As gas and oil reserves diminish, there will be a gradual shift in jobs that produce and transport energy. It is expected that scientists will be in demand in the oil, coal, gas, and conservation in-

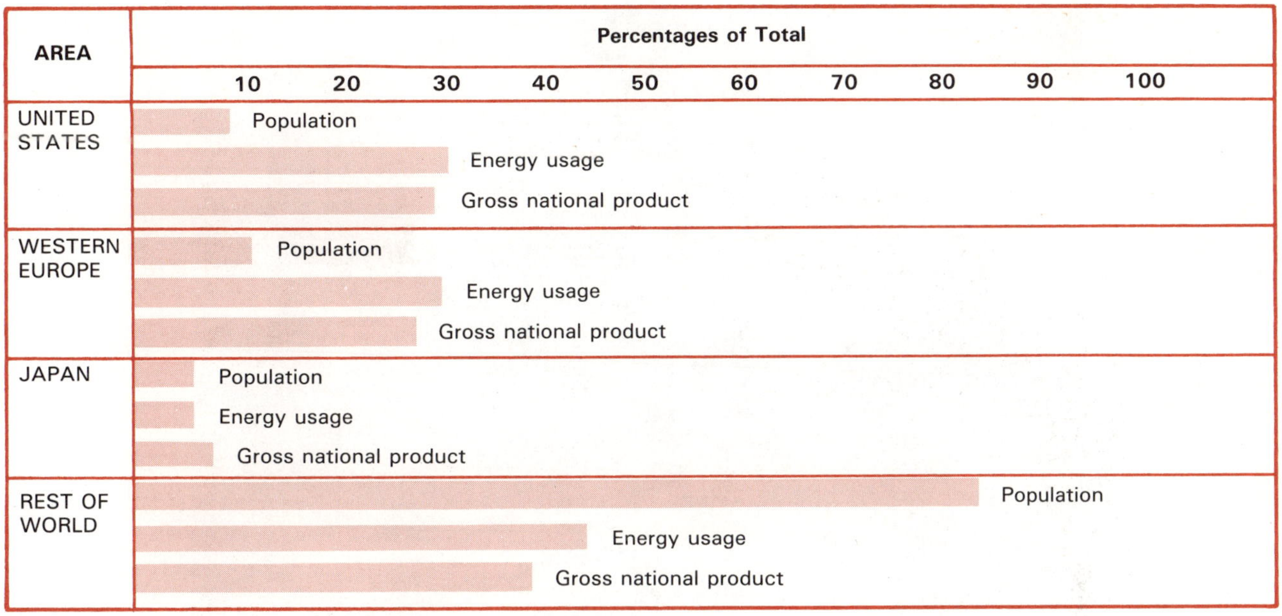

Fig. 14-2. Comparison of U.S. population and energy consumption with the rest of the world. With six percent of world population, U.S. uses a third of the world's energy.

dustry. They will be concerned with new technology for synthetic fuels and the alternatives to fossil fuels. They will also be in demand to solve the problems of waste disposal and handling of environmental hazards.

JOBS IN ENERGY EXPLORATION

A number of skills are required for exploration of energy resources under the ground. One group of specialists are called GEOPHYSICISTS. Their jobs are to explore solid earth for the presence of oil, coal, and other mineral deposits.

Exploration GEOPHYSICISTS use seismic prospecting techniques to locate the oil or minerals. They use a seismograph to send sound waves into the earth and record the echoes that return from rock layers below. See Fig. 14-3. This information can be studied to see if rock conditions are favorable to the presence of oil.

SEISMOLOGISTS work with GEOPHYSICISTS. They study and interpret the seismic data.

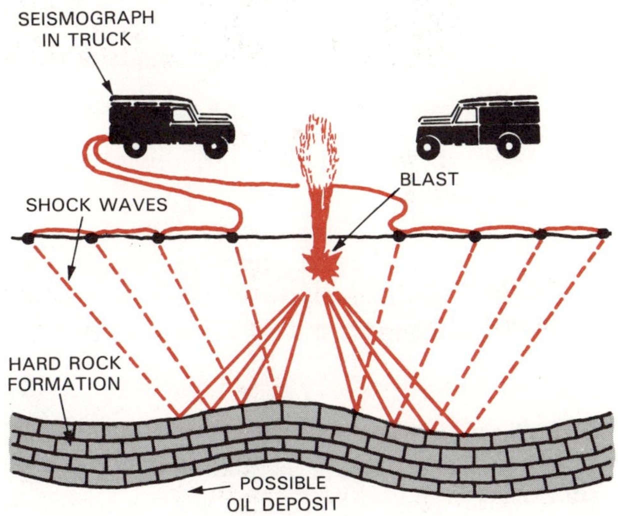

Fig. 14-3. Teams consisting of geophysicists and seismologists use charges of explosives to learn about possible oil, coal, or gas deposits underground. (Shell Oil Co.)

Geology is another field related to the oil industry. GEOLOGISTS study the makeup of the earth, identify rocks and minerals, make surveys, take measurements, and record data. PETRO-

Fig. 14-4. Drilling operations in an oil field. (Texaco Inc.)

LEUM GEOLOGISTS work with teams exploring for oil.

MINING ENGINEERS find, dig out, and prepare minerals for use. They design and construct pit and underground mines. Some develop special mining equipment. Many specialize in coal mining and some may work only on solving pollution problems resulting from coal mining activities.

PETROLEUM ENGINEERS work mostly at exploring and drilling for gas and oil. They design and develop pumps and other devices to force the oil to the surface. Some supervise drilling operations.

SEISMIC OBSERVERS work with seismologists. An electrical prospecting observer uses electrical apparatus for measuring earth resistance to electrical charges. Gravity prospecting observers record readings of meters and other gravity-measuring equipment for purpose of finding oil. They are assisted by observer helpers.

MINE SURVEYORS do surface and underground surveys at coal mines to control the direction and extent of mining. They also estimate the volume of coal in parts of the mine using survey information. Sometimes they draw up maps of mine workings.

OIL WELL DIRECTIONAL SURVEYORS measure sonar, electrical, or radioactive characteristics of earth formations in oil and gas wells. They bore test holes to determine how productive the well will be.

JOBS IN ENERGY EXTRACTION, REFINING, AND DISTRIBUTION

Once oil, gas, or coal deposits are discovered, extraction and refining processes are required. Also, distribution of the energy products must be done.

The DERRICK OPERATOR works on oil drilling crews rigging the derrick equipment and operating mud pumps. See Fig. 14-4.

ACIDIZERS operate equipment for treating oil wells with acid to make them produce more oil. They supervise the ACIDIZER HELPER in loading and mixing chemicals at the well site.

INDUSTRIAL TRUCK OPERATORS drive gasoline or electrically powered trucks or tractors equipped with a forklift, elevating platform, or hitch to move petroleum products in a warehouse.

A BLASTER plants and sets off charges of explosives in mining operations.

GASOLINE FINISHERS control equipment for blending natural gasoline with chemicals to improve its fuel qualities.

HEAVY TRUCK DRIVERS drive coal trucks in strip mining operations, haul gasoline, oil, and other petroleum products, Fig. 14-5. They usually lubricate their own trucks and perform minor repairs.

PIPELINE WORKERS are a classification of careers that are used in several industries. Many are employed in the petroleum industry. PIPELINE CONSTRUCTION INSPECTORS check materials and work during pipeline construction to see that they are up to specifications. They supervise or observe preparation of right-of-way, check trenches for depth and pitch, examine welds, coatings, and wrappings, and advises supervisor of corrections needed. A PIPELINER maintains and repairs pipelines, pumping stations, and tank farms. This person will drive equipment such as backhoes, bulldozers, and side booms to dig ditches. He or she will also lay pipe and backfill ditches. PIPELINE WORKER bolts together pipe sections working as a member of a team. She or he pours sealing compounds over joints and wraps asbestos coverings to prevent corrosion and leaks. See Fig. 14-6.

Fig. 14-5. Truck drivers are in demand to deliver fuel to gas stations throughout the United States.

Fig. 14-6. Pipeline workers' jobs may take them anywhere in the world. This 24 in. pipeline connects Beruk oil field in central Sumatra with three new oil fields. (Texaco Inc.)

MINE-CAR REPAIRERS are skilled at metalworking. They repair and replace damaged parts of underground mine cars. The work often includes welding, straightening parts, and refitting them.

MINE FOREMEN supervise and direct the work of other mine workers. They also supervise the opening of new cuts, pits, or underground rooms and passageways. Another responsibility is the construction and installation of mining equipment.

MINE MACHINERY MECHANICS repair and maintain mine machinery such as stripping and loading shovels, drilling and cutting machines, and continuous mining machines.

FIRE BOSSES inspect underground mines for fire hazards, presence of toxic gases, and inadequate ventilation. They tour hallways, shafts, and work areas looking for hazards. They also keep logs of accidents.

MINER is a term given to all who work in mines. Jobs are classified according to the work performed. For example, CUTTER OPERATORS control self-propelled machines that saw channels along the bottom or side of the working face of the coal mine.

STRIPPING SHOVEL OPERATORS operate the shovels that remove overburden at strip mines prior to actual coal mining operations. Some also repair and replace shovel parts.

TIMBER FRAMERS cut, fit, and install supporting timbers and other framework in underground mines, using carpentry tools. Some may build forms and pour concrete supports.

FILLING STATION ATTENDANTS are one type of salesperson. They service vehicles of all types with fuel and lubricants, pump the fuel into the fuel tanks, check and change oil, and lubricate the vehicles. Many also perform light maintenance and repair.

SALESPERSONS are responsible for selling energy products such as coal, gasoline, and various petroleum products. Some sell the products directly to the consumer. Some travel to call on customers. They describe their products to customers, take orders, and keep records of their sales calls.

These are but a small sample of the occupations within the fossil fuel industry. You can find others through consulting the Dictionary of Occupational Titles, or the Occupational Outlook Handbook. Both are published by the U.S. Labor Department.

CAREERS IN NUCLEAR POWER

There are 77 nuclear power generating stations in operation, Fig. 14-7. Because of public concern over the dangers of radiation and the problems of disposing of nuclear waste, construction on nuclear reactors has slowed. The growth of nuclear power and the potential of nuclear power for careers is uncertain. Also see Fig. 14-8.

However, trained personnel are needed for operation of the nuclear power stations now in use. Trained technical workers are needed for the following jobs:

NUCLEAR MATERIALS PROCESSING TECHNICIAN. These technicians help to fabricate (make), handle, reprocess, transport, and store nuclear materials and fuels. Candidates are selected with great care by the firms who employ them. Two years of study in technical schools is a good background for a trainee. They are employed by nuclear facilities that prepare nuclear fuels and by research organizations and chemical industries.

NUCLEAR INSTRUMENTATION AND CONTROL TECHNICIAN. These persons are specialists in maintaining and repairing instruments and control systems. They must have a broad background in electronics, mechanics, and mathematics. They find employment with electric utilities, manufacturers of nuclear instruments and controls, handlers and processors of nuclear fuels, reactor research, nuclear power plant builders, and many electronic and instrument industries.

Fig. 14-7. Nuclear power generating station. Built in 1976, this plant has a capacity of 965,000 kW. (Power Authority of the State of New York)

Fig. 14-8. Scientists and technicians are required for developing of fusion power reactors. Pictured is the ELMO Bumpy Torus fusion chamber. It is an experimental reactor which the industry hopes will lead to the design and construction of more efficient and safer nuclear power. (Oak Ridge National Laboratory and Union Carbide Corp., Nuclear Div.)

RADIATION PROTECTION TECHNICIAN. These men and women use special instruments to check radiation levels and keep records on intensity and duration (how long it lasts) of radioactivity. They also identify types of radiation. They may also train others in the proper use of instruments. Employers are the same as for nuclear instrumentation and control technicians.

NUCLEAR REACTOR (plant) OPERATOR TRAINEE. This is an entry level job and requires special training in the operation of a nuclear power plant or a reactor used in research. The trainees may learn the fundamentals of reactor operation at a two-year vocational-technical school. Employment and additional training comes from nuclear power stations, nuclear reactor research facilities, or from federal agencies that regulate nuclear activities.

JOBS IN SOLAR ENERGY

New jobs in the solar energy industries will be closely tied to conservation and the construction industry. Some of the job areas are:

1. Insulating older buildings to make them more energy efficient.
2. Building and remodeling houses to use solar energy and cogeneration for space and water heating. (Cogeneration is using waste energy from one operation to do another job. For example, spent steam from a power generator may be used to heat buildings.)
3. Legal and engineering jobs related to solar

Fig. 14-9. Solar age skills. Workers fabricate tubes for boiler of solar thermal electric pilot plant. Tubes 45 ft. long are welded at the rate of 60 in. per minute. (Rocketdyne Div., Rockwell International)

construction and conservation.

4. Collecting and recycling metals and organic wastes which still have some energy in them. New solar and conservation industries will provide jobs for sheet metal workers, carpenters, plumbers, pipefitters, construction workers, insulation workers, painters, welders, air conditioning, heating, and refrigeration technicians, masons, and electricians. See Fig. 14-9.

Careers will be available in law, sales, solar engineering, architecture, waste management, zoning, and consumer protection. Development of equipment for solar and conservation technology will create jobs for process, production, design, and chemical engineers.

It could be that in 20 years solar energy systems in homes and businesses could save 3 million barrels of oil a year. The solar industry has been growing. Sales and installation of solar heating products have grown from $25 million to $260 million in just two years.

Many of the persons needed to install solar heating systems will be trained by two unions. The Sheet Metal and Air Conditioning Contractors National Association and the Sheet Metal Workers International Association have set up 150 training programs all over the United States. They could train up to 11,000 solar installers a year.

JOB CATEGORIES

The following job descriptions are based upon solar careers listed in the Occupational Outlook Handbook, a publication of the U.S. Labor Department:

1. Solar energy systems installer plans the systems installation and tests the completed system.
2. Solar energy systems installer helper helps install and repair solar energy systems.
3. Solar technician performs all functions of a solar mechanic and can also design active solar systems for dwellings and light commercial use.
4. Solar engineer designs all types of active solar systems; must have a degree from an accredited four-year engineering program.

CAREERS IN KINETIC ENERGY

Between 1850 and 1930 the windmill industry provided many jobs. It died out with rural electrification.

It is about to be reformed again. Large companies have begun serious work in research and development on generators capable of large-scale production of electricity. Small wind generators that supply 5 to 10 kW are already working. They are big enough for ranches, farm, and home use. The government is looking at the possibility of typing many such small units together. Then large amounts of electricity could be generated for whole towns and cities.

Another plan being researched by Westinghouse would create whole "farms" of large wind generators in the ocean to use the breezes that are always present off shore.

Many problems must be solved. Among them are the design of floats to support the towers and power lines to carry the electricity to shore. Other questions to be worked on are what effect the towers will have on recreation, shiping, and fishing. Besides engineers and technicians, this development will require welders, steel workers, operating engineers, machinists, and electricians.

WATER POWER JOBS

The water power industry in the U.S. has been developed to the point where hydroelectric plants supply 14 percent of our electricity. Many workers will be needed to maintain and operate present plants.

Additional jobs will be available as we begin

Fig. 14-10. Highly skilled workers install generator rotor into stator at new hydroelectric power station. (Power Authority of State of New York)

to expand hydroelectric facilities. Three-fourths of the nation's water power resources have not been tapped. A survey by Army Corps of Engineers has marked 50,000 locations where hydroelectric plants could be built. This will require engineers, technicians, environmentalists, and surveyors. Construction will require a variety of skilled workers such as welders, concrete workers, carpenters, and pipefitters. See Fig. 14-10. Once these power plants are built, many technical and highly skilled jobs will open up.

BIOMASS CAREERS

Careers in biomass industries are expected to grow slowly at first. Rapid growth of this area is unlikely because of the need for better technology. At the same time, the problems of collection need to be solved. However, small industries are already at work.

WASTE PRODUCT INDUSTRIES

At least 20 companies are involved in bioenergy. Products include:

1. Entire systems for collecting fuel derived from municipal waste.
2. Silvicultural biomass (tree) farm.
3. Plants to produce methane by anaerobic digestion.
4. Plants to convert wastes into fuel by pyrolysis process.

FORESTRY

Forestry technicians or forestry aids assist foresters in care and management of forest lands and forest resources. Their work involves forest protection, improvement of forests, and harvesting of trees for forest products.

Duties can be varied. For example, in forest production they may estimate yields in certain areas. Other tasks could include supervising, surveying, and building of roads.

Most foresters work for either lumber companies or the federal government in the Forest Service of the U.S. Department of Agriculture. Some work for state governments. A few are employed by mining and oil companies on reforestation projects.

OTHER CAREER AREAS

There are many other professional and skilled people whose services are required in energy fields. For example, certain engineering specialties common to other industries are also employed in energy-related jobs. In general, engineers apply science and mathematics to solve technical design problems. They design industrial machines and equipment. They work at testing, production operations, or maintenance. They may supervise manufacturing processes in refining of fuels. Some will work only in sales.

Energy related engineering includes:

1. Agricultural engineering. Part of their work is to design systems to improve conservation of energy and soils.
2. Chemical engineering. Since petroleum is the raw material for many chemical products, this type of engineering is important in the oil industry. Chemical engineers may help design synthetic fuel plants or systems to prevent pollution.
3. Electrical engineering. These engineers may design electrical equipment needed for power generating and transmission.
4. Mechanical engineering. Many engineers of this type are engaged in the design and development of energy producing machines. Some may specialize in producing machines for the petroleum industry.

Other skilled occupations important to the energy industry include carpenters, electricians, machinists, auto mechanics, plumbers and pipefitters, concrete workers, and masons. Many are employed by the energy companies for maintenance work, Fig. 14-11.

Fig. 14-11. Electric power production requires careful maintenance. (Power Authority of State of New York)

DO YOU KNOW THESE TERMS?

Acidizer, agricultural engineer, blaster, chemical engineering, consumer protection, derrick operator, fire boss, forestry technician, gasoline finisher, geologist, geophysicist, mining engineer, nuclear instrumentation and control technician, nuclear materials processing technician, nuclear reactor operator trainee, petroleum engineer, radiation protection technician, seismic observer, seismologist, silvicultural biomass farm, timber framer.

SUGGESTED ACTIVITIES

1. Interview someone in your community who works for an energy company. Report to your

class on the skill requirements, educational requirements, duties, and responsibilities of the job.
2. Study the publications in the guidance counselor's office or library on career opportunities. Select an energy-related career and research it. Prepare a report for your class.
3. With your instructor's permission and assistance, invite a speaker from the energy industry to talk to your class.

TEST YOUR KNOWLEDGE

1. Congress predicted in 1979 as many as ____________ jobs could be generated in energy-related fields.
2. There (will, will not) continue to be jobs in traditional energy fields.
3. What do exploration geophysicists do?
4. Name one source of additional information on energy careers.

Giant earthmovers are controlled by heavy equipment operators. Training is usually provided on the job. This unit is removing overburden to get at a coal seam in a strip mining operation. (Consolidation Coal Co.)

DICTIONARY OF TECHNICAL TERMS

A

ABSORBER: Flat surface of a solar collector which absorbs the heat of solar radiation.

ACCELERATION: Change (speed up) in velocity over a period of time.

ACCUMULATOR: Another name for a storage battery. Electrical energy is stored as chemical energy.

ADVANCED GAS-COOLED REACTOR (AGR): Type of nuclear reactor in which heat is extracted by carbon dioide gas under pressure of 40 atmospheres at 634°C. Heat from the gas is absorbed by water in a heat exchanger and converted to steam for a turbogenerator.

ALGAE: A variety of plant life including seaweed, rockweed, and green scum on ponds.

ALTERNATIVE ENERGY SOURCES: Heat and power sources which do not rely on fossil fuels or nuclear fission. Includes solar, hydroelectric, wind, wave, tidal, biomass, nuclear fusion, geothermal and ocean thermal energy.

AMMETER: Instrument for measuring electrical current.

ANAEROBIC: Decay of biomass which takes place in the absence of air.

ANAEROBIC DIGESTER: An insulated, airtight container where gases from decaying organic waste are trapped. The gases are mostly methane.

ANAEROBIC DIGESTION: Controlled decay or organic matter without oxygen.

ANAEROBIC FERMANTATION: A breakdown of organic substances (including sugars and proteins) by microorganisms in an atmosphere which has no oxygen. End result of breakdown is alcohol and carbon dioxide or other simple molecules. This process is known as digestion and goes on in sewage disposal systems. Has application not only to use of manure as fuel source but to energy farming using carbohydrate-rich plants as an energy-food source (i.e. sugar cane and cassava).

ANEMOMETER: Device for measuring the velocity (speed) of wind.

ANTHRACITE: A coal older and much harder than bituminous.

ATOMIC NUMBER: Refers to the number of protons in the nucleus of an atom equal to the number of electrons around the nucleus of the atom in a neutral state.

ATOMIC WEIGHT: Ratio of the statistical average weight of the atoms of the element to 1/12th of the weight of one atom of carbon-12.

B

BARREL: Unit of measure for crude oil. A barrel equals 42 gallons.

BENCH: Leveled surface produced over an open-pit mine by a bulldozer so that heavier machines can be moved in to remove the overburden.

BIOFUELS: Fuels made from plant life.

BIOGAS: Any gas, usually combustible, produced as a result of the breakdown of organic materials by enzymes. Methane is produced from sewage in this way.

BIOLOGICAL CONVERSION: Also called bioconversion. A general term meaning the changing of one form of energy to another by plants or microorganisms. For example, plants change solar energy into chemical energy.

BIOLOGICAL ENERGY SOURCES: Animal wastes which can be converted by digesters to methane gas.

BIOLOGY: Science of living organisms.

BIOMASS: Organic material (trees, plants, and algae) as well as wastes like manure, dead plant life, sewage, and garbage.

BITUMINOUS COAL: An older form of coal. It is black and much harder than lignite.

BLACK BODY RADIATION: Reverse radiation. The earth and other substances give up their heat to the atmosphere. This principle can be used to provide cooling where there is no electric power.

BOILING WATER REACTOR: A light water nuclear reactor in which steam is generated directly in the reactor core under pressure of about 68 atmospheres.

BORON: Chemical element used in nuclear reactors being mixed with the steel used in the control rods. One of the chemical elements, it is a light, hard nonmetal. It is yellowish brown in solid state and a greenish brown color as a powder. It is obtained chiefly from ore but is also one of the elements found in geothermal steam, especially in Italy.

BREEDER REACTOR: A nuclear reactor designed to produce more fuel that it uses. Neutrons released by the fission process join other isotopes to make them fissionable.

BRIQUETTING: Method of densifying biomass by pressing it between rollers with depressions in them. The depressions produce shapes like charcoal briquettes.

BTU: Unit of measure for heat. It is the amount of heat needed to raise the temperature of 1 lb. of water 1°F.

BY-PRODUCTS: Materials produced or left over after the making of another product. Some by-products are useful; others are not.

C

CADMIUM SULPHIDE CERAMIC SOLAR CELL: A type of solar cell used mostly in space craft.

CARCINOGENS: Chemical elements which produce high risk of cancer for those exposed to the chemicals.

CATALYTIC CONVERSION: A process for changing coal into a liquid fuel.

CATHODE: Negative terminal of a battery.

CELLS: With reference to world wind patterns, areas of the earth where wind patterns differ. There are three distinct areas of patterns.

CENTRIFUGAL FORCE: Tendency of an object traveling in a circular path to move outward from the center of the circular path.

CENTRIFUGE: A device which spins containers of materials to be separated at high speed. Heavier elements settle to the bottom.

CHAIN REACTION: The splitting of an atom in nuclear fission which, in turn, splits atom after atom.

CHANGEABLE ENERGY SOURCE: Energy source which is not always available. For example, wind energy is a changeable source since it is not always blowing.

CHAR: Porous solid fuel which is nearly pure carbon. Results from incomplete burning of organic material. When made from coal it is called coke. That made from wood is called charcoal.

CHEMICAL ENERGY: Energy locked away and stored in elements such as growing plants, coal, oil, or other compounds.

CHLOROPLASTS: Small disc-shaped bodies found in leaves, young stems of plants, and in algae. They contain chlorophyll and are being studied as a source of hydrogen gas.

COAL GASIFICATION: Converting coal to a gas fuel, a substitute for scarce natural gas.

COAL HYDROGENATION: Use of hydrogen to convert coal into liquid and gas products.

COGENERATION: Use of an energy source to produce two or more types of energy needed for heat or power. For example, spent steam from an electric power plant can be used as process heat for manufacturing or to heat buildings.

COMPOUND TURBINE: Steam turbine consisting of a series of turbines on one shaft.

CONDUCTION: Heating of matter through molecular activity. As heat is applied, the atoms move faster, bump into other atoms, and cause them to agitate as heat is passed along.

CONSTANT ENERGY SOURCE: An energy

source which is always available when needed.

CONTINUOUS MINING: Use of special mobile machines to mine and load coal in a nonstop operation.

CONTROL ROD: A metal rod which can be lowered into fissioning material to slow down the fission rate.

CONVECTION: Currents or movement within fluids caused by heating.

CONVECTIVE LOOP: A closed loop of duct work or piping connecting a flat plate collector and a storage device. Sun heats the fluid (water or air) in the collector causing warm fluid to flow to storage and cooler fluid to move into the collector.

CUBING: Compressing biomass into large pellets 1 to 2 in. in diameter.

D

DARRIEUS ROTOR: Vertical shaft rotor in which the narrow blades are curved to catch the wind from all directions. Its appearance is somewhat like an egg beater.

DEGRADED: Made less useful.

DENSIFICATION: Compressing matter so it takes up less space and delivers more heat per unit of weight.

DEUTERIUM: An isotope of hydrogen whose nucleus contains one neutron and one proton. It is twice as heavy as the nucleus of normal hydrogen. Sometimes called "Heavy Hydrogen."

DIESEL ENGINES: Type of internal combustion engine which runs on diesel fuel and does not depend on electric spark for fuel ignition.

DIGESTER: A mechanism into which can be placed organic waste where it will ferment anaerobically (without oxygen). As the waste breaks down it releases methane which then is collected in a connected storage vessel.

DIRECT ENERGY CONVERSION: When one form of energy is converted into another form ready to do work. As an example, a flashlight battery converts chemical energy into electrical energy which produces light.

DIRECT GAIN SOLAR HEATING: Space to be heated receives sunlight directly into the space. Heat is stored in floor, wall, or other structures in the room.

DOWN WIND ROTOR: Rotor design in which the wind blows over the rotor housing to reach the rotor.

DRIFT MINE: Mine in which the shaft is dug horizontally into a hill.

DRY STEAM: Superheated steam with no associated liquid. Used to describe geothermal systems.

E

EFFICIENCY: Ratio between the amount of energy put into a power system and the energy output.

ELECTRODE: Either terminal of an electrical source.

ELECTROLYSIS: Breakdown of a compound into its elements by passing an electric current through an electrolyte (such as a sulphuric acid solution). The process is used to store electricity.

ELECTRON: Negatively charged particle which orbits the nucleus of an atom.

ELECTRON VOLT: A tiny amount of energy equal to fraction of a calorie.

ELECTRICAL ENERGY: Energy resulting from the displacement and flow of free electrons.

ELECTROMAGNETIC RADIATION: Theory of energy transmission in which energy released by solar fusion travels through space in waves of varying lengths.

ELECTROMAGNETIC WAVES: The traveling of electromagnetic radiation through space or through conducting material. Light from the sun is a visible ray of this type. Others are not visible.

ELECTRON: Negatively charged particle found orbiting all atoms.

ENERGY: The ability to do work.

ENERGY CONVERSION: Changing of one form of energy to another. For example, burning of oil converts chemical energy into thermal (heat) energy.

ENERGY CROPS: Plant life cultivated solely for use as an energy source.

ENERGY WIND: Wind which blows at a speed

of 10 to 25 mph, suitable for operating a wind generator.

ENTROPY: The principle that all energy, as it is used, becomes so random (scattered or diffused) that it loses the ability to do work.

ENZYME: A type of protein which is responsible for producing chemical action within plant and animal cells.

ESTUARY: Place where a river flows into a sea or ocean.

ETHANOL: Also called ethyl alcohol. A type of alcohol produced by fermentation of grains, fruits, and other biomass.

ENTRIES: Long parallel horizontal shafts or hallways cut in a coal vein for longwall mining. The mining machine works back and forth removing the coal between the two entries.

EVACUATED TUBE COLLECTOR: A solar collector design consisting of a tube within a tube. The tubes are separated by a vacuum. Radiation travels easily through the vacuum but heat does not escape easily through a vacuum. This makes for a very efficient collector.

EXTERNAL COMBUSTION ENGINE: Engine in which the fuel is burned outside the cylinders.

EXTRUSION: A continuous forming process in which the product is shaped by forcing it through a die.

F

FERMENTATION: A process for making alcohol which uses anaerobic action. Enzymes act on starches and sugars in fruits, grains, and other biomass forms.

FERROELECTRIC CONVERSION: Nonconducting materials are made to absorb heat by producing change in molecules of the materials.

FISSILE: A uranium isotope able to go into a chain reaction to create fission.

FISSION: The basis of nuclear power. The splitting of the nucleus of an atom which produces two nuclei and releases tremendous amounts of heat energy.

FLASHING: In geothermal energy, the conversion of high-temperature water under high pressure to steam by lowering the pressure.

FLAT PLATE COLLECTOR: Device designed to absorb solar radiation as heat and transfer it through air or liquid to storage or living space.

FLUIDIZED COMBUSTION: New burning process which cuts down the sulfur, nitrogen oxide, and fly ash emissions or coal.

FOOT-POUND: Unit of measure for work. A foot-pound is the amount of force necessary to move a 1 lb. load a distance of 1 ft.

FREE PISTON ENGINE: An external combustion engine which combines elements of both reciprocating and gas turbine engines. Pistons are called free because they are not connected to anything. The pistons direct hot expanding gases onto a turbine.

FUEL ASSEMBLIES: Collection of rods, tubes, plates, or other shapes into which nuclear fuel is placed for use in a reactor.

FUEL CELL: An energy conversion device which combines fuel and oxygen to produce electrical energy without combustion.

FUSION: A nuclear reaction which fuses two atoms and releases energy.

FUTURE RESERVES: Supplies of energy which are thought to exist with less certainty than proven reserves and which may one day be recoverable as technology advances.

G

GAMMA RAYS: One of the invisible waves in the electromagnetic spectrum. Its wavelength is one-two hundred-fifty-trillionth of an inch.

GASIFICATION: Name for a number of processes which convert solids such as coal and organic material to a liquid or gaseous fuel.

GENERATOR: Device which, when energized by an outside force, spins and creates an electric current.

GEOTHERMAL ENERGY: Natural heat energy of the earth's interior. This heat is often transferred to water underground and, thus, becomes recoverable.

GEOTHERMAL FRACTURING: Creation of a heat reservoir deep beneath earth's surface.

Can be done artifically to induce flow of geothermal well.

GONDOLA CAR: Rail car with an open top designed for hauling coal. Some have a hopper bottom.

GRAVITATIONAL STORAGE: The storage of water as behind a dam or in a holding pond or water tower where it can provide energy as needed.

GREENHOUSE EFFECT: A heat trapping effect based upon the fact that solar radiation will readily pass through transparent materials but the heat it creates will not readily pass back through the transparent material.

H

HEAD: Distance water falls to where its energy is used. In a hydroelectric plant it is the distance from the inlet at the top of the dam to the turbine at the bottom of the dam.

HEAT ENGINE: Machine which can harness the energy of heat to do work.

HEAT EXCHANGER: Device for transferring heat from one fluid to another.

HEAVY HYDROGEN: (See deuterium.)

HELIOSTAT: A movable mirror linked to a tracking device to keep it turned to the sun. It reflects sunlight to a certain spot.

HELIUS™ ROTOR: Vertical-shaft rotor similar to the Savonius rotor but with blades curved and angled to collect more wind energy.

HORIZONTAL AXIS WIND MACHINE: Wind generator whose rotor shaft is horizontal while the propeller revolves in a vertical plane.

HORSE LATITUDES: High pressure belts north and south of the equator where there is little or no wind.

HORSEPOWER: The unit of measure for power. 1 hp equals energy needed to lift 33,000 lb. 1 ft. in 1 minute.

HYDROCARBONS: A compound of hydrogen and carbon which makes up the whole family of fossil fuels.

HYDROGENATION: Also called liquefaction. Process for producing gaseous and liquid fuels from coal and waste. Heat, steam, and carbon monoxide are used to carry on the process.

HYDROX CELL: A type of fuel cell. It uses hydrogen as a fuel and oxygen as an oxidizer.

I

IMPULSE TURBINE: A type of steam turbine in which the steam is piped to the turbine blades. It is released through nozzles and strikes the turbine vanes at high speed.

INDIRECT ENERGY CONVERSION: Changing the form of energy several times in order to produce a form which will preform work. For instance, a gasoline engine must convert chemical energy into heat energy, and the heat energy into mechanical energy, before the engine can do work.

INDIRECT GAIN SOLAR HEATING: Passive solar heating which uses a masonry structure or other device to collect solar radiation. Then the storage structure passes the heat on to the living space.

INDUSTRIAL SECTOR: Energy-using segment of United States which includes all types of manufacturing and extraction industries.

INERTIAL STORAGE: Storage of kinetic energy as in a flywheel.

INFRARED RAYS: One of the invisible electromagnetic waves in solar energy.

IN SITU: A method of extracting uranium and some fossil fuels. Part of the refining or separating process takes place underground. The phrase means "in the natural place."

INSOLATION: Shortened form of "Incident Solar Radiation." It means the amount of radiation falling on the ground at any given time. It is expressed in Btu per square foot per hour.

INTERNAL COMBUSTION ENGINE: A heat engine designed to burn the fuel inside the engine.

INVERTER: Device which steps up voltage and converts direct current to alternating current.

IONIZE: To remove some or all electrons from an atom or molecule. This leaves the atom with a positive charge. One or more electrons could also be added giving the atom a negative charge.

ISOLATED GAIN SOLAR HEATING: Solar heat is collected in one space before it is transferred by natural convection to the space where the heat is needed. Amount of heat admitted to living space can be controlled by vents.

ISOTOPE: Atoms of the same element which have different numbers of neutrons.

J

JET ENGINE: Engine producing motion with forceful stream of heated air and exhaust gases.

JOULE: Metric unit of power.

K

KEROGEN: A brown-gray organic substance in oil-bearing shale (marlstone). When heated it reacts to form oil and gas.

KINETIC ENERGY: Energy of motion. The ability of moving objects or matter to do work.

L

LANGLEY: Metric unit of measure for solar energy. A Langley equals 1 calorie per sq. centimeter.

LASER: A device for producing an intense beam of light.

LEACHING: Method of separating certain materials by allowing them to be dissolved in a liquid such as water.

LEG: Low energy gas, a fuel produced when small amounts of air and steam act upon char.

LIGNITE: Also known as "young coal." Soft, brown, fibrous materials which is sometimes used as fuel.

LIQUEFICATION: A process for liquefying coal; that is, turn it into a liquid fuel. The process adds hydrogen to the coal to change the carbon-hydrogen ratio from 16-1 ro 6-1. Also called hydrogenation.

LOCAL WINDS: Winds caused by local temperature differences and topography.

LONGWALL MINING: Mining system based on use of a continuous mining machine. It works back and forth on the face of the seam between two parallel halls. The ceiling behind the mining machine is allowed to collapse as the machine advances.

M

MECHANICAL ENERGY: Kinetic energy or the energy of motion.

METABOLIC CONVERSION: Use of enzymes, molds, or bacteria to convert organic wastes to gas fuels.

METHANOL: An inflammable natural gas which becomes explosive when mixed with air. It can be manufactured from crude petroleum or other organic material.

MHD GENERATOR: Device for generating electricity without moving parts. Hot gases at 4000°F are passed across a magnetic field. Electrodes capture free electrons that flow through conductors connected to the electrodes.

MICROSPHERES: The tiny particles of isotopes which make up the fissionable material in the High Temperature Gas Reactor.

MICROWAVE: One of the longer electromagnetic waves emitted by the sun.

MISTRAL: A local wind, strong and cold, in the Rhone Valley of France.

MODERATOR: A material, such as graphite, water, deuterium, or beryllium, used in a nuclear reactor to slow down the rate of fission.

MOLECULES: Smallest particle of an element and without which it would not be that element.

MOMENTS OF FORCE: Metric unit for twisting force. It is measured in newton meters.

N

NEAP TIDE: A high tide which occurs while the gravitational pull of the moon and the sun are working against each other.

NEUTRON: Neutral particle in an atom which has neither positive nor negative charge.

NONENERGY SECTOR: An energy using part of our society which uses fossil fuels as raw material for products.

NUCLEAR ENERGY: Energy which is released by splitting of an atom of matter.

NUCLEUS: The center of an atom which is made up of neutrons and protons.

O

OPEN PIT MINING: Surface mining of coal in which large scoops work in open pits.

ORBIT: Circular path of an electron around the nucleus of an atom.

OVERBURDEN: Rock and soil lying over the top of a coal bed.

OVERSHOT WATER WHEEL: Water wheel designed so that the water flows over the top. Thus, the weight of the water is added to its motion. The overshot wheel has more power than the undershot.

P

PARABOLIC COLLECTOR: A dish-shaped solar collector whose curved shape concentrates the solar rays at a certain spot.

PASCAL: Metric unit of pressure.

PASSIVE SOLAR SYSTEMS: Solar heating systems which use no mechanical methods for bringing in heat and storing it for conditioning living space temperatures.

PEAT: An early stage in the formation of coal from decaying vegetation.

PELLETIZING: To compress matter, such as biomass, into small particles or pellets.

PHASE-CHANGE MATERIALS: Materials which change from solids to liquids as they collect and store heat.

PHOTON: Bundle of energy which, in theory, travels from the sun in an unceasing bombardment accounting for solar energy.

PHOTOSYNTHESIS: Chemical process by which plants absorb and store solar energy.

PHOTOVOLATIC CELL: Device which can convert solar energy directly into electricity.

PLANETARY WINDS: Mass movement of air on earth caused by the great temperature differences between the equator and the poles.

PLASMA: Gaseous mixture of positive and negative ions. High-temperature plasmas are used in controlled fusion experiments.

PLENUM: An enclosed space connected to a heat source where heated air collects to be distributed by natural convection or by blowers.

PLUTONIUM: A heavy, radioactive manufactured metallic element. One of its isotopes is used to produce a fissionable material known as PU 239.

POLYCYCLIC AROMATIC HYDROCARBONS: A carcinogen produced by the high temperatures used in converting coal to synthetic fuels.

POST MILL: Early windmill design. A large, round, wooden shaft, called a post, supported the vane and kept the mill turned into the wind.

POND: In passive solar heating, water stored on the roof to collect the heat from sunlight. At night the heated water warms the living space below.

POTENTIAL ENERGY: Energy at rest or in storage which is not working but is capable of doing work.

POWER: Amount of work done in a given period of time. $(\frac{\text{work}}{\text{time}})$

PRESSURE: Force exerted by a contained gas or liquid.

PRESSURE CYLINDER: Cylinders containing compressed air or carbon dioxide. Coal miners place them in drilled holes. When expanded by compressed gas, they break up the coal.

PROTON: A positively charged particle found in any atom. One or more protons, along with neutrons, form the atom's center.

PROVEN RESERVES: Supplies of fossil fuels which are thought to exist with a great deal of certainty and which can be recovered with current technology.

PYROLYSIS: Heating of coal or other organic material without oxygen, to convert coal into other fuels and organic wastes into liquid and gaseous fuel.

Q

QUAD: Unit of measure for large quantities of energy. It stands for 1 quadrillion Btu.

QUANTUM THEORY: A theory of radiation suggesting that its cause is particles or bundles of energy which travel through space and give up their energy upon striking a surface through which they cannot pass.

R

RADIANT ENERGY: Energy of the type found in sunlight which travels in waves from its source. It creates light and heat.

RADIATION: Traveling of heat and light energy in waves through space.

RANKINE ENGINE: Reversible heat engine which uses hot water to vaporize volatile fluids such as Freon.

REACTION TURBINE: Steam powered turbine in which the steam passes through several sets of blades which are set on the same shaft. Each set is at a different angle.

REACTOR: A strong, closed container in which nuclear fission takes place.

REACTOR VESSEL: Steel pressure tank in a nuclear reactor which contains the fuel rods. Its walls are 9 in. thick.

REINJECTION WELL: In geothermal installations, a well used to return condensed liquid to the rock formation where it may be reheated thus extending the life of the well.

RESIDENTIAL/COMMERCIAL SECTOR: A segment of the energy-using activities in the United States. It includes offices, schools, and individual homes which consume energy for heat, light, and operation of various machines and appliances.

RESIDUES: That which remains after a part has been removed. In agriculture, lumbering, or food processing, the rejected plant matter such as leaves, husks, bark, or pits.

ROCKET: Any heat engine that is propelled (moved) by the reaction to the high speed exhausting of gases. Fuel is either solid or liquid and is carried inside the rocket.

ROOM AND PILLAR MINING: Coal mining system in which coal is removed by making a series of "rooms" with large pillars of coal left at intervals to support the roof.

ROTARY ENGINE: Internal combustion engine with a triangular rotor replacing the conventional piston.

ROTOR: In a wind generator, the propeller which is turned by the force of the wind and which, in turn, rotates the generator.

S

SANTA ANA WIND: A hot dry wind that comes out of the Santa Ana mountains. Air on the leeward side is compressed, heats up, and expands across the plain in a northwesterly direction.

SAVONIUS ROTOR: A vertical shaft rotor with curved blades designed to catch the wind from any direction.

SEISMIC: Related to shock waves induced in the ground to study formations underground.

SEMICONDUCTOR: An electronic part which sometimes acts like a conductor and sometimes like an insulator.

SHAFT MINE: Mine whose entry shaft is sunk straight down into the earth.

SHEARERS: A mining machine which cuts away the coal during a longwall mining operation.

SHOOTING: Use of explosives to break up coal deposits during open pit mining operations.

SINTERING: A baking process done in a furnace.

SLOPE MINE: Mine is which the main shaft slopes gently downward.

SLURRY: Coal which has been crushed to consistency of granulated sugar and mixed with water or some other fluid so it can be transported via pipeline.

SMOCK MILL: Large mill of an early design which used a second wheel or rotor to keep the main wheel into the wind.

SODIUM: Soft, silvery-white metallic alkali. It is valuable as a coolant for a nuclear reactor because it can absorb more heat than water or helium, two other common reactor coolants.

SOLAR CONSTANT: The rate at which solar radiation falls on the outer limits of earth's atmosphere. This constant has been set at 1.36 kilowatts per square meter.

SOLVENT REFINING: Making solid fuel from coal. It is like a solid tar. Its heat value is higher than coal and much cleaner burning.

SPOIL: Another term for loosened overburden, the dirt and rock that is removed from a pit mine.

SPOIL AREAS: Place where overburden is

stored during open pit mining.

SPRING TIDE: Tides which are caused when the gravitational pull of the sun and the moon are working together.

STEAM ENGINE: Heat engine which turns the heat energy in steam to mechanical energy. It is a type of external combustion engine.

STEMMING MATERIAL: Drill cuttings which are tamped over dynamite to keep explosive force from simply going upward out of the hole.

STIRLING ENGINE: External (outside) combustion engine in which air or hydrogen is first heated and then cooled to drive the piston up and down.

STORAGE BATTERY: Device made up of an electrolyte and plates or terminals. Its purpose is to provide chemical storage of electricity.

SYNFUELS: Name given to natural fossil liquid fuels which are gotten from coal and an oil bearing shale called marlstone.

SYNTHESIS GAS: Also known as syngas. A gas made by burning biomass in the presence of oxygen. It is a mixture of carbon monoxide and hydrogen.

T

THERMOELECTRIC CONVERTER: Device which generates electricity directly from heat applied to the junction of two unlike metals.

THERMIONIC CONVERTER: Device consisting of two metallic plates. When one is heated, electrons "boil" off and create an electric current if the plates have conductors to create an electrical circuit.

THERMOMAGNETIC CONVERSION: Device which uses heat to make a change in the magnetic field of a magnet thus displacing electrons and creating an electric current.

THERMOSIPHONING: Natural circulation of a heated fluid without aid of fans or pumps. Heated fluid drives the system as it rises, pulling in descending cooler fluids.

THORIUM: A naturally radioactive element which is found in nature. It can be made to yield a fissionable uranium isotope.

TIDAL POWER: Use of tides as an energy source to produce electrical power.

TIDAL RANGE: Difference between low tide and high tide.

TIPPLE: Building or plant where bituminous coal is prepared for use or shipping.

TOKAMAK: The doughnut shaped chamber used in a fusion reactor to contain the superheated fusion material called plasma.

TOPOGRAPHY: The condition of the land surface; that is, mountainous, hilly, flat, or rolling.

TOROIDAL: In the shape of a doughnut.

TORQUE: Turning force measured in pound-feet (force times distance).

TOWN GAS: A gas made from a coal gasification process which used to be supplied to cities for cooking and light.

TRADE WINDS: Planetary winds that move toward the equator usually at a constant speed and direction.

TRANSLUCENT: Quality of some solid matter by which it transmits some light but objects beyond are not visible.

TRANSMUTATION: A process which separates radioactive chemicals into materials with a short radioactive life. Scientists are exploring its use as a method of reducing storage time of radioactive waste.

TRANSPARENT: Matter which transmits light so that objects can be seen through it.

TRANSPORTATION SECTOR: One of the major users of energy. It includes all types of vehicles used to move people and materials.

TRITIUM: Radioactive isotope of hydrogen. Its nucleus contains one proton and two neutrons. It may be used as fuel in the early fusion reactors.

TROMBE WALL: Storage wall located behind glazing so it can receive and store heat generated by sunlight. Stored heat is later passed on to living space.

TURBINE: Device somewhat like a water wheel with curved vanes attached to a central spindle. It is used to turn a generator in modern hydroelectric plants.

TURBOFAN ENGINE: Internal combustion engine which is like the turbojet engine except that a fan draws extra air into the engine for greater power.

TURBOJET ENGINE: Internal combustion engine used in aircraft, boats, and locomotives. Like the reciprocating engine it has intake, compression, combustion, and exhaust. However, these actions take place continouously rather than as a cycle.

U

U-235: The isotope of uranium which is used as fuel for the nuclear reactor.

ULTRAVIOLET RAYS: Electromagnetic radiation of a wavelength which is not visible.

UNDERGROUND MINE: A mine deep in the earth which is reached by digging a vertical shaft. Horizontal shafts are then excavated in the work area.

UNDERSHOT WATER WHEEL: A water wheel designed to let the water flow under it and push against its paddles.

UPWIND ROTOR: A rotor which heads into the wind with the generator and housing behind it.

URANIUM: A radioactive element used to produce nuclear fission. It is a heavy, hard, nickel-white metal.

URANIUM HEXAFLUORIDE: A gas produced from yellow cake.

V

VELOCITY: Speed of any object.

VERTICAL AXIS WIND MACHINE: Wind generator whose rotor spins on a vertical axis.

VOLTAGE REGULATOR: Device to keep voltage to a system at a certain level.

VOLUMETRIC HEAT CAPACITY: Amount of heat-holding capacity of a cubic foot of material.

W

WALKING DRAGLINE: Large bucket suspended by cable from the end of a boom. Used for stripping in open pit mining.

WALKING PROPS: In coal mining, hydraulic roof supports which advance automatically as mining proceeds.

WATT: Metric unit of power. One watt equals 1 joule of work per second.

WHEEL EXCAVATOR: Large machine consisting of a boom, conveyor, and a wheel with scoops on it. It is used to strip coal from a coal bed.

WINDMILL: A machine which uses the energy of wind to do work.

WINDOW COLLECTOR: Small flat plate collector which delivers solar heated air through a window.

WORK: Application of energy or force which produces motion or change.

Y

YELLOW CAKE: Uranium ore which has been ground to bright yellow slurry (soupy mixture) or powder.

ACKNOWLEDGEMENTS

I wish to thank those whose help or influence made this book possible:

My father, Hilar, 101 years old at this writing, whose example first taught me the importance of hard work and finishing what I had started.

My wife, Leona, for her encouragement and assistance in research, correspondence, and final manuscript preparation.

Jan Witt, artist and friend, for many of the line illustrations.

The many teachers who influenced me and helped me along the way to my chosen work.

Co-workers who critiqued the text and shepherded it through production.

The many manufacturers and other persons who supplied illustrations and information.

INDEX

D

E